Physics
for CCEA A2 Level

COLOURPOINT
EDUCATIONAL

Pat Carson and Roy White

ISBN: 978 1 904242 65 9

First Edition
Third Impression, 2015

Layout and design:
April Sky Design, Newtownards

Printed by:
GPS Colour Graphics Limited, Belfast

Colourpoint Creative Limited
Colourpoint House
Jubilee Business Park
21 Jubilee Road
Newtownards
County Down
Northern Ireland
BT23 4YH

Tel: 028 9182 6339
Fax: 028 9182 1900
E-mail: info@colourpoint.co.uk
Web site: www.colourpointeducational.com

The Authors

Roy White has been teaching Physics to A-level for over 30 years in Belfast. He is currently Head of Department and an enthusiastic classroom practitioner. He works for CCEA as Chair of Examiners for Double Award Science, Principal Examiner for GCSE Physics and as Chair of Examiners for Applied Science. In addition to this text, he has been the author or co-author of three successful books supporting the work of science teachers in Northern Ireland.

Pat Carson has been teaching Physics to A-level for over 30 years in Belfast and Londonderry. He is currently Vice-Principal in a Londonderry Grammar school. He works for CCEA as Chief Examiner for GCSE Physics.

Note to Teachers

The content of the CCEA A2 specification is divided into three main sections named A2 1, A2 2 and A2 3. However, in order to avoid confusion with the section numbers in the first book (*Physics for CCEA AS Level*), and in order to match the numbers used in the details of the specification, these units are named 4, 5 and 6 in this book. Therefore:

- Unit A2 1 in the specification = Unit 4 in the book.
- Unit A2 2 in the specification = Unit 5 in the book.
- Unit A2 3 in the specification = Unit 6 in the book.

In addition, chapters in Unit 4 and Unit 5 are named after the equivalent sections of the CCEA specification. For example the chapter entitled "5.5 Magnetic Fields" corresponds to section 5.5 of the CCEA specification on Magnetic Fields. Note that due to the nature of the material, chapters in Unit 6 are not broken down in this way.

Finally, this book also includes a section giving advice on the Synoptic Assessment aspect of A2 Physics. This section does not cover actual content from the specification, but rather deals with the manner in which students will be expected to display their knowledge through written examination and experimental tests. This section has been named Unit 7 for consistency.

CONTENTS

Unit 4 (A2 1): Momentum, Thermal Physics, Circular Motion, Oscillations, Atomic and Nuclear Physics

Unit 5 (A2 2): Fields and their Applications

Unit 6 (A2 3): Practical Techniques

Unit 7: Synoptic Assessment

Unit 4 (A2 1)

Momentum, Thermal Physics, Circular Motion, Oscillations, Atomic and Nuclear Physics

4.1 Momentum

Students should be able to:

4.1.1 Define momentum;

4.1.2 Calculate momentum;

4.1.3 Demonstrate an appreciation of the conservation of linear momentum;

4.1.4 Perform calculations involving collisions in one dimension;

4.1.5 Use the terms 'elastic' and 'inelastic' to describe collisions;

You were first introduced to momentum in the context of the de Broglie equation in AS Module 1. You will recall that **the momentum of a body is defined as the product of its mass and its velocity.** In symbols this is often written:

$$p = mv$$

where p is the momentum in Ns (or kg ms^{-1})
m is the mass in kg
v is the velocity in ms^{-1}

Note that momentum and velocity are vectors and take place in the same direction.

This kind of momentum is sometimes called linear momentum, because it relates to the momentum of a body moving in a straight line, as opposed to the momentum of a body rotating, like a spinning top.

Worked Examples

Example 1

Show that the unit Ns is equivalent to the unit kg ms^{-1}.

Solution

From F = ma, we see that the unit of force, the newton, is equivalent to the kg ms^{-2}

So, the Ns $= kgms^{-2} \times s = kgms^{-1}$

Example 2

Calculate the momentum of a car of mass 800 kg travelling due North with a speed of 15 ms^{-1}.

Solution

$p = mv$

$= 800 \times 15 = 12000$ Ns due North

Principle of Conservation of Linear Momentum

It can be shown experimentally that the total momentum of a **closed system** remains constant, even during collisions. By a **closed system**, we mean one where no external forces are acting. A closed system is therefore one in which the only forces which contribute to the momentum change of an individual object are the forces acting between the objects themselves.

So, for example, in a system of colliding neutrons, far away from any other particles, the total momentum is conserved. Similarly, when two trolleys collide on a linear air track, the friction force can be neglected and the total momentum before the collision is equal to the total momentum afterwards.

Consider the collision of two balls on a snooker table. The collision occurs in an isolated system as long as friction is small enough that its influence upon the momentum of the balls can be neglected. In this case, the only unbalanced forces acting upon the two balls are the contact forces which they apply to one another. These two forces are considered **internal** forces since they result from a source within the system – that source being the contact of the two balls. For such a collision, total system momentum is conserved.

However, where cars collide on a road where friction is large, then friction is considered an **external** force. The system of colliding objects is not closed and linear momentum is not conserved.

This allows us to state the **Principle of Conservation of Momentum** for bodies in collision.

If no external forces are acting, the total momentum of a system of colliding bodies is constant.

Collisions

When we apply this principle to collisions, it can be simply stated as

Total momentum before collision = Total momentum after collision

As we shall see in the examples, we must always remember that **momentum is a vector** and to **assign one direction as positive and the opposite direction as negative.** Thus if a car of mass 1000 kg moving at 5 ms^{-1} **to the right** has a momentum of +5000 kg ms^{-1}, then the same car moving at 10 ms^{-1} **to the left** has momentum of –10000 kg ms^{-1}. The choice as to which direction is positive is entirely arbitrary.

Worked Examples

Example 1

A toy truck of mass 400 g, moving to the right with a speed of 4 ms^{-1} collides with and sticks to a toy tricycle of mass 1600 g moving to the left with a speed of 3 ms^{-1}. Calculate

(a) the momentum of each toy prior to the collision and

(b) the velocity of the combination after the collision

Solution

(a) Taking motion to the right as positive: ⟶ Positive

and motion to the left as negative: Negative ⟵

Momentum of truck before collision = mv

$mv = 0.4 \text{ kg} \times +4 \text{ ms}^{-1} = +1.6 \text{ kgms}^{-1}$

Momentum of tricycle before collision = mv

$mv = 1.6 \text{ kg} \times -3 \text{ ms}^{-1} = -4.8 \text{ kgms}^{-1}$

(b) By the Principle of Conservation of Momentum

Total momentum before collision = Total momentum after collision

$$\{1.6 + (-4.8)\} \text{ kg ms}^{-1} = \text{Mass of combination} \times \text{Velocity of combination after collision}$$

$$-3.2 = (0.4 + 1.6) \times v_{after}$$

$$v_{after} = -3.2 \div 2.0 = -1.6 \text{ ms}^{-1}$$

The minus sign shows that the **combined truck and tricycle is moving to the left,** that is, it is moving in the same direction as the tricycle was moving originally.

Example 2

Two girls of masses 45 kg and 60 kg stand facing one another on light frictionless trolleys holding the ends of a strong taut rope between them. The lighter girl tugs the rope and starts to move towards her neighbour with a velocity of 2 ms^{-1}. Calculate the initial velocity of the heavier girl.

Solution

Assume after the rope is pulled the 45 kg girl moves in the direction of positive velocity, and the 60 kg girl moves with a velocity v.

Total momentum before motion begins = 0

Total momentum after rope is pulled = $45 \times 2 + 60 \times v$

By the Principle of Conservation of Linear Momentum, $0 = 45 \times 2 + 60 \times v$

Hence, $v = -90 \div 60 = -1.5 \text{ ms}^{-1}$. The minus sign shows that the 60 kg girl moves in the direction of negative velocity. So the 60 kg girl moves towards her neighbour at a speed of 1.5 ms^{-1}

Example 3

A bullet of mass 6 g is fired from a pistol of mass 0.5 kg. If the muzzle velocity of the bullet is 300 ms^{-1}, calculate the recoil velocity of the gun.

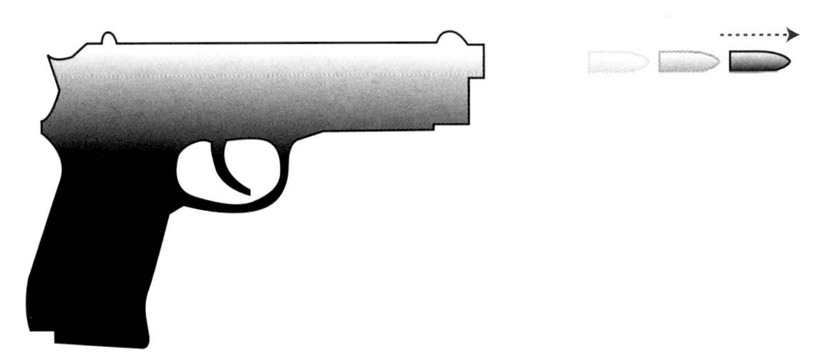

Solution

$$\text{Momentum after firing} = \text{Momentum before firing}$$

$$(0.006 \times 300)_{bullet} + (0.5 \times v_{recoil})_{pistol} = 0$$

$$\text{So, } v_{recoil} = -(1.8 \div 0.5) = -3.6 \text{ ms}^{-1}$$

The minus sign indicates that the velocities of the bullet and the recoil of the pistol are in opposite directions.

Collision Classification

Collisions may be classified as elastic or inelastic.

Elastic collisions are those in which kinetic energy is conserved. These only occur on an atomic scale, such as the collision of two molecules of an ideal gas.

Inelastic collisions are those in which kinetic energy is not conserved. Frequently inelastic collisions involve kinetic energy being converted to thermal energy or sound. An example is the collision of a tennis ball with a racquet. A **completely inelastic** collision is one in which two bodies stick together on impact. Here the loss of kinetic energy is very large, though not complete. An example is a rifle bullet embedding itself in a sandbag.

Suppose a car of mass 1200 kg moving at 10 ms^{-1} collides with a stationary car of mass 800 kg and the two cars stick together. To calculate the speed of the two cars after the collision we apply the Principle of Conservation of Momentum.

$$\begin{aligned} \text{Momentum before collision} &= \text{Momentum after collision} \\ 1200 \times 10 + 800 \times 0 &= 2000 \times v \\ v &= 12000 \div 2000 \\ &= 6.0 \text{ ms}^{-1} \end{aligned}$$

Before the collision the total kinetic energy was $\frac{1}{2}mv^2 = \frac{1}{2} \times 1200 \times 10^2 = 60\ 000$ J.

After the collision the total kinetic energy was $\frac{1}{2}mv^2 = \frac{1}{2} \times 2000 \times 6^2 = 36\ 000$ J.

The collision has resulted in **a reduction of 24 000 J of kinetic energy.**

The kinetic energy is therefore not conserved.

Now the **Law of Conservation of Energy** tells us that you cannot just "lose" energy. This 24 kJ of **kinetic** energy has changed into other forms. When the cars collide, a lot of sound and heat is converted from kinetic energy. The missing kinetic energy has been changed into these forms and work is done in changing the shape of the cars. But the **total** energy of the system **does not change.**

We can sum up these ideas in a table.

	Momentum	Kinetic energy	Total energy
Inelastic Collisions	is conserved	is NOT conserved	is conserved
Elastic Collisions	is conserved	is conserved	is conserved

Exercise 1

Examination Questions

1 A bullet of mass 15 g is fired horizontally from a gun with a velocity of 250 ms^{-1}. It hits, and becomes embedded in, a block of wood of mass 3000 g, which is freely suspended by long strings as shown in the diagram below. Air resistance can be neglected.

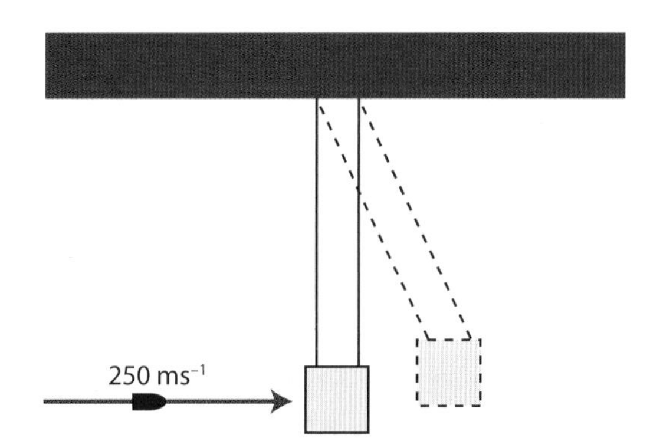

(i) Calculate the magnitude of the momentum of the bullet as it leaves the gun.

(ii) Calculate the speed of the wooden block when the bullet strikes it.

(iii) Use your answer to part (ii) to calculate the kinetic energy of the wooden block and the embedded bullet immediately after the impact.

(iv) Hence calculate the maximum height above the equilibrium position to which the wooden block, with the embedded bullet, rises after impact.

2 In an experiment to verify the Principle of Conservation of Linear Momentum, two bodies are caused to collide on a horizontal, friction-free track. Before the collision, one of the bodies, of mass 1.5 kg, is stationary, while the other, of mass 0.5 kg, moves with a speed of 0.18 ms^{-1} to the right. Following the collision, the two bodies remain joined together, and move towards the right.

(i) Why is the experiment conducted on a friction-free track?

(ii) This collision is an example of an inelastic collision. What is the difference between an inelastic and an elastic collision?

(iii) Calculate the speed of the combined masses following the collision.

(iv) Calculate the constant force F needed to stop the bodies in a time of 0.15 s after the collision.

3 Two frictionless trolleys, A and B, of mass m and 3m respectively, are on a horizontal track.

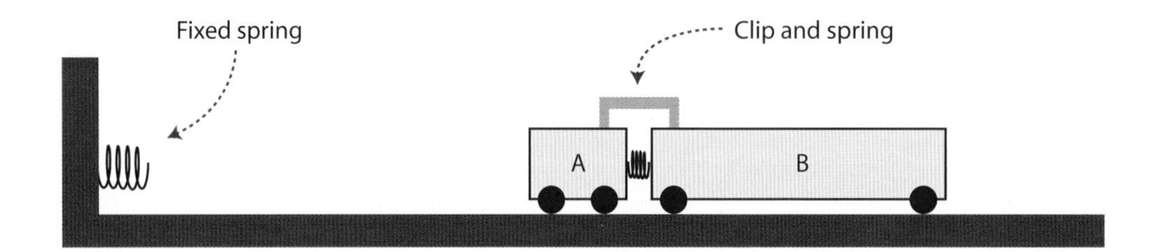

Initially they are clipped together by a device which incorporates a spring, compressed between the trolleys. At time t = 0, the clip is released and the trolleys move apart, the spring falling away. The time during which the spring expands is negligible. The velocity of B is then u to the right.

(a) Show that, as the trolleys move apart, the magnitude of the velocity of trolley A is 3u.

(b) At time $t = t_1$, trolley A collides elastically with a fixed spring and rebounds. The compression and expansion of the fixed spring takes place in a negligibly short time. Trolley A catches up with trolley B at time $t = t_2$.

(i) What is the velocity of trolley A between $t = t_1$ and $t = t_2$?

(ii) Find an expression for t_2 in terms of t_1.

(c) When trolley A catches up with trolley B at time $t = t_2$ the clip operates so as to link them again, this time without the spring between them, so that they move together with a velocity v. Calculate the common velocity v in terms of u.

(d) Initially, before the clip was opened, the trolleys were at rest and the total momentum of the system was zero. However, your answer to (c) should show that the total momentum after $t = t_2$ is not zero. Discuss this result with reference to the Principle of Conservation of Linear Momentum.

4. (a) Collisions between bodies can be classified as elastic or inelastic. In which of these types of collision is (i) linear momentum, (ii) kinetic energy, (iii) total energy, conserved?

(b) Particle A of mass **m**, moving with velocity **u** makes a head on elastic collision with particle B of mass **M** which is initially at rest. After collision the velocity of A is **v** and the velocity of B is **V**. The directions of these velocities are defined in the diagram below.

For this collision express the conservation of momentum and kinetic energy in the form of equations, using the symbols used above for mass and velocity.

(c) For the collision shown above, it can be shown that

$$v = \frac{(m-M)}{(m+M)}u$$

Use this result to find an expression for the ratio R of the kinetic energy of particle A after collision to the kinetic energy of A before collision.

R = kinetic energy of A after collision ÷ kinetic energy of A before collision.

(d) For the special case in which the two particles are of equal mass (M = m) use the equation above to describe what the motion of the two particles after collision. What is the value of R in this case?

(e) For the special case in which particle B is of infinite mass, use the equation above to describe the motion of the particles after collision. What is the value of R in this case?

(f) In a nuclear reactor uranium nuclei undergo fission when they absorb neutrons. To slow the neutrons down, to allow this absorption to take place, a moderator is used. The neutrons are slowed by head-on collisions with the atoms of the moderator. Use the equation above to show that graphite (carbon) is a better moderator than a heavy element such as lead.

Mass of a carbon atom = 12, the mass of a lead atom = 206 and the mass of a neutron = 1

4.2 Thermal Physics

Students should be able to:

4.2.1 Describe simple experiments on the behaviour of gases to show that $pV = \text{constant}$ for a fixed mass of gas at constant temperature, and $\frac{p}{T} = \text{constant}$ for a fixed mass of gas at constant volume, leading to the equation $\frac{pV}{T} = \text{constant}$;

4.2.2 Recall and use the ideal gas equation $pV = nRT$;

4.2.3 Recall and use the ideal gas equation in the form $pV = NkT$;

4.2.4 Use the equation $pV = \frac{1}{3}Nm\langle c^2 \rangle$;

4.2.5 Demonstrate an understanding of the concept of absolute zero of temperature;

4.2.6 Demonstrate an understanding of the concept of internal energy as the random distribution of potential and kinetic energy among molecules;

4.2.7 Use the equations for average molecular kinetic energy $\frac{1}{2}m\langle c^2 \rangle = \frac{3}{2}kT$

4.2.8 Perform and describe an electrical method for determination of specific heat capacity;

4.2.9 Use the equation $Q = mc\Delta\theta$

The Behaviour of Gases

Volume, pressure and temperature are the **macroscopic** (large-scale) properties of a gas. They are the properties of the gas that you can observe in the laboratory. The pressure, volume and temperature of a gas are all inter-related. If you change one, at least one of the others changes. These properties also depend on the amount of gas present. If you want to find out the connections between the macroscopic properties of a gas, you need to carry out experiments with a fixed amount of gas.

A sealed balloon contains a fixed amount of gas. Its volume increases when it is placed in warm water and the temperature rises. But the pressure changes as well. To investigate gas behaviour, as well as keeping the amount of gas constant, you also need to keep one of the other properties constant (either the pressure or the volume or the temperature) and investigate how the remaining two properties depend on each other.

Volume and Pressure Changes Leading to Boyle's Law

In the diagram below the oil in the closed tube traps a **fixed mass** of air above it. If the column of air has a constant cross-section, **its length is proportional to its volume**. This experiment involves measuring the length of this column and recording the corresponding pressure on the Bourdon gauge. The oil used must have a low vapour pressure, otherwise we are compressing an oil vapour and air mixture, whose oil content will change with pressure. This in turn would cause the mass of the material being compressed to change.

Using the hand pump (some types of apparatus incorporate a foot pump) we can increase very slowly the pressure acting on the trapped air. Compressing the gas warms it slightly, so after every compression we need to wait a few moments for the temperature of the trapped air to return to room temperature. Then we record the new volume and pressure readings in a table.

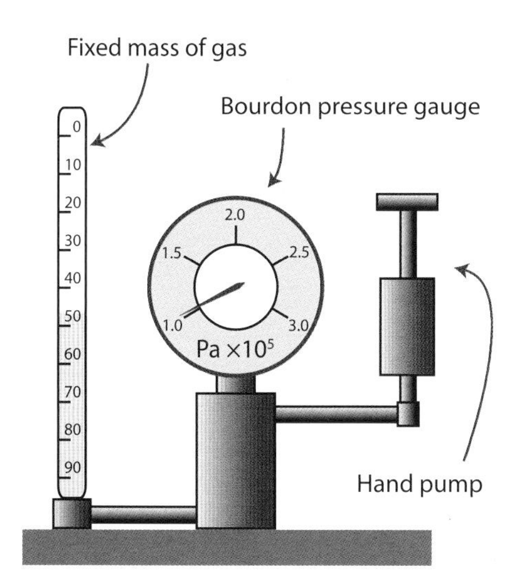

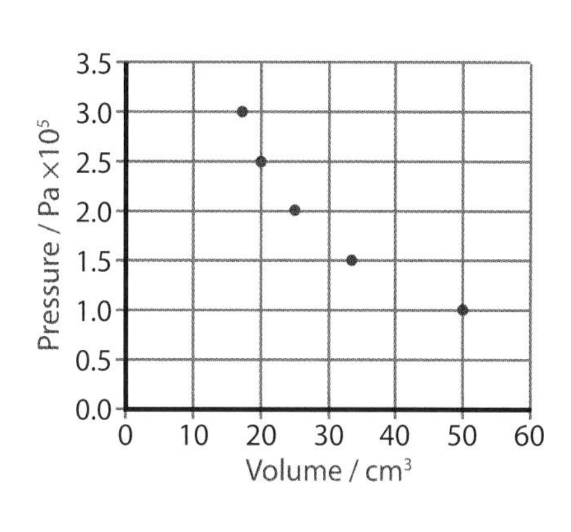

We can repeat this for several more values of pressure and then plot a graph of pressure against volume. The graph is called an **isothermal**, because it shows how pressure changes with volume at constant temperature. ("iso" means "the same" and "thermal" means "temperature".)

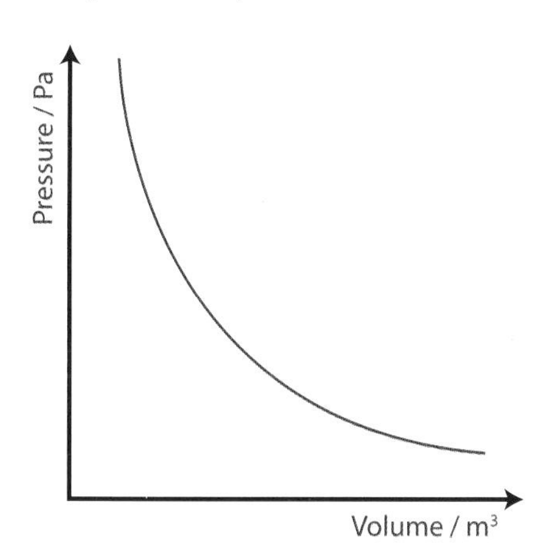

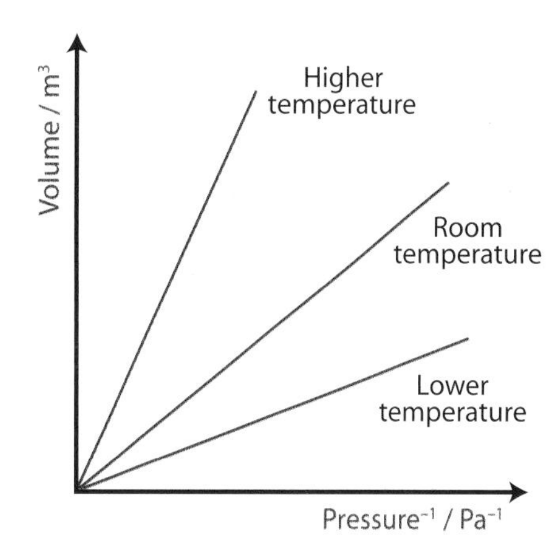

The graph of pressure against volume is a curve which shows that volume decreases as pressure increases – there is some kind of inverse relationship between pressure and volume. To determine the nature of this relationship, we plot a graph of volume against 1/pressure as shown above. This graph is a straight line through the origin, confirming that volume is directly proportional to the inverse of the volume. More succinctly, the volume is inversely proportional to the pressure. This is formally expressed as **Boyle's Law** which states:

For **a fixed mass** of gas **at constant temperature**, the volume is **inversely proportional** to the applied pressure.

Boyle's Law can be expressed as an equation:

$$\text{pressure} \times \text{volume} = \text{constant} \quad \text{or} \quad PV = \text{a constant}$$

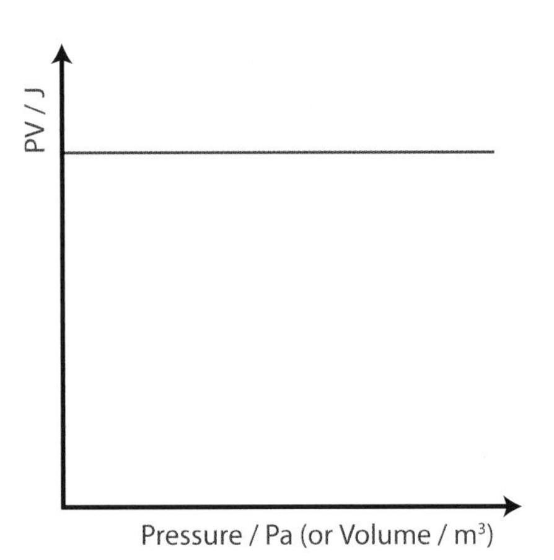

Boyle's Law can be demonstrated graphically in a number of different ways. The graph of pressure against volume is a hyperbolic curve as shown above; the graph of the product PV against V (or P) is a horizontal straight line. The reader is encouraged to attempt question 3 in Exercise 2 to demonstrate these graphical relationships.

For a **virtual experiment** on Boyle's Law, the material at the web address below may be useful:

http://tinyurl.com/ldrjn3

Its just a thought...

Robert Boyle (1627-1691) was born in Lismore Castle, Ireland, the seventh son of the Earl of Cork. He could speak French and Latin fluently as a child and was sent to Eton at the age of eight. He spent his early life at Gresham College London where he joined a group called *The Invisible Society*, which had the purpose of promoting scientific study and thinking. Later it was to become *The Royal Society* and Boyle was elected its President in 1680. However, Boyle spent much of his life in Oxford carrying out scientific research, some of it with an assistant called Robert Hooke (of Hooke's Law fame!). He published a book with the (abbreviated) title *The Spring of Air* and gave us the modern meaning of the word "elastic". Boyle's studies took him into many different fields. He was interested in bioluminescence, hydrometers, electricity, colour, combustion and theology (he learned Greek, Hebrew and Syriac).

Boyle is called **The Father of Chemistry**, because he was the main agent responsible for changing its outlook from alchemy (which was primarily interested in being able to change base metals like lead into precious metals like gold) to modern chemistry.

Exercise 2

Questions on Boyle's Law

1 One form of Boyle's Law is PV = a constant. Show that the unit for the constant is the Joule.

2 A fixed mass of gas has a volume of 24.0 litres when the pressure is 100 kPa.

(a) Calculate the volume when the pressure is (i) 50 kPa (ii) 150 kPa.

(b) Calculate its pressure when its volume is (i) 4.8 litres (ii) 8 litres (iii) 12 litres

3 The data in the table below were obtained in a Boyle's Law investigation. Copy and complete the table. Remember to provide the units in the table for the product PV and the quantity P^{-1}.

(a) Plot graphs of (i) P against V (ii) PV against P (iii) V against P^{-1}

(b) Sketch on your graphs what you would expect to obtain if the experiment had been carried out at a higher temperature.

Pressure, P in MPa	0.5	1.0	1.5	2.0	2.5	3.0
Volume, V in cm^3	80.0	40.0	26.7	20.0	16.0	13.3
$P \times V$ in						
P^{-1} in						

4 An experiment to measure the variation of the volume of a fixed amount of air with pressure is conducted using a sample of air trapped in a vertical glass tube by oil of low vapour pressure. Why is low vapour pressure necessary?

Explain why it is necessary to keep the temperature of the air constant throughout the experiment. Describe the experimental techniques you would consider using to achieve constant temperature.

[CCEA Module 4 May 2003]

Volume and Temperature Changes leading to Charles's Law

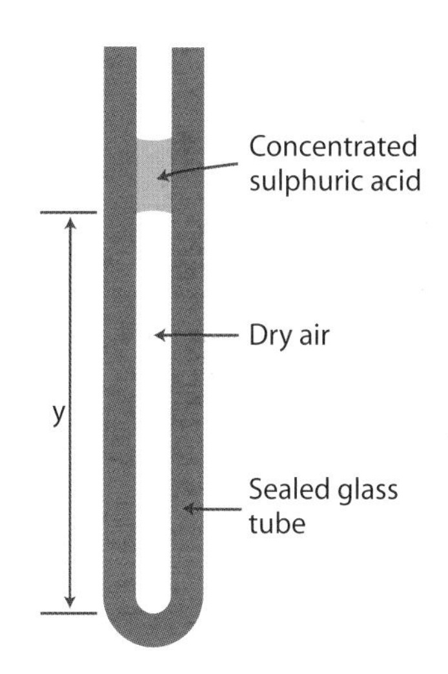

The traditional experiment involves investigating how the volume of a fixed mass of air at constant pressure varies as the temperature changes.

In the usual apparatus, the air is held inside a glass capillary tube by a short length of concentrated sulphuric acid (the concentrated acid traps and **dries** the air to give better results). The length y of the trapped air is a measure of the volume of the air **because the area of cross section of the capillary tube is constant.**

The glass tube is attached to a ruler with rubber bands. The position of the tube is adjusted until the bottom of the trapped air is opposite the zero mark of the ruler.

The experiment is carried out at constant pressure – the pressure on the gas is that due to the atmosphere and the thread of concentrated sulphuric acid.

The apparatus is then placed in a tall beaker of cold water with a thermometer. Throughout the experiment the water is stirred regularly so that the trapped air is the same temperature as the water. The "volume" of the trapped air and the temperature are then recorded in a results table. The water is then heated until it is about 10°C hotter and another pair of readings of volume and temperature recorded.

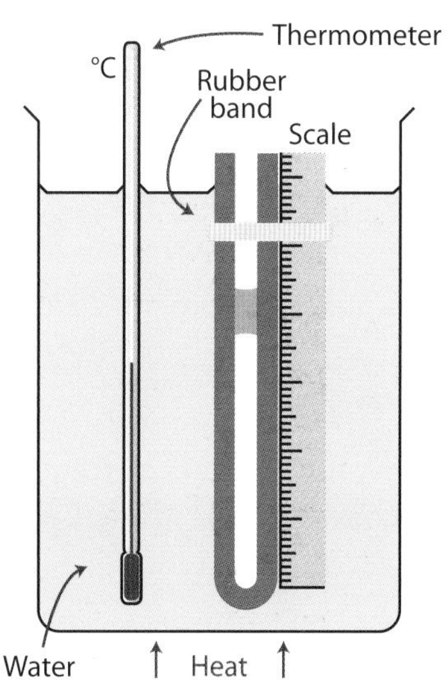

This process is repeated, increasing the temperature of the water until it boils. A graph is then plotted of volume (vertical axis) against temperature in degrees Celsius (horizontal axis).

Below is a typical set of results.

Length of trapped air thread in mm	65	67	69	72	74	76	78	81	83
Temperature in °C	20	30	40	50	60	70	80	90	100
Temperature in K	293	303	313	323	333	343	353	363	373

The reader is encouraged to use the above data to plot (i) a graph of length of air thread in mm (proportional to volume) against temperature in °C and (ii) a graph of length of air thread in mm against temperature in kelvin.

Although air is used in school laboratory experiments, other gases give similar results. Below is a sketch of a graph, typical of the type obtained in this experiment.

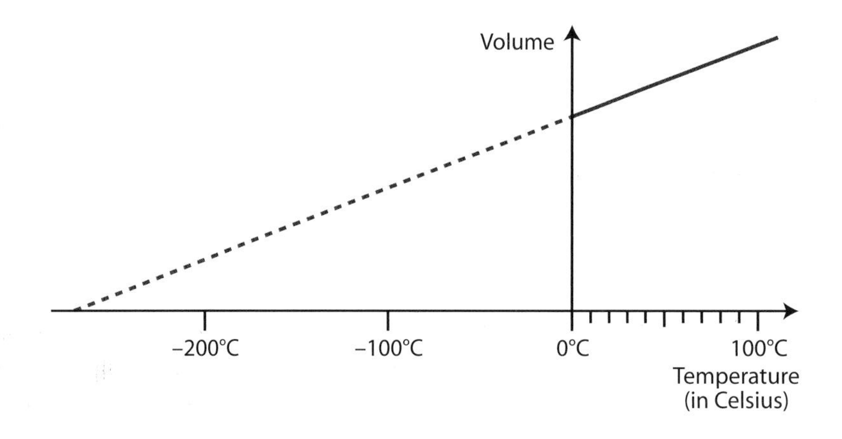

The graph of volume against the **Celsius** temperature is a straight line of positive slope. This shows that **the volume of a fixed mass of gas at constant pressure increases uniformly with temperature.** However as the graph does not pass through the origin, it does not illustrate proportionality.

However, the graph extended backwards touches the horizontal axis at –273°C. Suppose now we define a new scale of temperature, the Kelvin scale, T, by the equation:

$T = \theta + 273$ where T is the temperature in Kelvin and

θ is the temperature in degrees Celsius

In science, we often measure temperature on the **Kelvin** scale. This begins at zero Kelvin (0 K) and increases just like the Celsius scale. This means that **0 °C becomes 273 Kelvin** (273 K) and **100 °C becomes 373 K** and so on. You were first introduced to this in Module 1 of the AS course.

Now, the graph is a straight line through a (0,0) origin when temperature is measured on the Kelvin scale. The dotted vertical line shows the previous axis when temperature was measured on the Celsius scale.

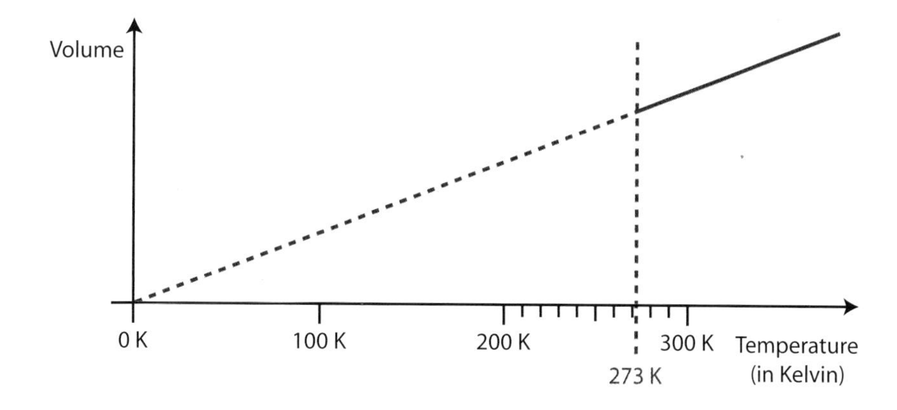

Now that the graph is a straight line **passing through the origin,** it certainly does demonstrate proportionality. It tells us that

For a **fixed mass of gas at constant pressure** the volume (V) is directly proportional to the **Kelvin** (or absolute) temperature (T).

This is called **Charles' Law.**

Charles' Law can be expressed as an equation:

$$V = \text{constant} \times T \quad \text{or} \quad \frac{V}{T} = \text{constant}$$

where V is the gas volume and T is the temperature in Kelvin

The graph suggests that 0 K is the temperature at which the volume **would become zero.** But in fact this is not so. Every gas would liquefy before reaching this temperature. However the graph also suggests that **there is a limit to how cold objects can be.** This temperature (0 K) is called the **absolute zero** of temperature. Absolute zero is the temperature at which all molecular motion stops and is approximately –273.16°C, although for purposes of calculations, it is sufficient to use –273°C. The temperature has never been reached because the methods of measuring such temperatures change the temperature of the system. However, physicists are confident that they have come within about one millionth of a degree of absolute zero.

Exercise 3

1 Why can we be confident that the volume of the gas is not directly proportional to the Celsius temperature?

2 A fixed mass of gas at 27°C has a volume of 12.0 litres. Find its volume when its temperature rises to 127 °C at constant pressure.

3 The volume of a fixed mass of gas rises from 25 litres to 40 litres when it is heated at constant pressure. The initial temperature is 77 °C. Find the final temperature in °C.

4 A fixed mass of gas has a volume of 12.0 litres and has a temperature of 100°C. To what temperature must the gas be cooled to reduce its volume to 1.2 litres?

Pressure and Temperature Changes Leading to the Pressure Law

The traditional experiment involves investigating how the pressure of a fixed mass of air at constant volume varies as the temperature changes.

The apparatus shown in the diagram is used to investigate how the pressure of air changes as the temperature rises. The pressure is measured using a Bourdon pressure gauge.

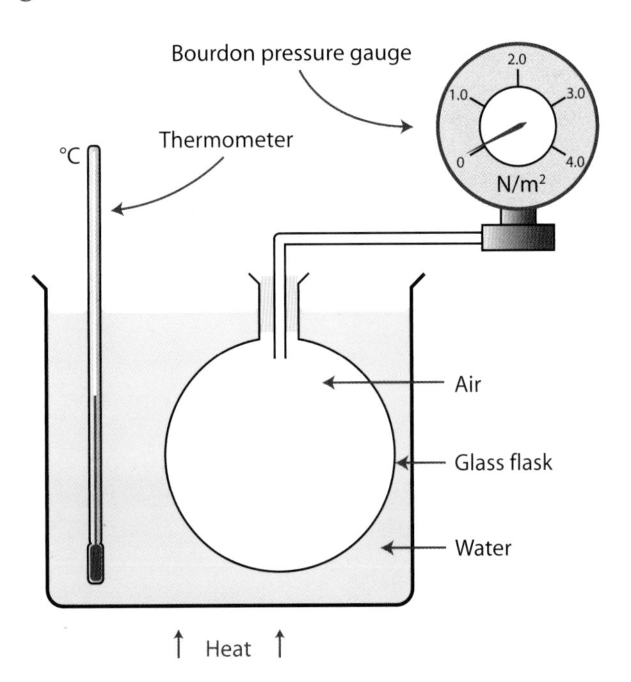

A large glass flask is placed in a tall beaker of cold water with a thermometer. Throughout the experiment the water is stirred regularly so that the trapped air is the same temperature as the water. The pressure of the trapped air and the temperature are then recorded in a results table. The water is then heated until it is about 10°C hotter and another pair of readings of pressure and temperature is recorded. This process is repeated, increasing the temperature of the water until it boils. A graph is then plotted of pressure (vertical axis) against temperature in degrees Celsius (horizontal axis).

Pressure × 10^6 (N/m²)	Temperature (°C)	Absolute temperature (K)

Below are some typical results.

Pressure in MPa	2.00	2.07	2.14	2.20	2.27	2.34	2.41	2.48	2.55
Temperature in degrees Celsius	20	30	40	50	60	70	80	90	100
Temperature in Kelvin	293	303	313	323	333	343	353	363	373

The reader is encouraged to use the above data to plot (i) a graph of pressure against temperature in °C and (ii) a graph of pressure against temperature in Kelvin.

Below is a sketch of the graph of pressure against temperature in Kelvin. Since it is a straight line through the (0,0) origin, it shows direct proportion.

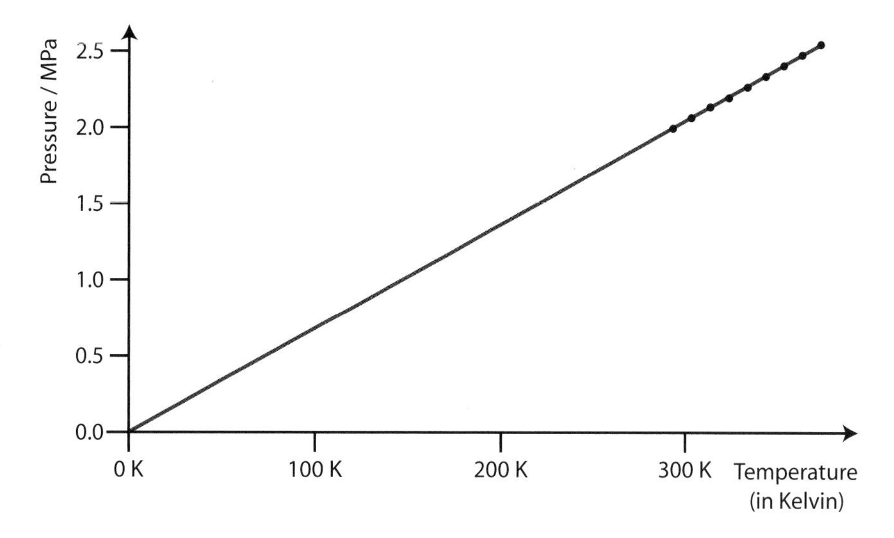

Since the graph of pressure against Kelvin temperature is a straight line passing through the origin, it tells us that:

For a fixed mass of gas, at constant volume, the pressure (p) is directly proportional to the Kelvin temperature (T).

This is known as the **Pressure Law.**

In symbols:

$$p = \text{constant} \times T \text{ (in Kelvin)} \quad \text{or} \quad \frac{p}{T} = \text{constant}$$

where p = the pressure of the gas
T = the Kelvin temperature

This apparatus can be used as a thermometer, called a **constant volume gas thermometer.** The dial of the pressure gauge can be marked out in °C.

The Ideal Gas Equation

From these three experiments, we have the equations:

- $pV = \text{constant}$ (Boyle's Law, for a fixed mass of gas at constant temperature)
- $\frac{V}{T} = \text{constant}$ (Charles' Law, for a fixed mass of gas at constant pressure)
- $\frac{P}{T} = \text{constant}$ (Pressure Law, for a fixed mass of gas at constant volume)

These three equations can be combined into one, called the **ideal gas equation:**

$$\frac{pV}{T} = \text{a constant}$$

If a fixed mass of gas has values p_1, V_1 and T_1, and then some time later has values p_2, V_2 and T_2, then the equation becomes:

$$\frac{p_1V_1}{T_1} = \frac{p_2V_2}{T_2}$$

Exercise 4

1 A deep-sea diver is working at a depth where the pressure is 3.0 atmospheres. He is breathing out air bubbles. The volume of each bubble is 2 cm³. At the surface the pressure is 1.0 atmosphere.

What is the volume of each bubble when it reaches the surface?

2. A cycle pump contains 70 cm³ of air at a pressure of 1.0 atmosphere and a temperature of 7°C. The air is compressed to 30 cm³ at a temperature of 27°C.

Calculate the final pressure.

$T_1 = 7°C$
$V_1 = 70\ cm^3$
$p_1 = 1.0\ atm$
Exit closed
$T_2 = 27°C$
$V_2 = 30\ cm^3$
$p_2 = ?$

Exercise 5

Examination Questions

1 The pressure in the tube of a bicycle tyre is 500 kPa above atmospheric pressure of 100 kPa. The initial volume of air at atmospheric pressure trapped in the barrel of the pump by the piston is 2.00×10^{-4} m³. The temperature of the air is 17°C. When the piston compresses the air the volume is reduced to 4.00×10^{-5} m³ and at that instant the one-way valve connecting the pump to the tube opens, allowing air to enter the tube. Calculate the temperature of the air in the pump at this instant in °C.

[CCEA Module 4 June 2005]

2 A syringe contains 40 cm^3 of neon gas at a temperature of 15°C and a pressure of 55 kPa. The temperature is increased to 350 °C. The gas is allowed to expand and the pressure rises to atmospheric (101 kPa) Calculate the new volume of gas in the syringe.

[CCEA Module 4 Jan 2004]

3 A student gives the following incomplete statement of one of the laws for an ideal gas: "The volume of an ideal gas is directly proportional to its temperature in Kelvin."

(i) Identify two important omissions from the correct and complete version of this statement.

(ii) Describe an experiment to investigate the law referred to above. Include a labelled diagram and indicate how you would process the results.

[CCEA Module 4 Jan 2004]

4 A student gives the following incomplete statement of one of the gas laws:

"The pressure of an ideal gas is directly proportional to its temperature."

(i) Identify three omissions from the complete and correct statement of the law.

(ii) On the axes sketch a graph showing the variation of pressure p of an ideal gas with temperature θ over the temperature range from 0°C to 100°C.

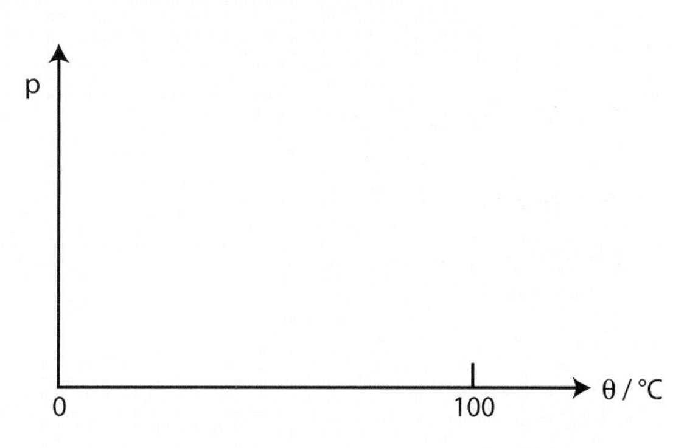

(iii) In an experiment carried out with hydrogen, which behaves approximately as an ideal gas, the measured pressure of the gas at 50°C is 1.3×10^5 Pa. A graph of pressure against Celsius temperature is drawn. Calculate the gradient of this graph and state the unit of the gradient.

[CCEA Module 4 May 2003]

The Mole

The experimental gas laws show that for a fixed mass of gas:

$$pV = \text{a constant} \times T$$

It turns out that the constant depends only on the number of molecules in the sample of gas being considered. But the number of molecules present in a sample of gas is enormous – and the best way to express this idea is in terms of the number of **moles** of gas.

You will recall that **amount of substance** is measured in moles and that the mole (abbreviated to **mol**) is one of the six SI base units introduced in your AS course. But what exactly is a mole? In 1960 physicists and chemists agreed to assign the value 12, **exactly**, to the relative atomic mass of the isotope of carbon with mass number 12 (carbon 12, ^{12}C) The unit of "amount of substance" is fixed in terms of the number of atoms in exactly 0.012 kg of carbon 12. This number of atoms is called Avogadro's Number (N_A).

The mole is the amount of substance which contains as many particles as there are atoms in 0.012 kilogram of carbon 12.

So, a mole of gas molecules is simply Avogadro's number of those molecules; a mole of electrons is Avogadro's number of electrons and so on. **Avogadro's number is the number of particles per mole.** Its numerical value is 6.02×10^{23} mol^{-1}.

Worked Examples

Example 1

A sample of hydrogen gas, H_2, has a mass of 12 g (0.012 kg). If the relative **atomic** mass of hydrogen is 1, calculate (i) the number of moles in the sample of gas and (ii) the number of hydrogen molecules in the sample.

Solution

(i) Hydrogen has 2 atoms in its molecule. If the relative atomic mass of hydrogen is 1, then

1 mole of hydrogen molecules has a mass of 2 g.

So 12 g of hydrogen contains 12÷2 = 6 moles.

(ii) 6 moles of hydrogen molecules contains $6\times N_A$ molecules

or $6 \times 6.02\times10^{23} = 3.612\times10^{24}$ molecules

Example 2

A nitrogen sample contains 2.107×10^{24} molecules and has a mass of 0.098 kg. Find

(i) the number of moles of nitrogen and

(ii) the relative molecular mass of nitrogen.

Solution

(i) $$\text{Number of moles} = \frac{\text{Number of particles}}{\text{Avogadro's number}} = \frac{2.107\times10^{24}}{6.02\times10^{23}} = 3.5 \text{ mol}$$

(ii) Relative molecular mass = mass of 1 mole in grams

$$= \frac{\text{mass of gas sample in grams}}{\text{number of moles}}$$

$$= \frac{(0.098\times1000)}{3.5} = \frac{98}{3.5} = 28 \text{ grams mol}^{-1}$$

Kinetic Theory and Ideal Gases

The kinetic theory attempts to explain the macroscopic behaviour of a gas by examining its microscopic properties i.e. behaviour of the molecules. In particular, the collisions of the molecules with the walls of the container produce an outward force or pressure.

To apply the kinetic theory we have to make some assumptions. These assumptions define the characteristics of what physicists call **an ideal gas.**

Ideal Gas Assumptions

- There are no intermolecular forces - the only time the molecules exert a force on each other is when they collide.
- The molecules themselves have a volume which is negligible compared to the volume of the gas.
- The collisions between molecules and between molecules and the walls of the container are elastic, so both kinetic energy and momentum are conserved.
- The duration of a collision is negligible compared with the time between collisions.
- Between collisions the molecules move with constant velocity.

Kinetic Theory and the Behaviour of Gases

1. How does the kinetic theory explain the pressure exerted by a gas on the walls of the container?

   Molecules collide elastically with the sides of the container.

   Each collision results in a momentum change for the molecules.

   Velocity of molecule before collision with wall = +v

   Momentum of molecule before collision with wall = +mv

   Velocity of molecule after collision with wall = –v

   Momentum of molecule after collision with wall = –mv

   Change in momentum as a result of collision = –mv – mv = –2mv

   A momentum change (by Newton's 2nd law) implies a force was exerted on the molecules by the wall and of course by the molecule on the wall (Newton's 3rd law).

   The total force on the wall is the sum of the forces exerted by all the colliding molecules.

   The pressure on the wall is ratio of this total force to the area of the wall.

2. For a fixed mass of gas at constant temperature Boyle's Law tells us that as the volume of a gas is reduced the pressure increases. How does the kinetic theory explain this?

   A reduced volume means that the molecules have a shorter distance to travel to the walls of the container. The momentum change per collision is the same but the shorter distance means that more collisions occur per second so the pressure increases.

3. **For a fixed mass of gas at constant volume the Pressure Law tells us that as the temperature is increased the pressure increases. How does the kinetic theory explain this?**

   An increase in temperature means that the kinetic energy, the momentum and velocity of the molecules increase.

   The momentum change per collision increases and so also does the number of collision per second; both of these contribute to an increase in pressure.

4. **For a fixed mass of gas at constant pressure Charles' Law tells us that as the temperature is increased the volume increases. How does the kinetic theory explain this?**

   An increase in temperature means that the kinetic energy, the momentum and velocity of the molecules increase.

   The momentum change per collision increases. To maintain the same pressure the number of collisions per second must decrease. Expansion ensures that the molecules have a greater distance to travel before they collide with the container. A greater distance means a greater time and so the number of collisions per second is reduced. This increase in one factor coupled with a decrease in the other ensures that the pressure remains constant.

The Ideal Gas Equation

We can now write down the equation for **an ideal gas.**

$$pV = nRT$$

where p is the gas pressure in Pa
V is the gas volume in m^3
n is the number of moles of gas in mol
T is the temperature of gas in Kelvin

R is a constant, known as the **universal gas constant,** having a value of 8.31 J mol^{-1} K^{-1}

R is a **universal** constant because **it applies to all gases** provided their behaviour is ideal.

This equation is not supplied in the CCEA formula sheet and must be remembered.

Exercise 6

Hydrogen gas consists of diatomic molecules. The mass of a hydrogen molecule is 3.34×10^{-27} kg.

(a) Calculate the mass of one mole of hydrogen gas.

(b) At a certain temperature and pressure the density of this hydrogen gas sample is 8.99×10^{-2} kg m^{-3}. Calculate the volume of the gas under these conditions.

(c) If the pressure of the gas is 101.5 kPa, calculate the gas temperature in degrees Celsius, giving your answer to the nearest degree.

[CCEA June 2002 (amended)]

The Boltzmann Constant

The Boltzmann constant, k, is defined by the equation

$$k = \frac{R}{N_A}$$

where R is the universal gas constant and N_A is Avogadro's number. The Boltzmann constant is therefore 8.31 J mol^{-1} K^{-1} / 6.02×10^{23} mol^{-1} and has a value of 1.38×10^{-23}J K^{-1}.

Now consider a gas containing N molecules. The number of moles in this gas, n is given by

$$n = \frac{N}{N_A}$$

Combining the two equations above gives

$$nR = \left(\frac{N}{N_A}\right) \cdot (k\,N_A) = Nk$$

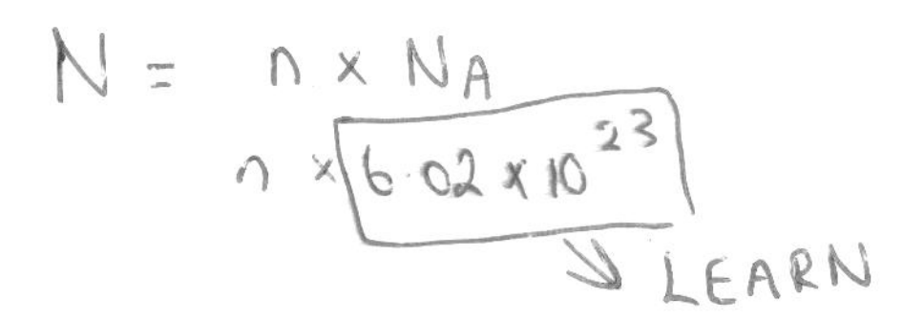

Hence, the ideal gas equation, pV = n R T, becomes

$$pV = NkT$$

This equation is important because it links the number of particles in the gas, N, with its macroscopic properties of pressure volume and temperature. It also tells us that equal volumes of all gases under the same conditions of temperature and pressure contain the same number of molecules, N. This is sometimes called **Avogadro's Law.** The equation is not supplied in the CCEA formula sheet and must be remembered.

The equation is important for another reason. It is the first where the product pV has been directly linked with the number of molecules in the gas. Following on from the gas laws, we are beginning to obtain information about the **microscopic** properties of gas molecules.

Linking the Equation of State to the Molecular Speeds of Gases

Heating a gas generally causes its molecules to move faster, collide more frequently and with greater force on the walls of the container and so increase the gas pressure. There is therefore some statistical relationship between the pressure, volume, temperature and kinetic energy of the molecules. Before stating that relationship, we should first define a few statistical terms.

Mean speed $\langle c \rangle$

If the speeds of the N molecules in a sample of gas are c_1, c_2, c_3, ..., c_N, then the mean speed, $\langle c \rangle$, is defined by:

$$\langle c \rangle = \frac{c_1 + c_2 + c_3 + \ldots c_N}{N}$$

Mean square speed $\langle c^2 \rangle$:

$$\langle c^2 \rangle = \frac{c_1^{\,2} + c_2^{\,2} + c_3^{\,2} + \ldots c_N^{\,2}}{N} = \frac{c_i^{\,2}}{N}$$

Root mean square speed c_{rms} :

$$c_{rms} = \frac{\sqrt{\langle c^2 \rangle}}{N}$$

Worked Example

Five particles have speeds (in ms^{-1}) of 2, 4, 5, 5 and 7. Find

(a) mean speed $\langle c \rangle$

(b) mean square speed $\langle c^2 \rangle$

(c) root mean square speed c_{rms}

Solution

(a) Mean speed $\langle c \rangle = \dfrac{2+4+5+5+7}{5} = 4.60\ ms^{-1}$

(b) Mean square speed $\langle c^2 \rangle = \dfrac{2^2+4^2+5^2+5^2+7^2}{5} = \dfrac{119}{5} = 23.8\ m^2s^{-2}$ (note the unit!)

(c) Root mean square speed $c_{rms} = \sqrt{\dfrac{2^2+4^2+5^2+5^2+7^2}{5}} = 4.88\ ms^{-1}$

It can be shown that the mathematical link between the speeds of the molecules and the gas pressure, p, is:

$$pV = \tfrac{1}{3}Nm\langle c^2 \rangle$$

where N is the number of molecules present in the gas
m is the mass of a single molecule
and the other symbols have their usual meaning

There is no need for A2 students to derive this equation although some older texts show where it comes from in great detail. CCEA candidates need only know how to **use** it – the equation itself is reproduced on the CCEA formula sheet.

Gas Density

The density of the gas, ρ, is defined by the equation:

$$\rho = \frac{\text{Mass of the gas, M}}{\text{Volume of the gas, V}}$$

and the mass of gas is the product of the total number of molecules present, N, and the mass of each molecule, m. It follows that:

$$\rho = \frac{M}{V} = \frac{Nm}{V}$$

and hence

$$p = \tfrac{1}{3}\rho\left\langle c^2\right\rangle$$

Exercise 7

1 A tyre contains a gas at a pressure of 150 kPa. If the gas has a density of 2.0 kg m^{-3}, find the root mean square speed of the molecules.

2 A cylinder contains gas at a temperature of 300 K and at atmospheric pressure. More gas at the same temperature is pumped into the cylinder until the pressure rises to 200 kPa above atmospheric pressure. If the volume of the cylinder is 0.015 m^3 calculate, in moles, the amount of extra gas pumped in. Take R = 8.31 J mol^{-1} K^{-1}

3 Helium is monatomic with a relative atomic mass of 4.0. Calculate

(i) the density of helium gas at a temperature of 0°C and a pressure of 101 kPa and

(ii) the rms speed of the helium atoms under these conditions.

4 The root mean square speed of five molecules (in ms^{-1}) is 306 ms^{-1}. Four of the molecules have speeds (in ms^{-1}) of 301, 301, 305 and 310. Find the speed of the fifth molecule.

Molecular Speeds and Temperature

Since $pV = \tfrac{1}{3}Nm\left\langle c^2\right\rangle$, and $pV = NkT$

It follows therefore that:

$$\tfrac{1}{3}Nm\left\langle c^2\right\rangle = NkT$$

Multiplying both sides by $\frac{3}{2N}$ gives:

$$\tfrac{1}{2}m\left\langle c^2\right\rangle = \frac{3kT}{2}$$

Thus,

$$\tfrac{1}{2}m\langle c^2\rangle = \tfrac{3}{2}kT$$

The equation above is important – it links the average kinetic energy of a collection of gas molecules with the Kelvin temperature. Note in particular that the **average kinetic energy of the molecules is directly proportional to the Kelvin temperature.**

The equation applies only to ideal gases, where the only energy possessed by the atoms is the kinetic energy of translational motion.

Worked Examples

Example 1

Calculate (i) the mass of a hydrogen molecule and (ii) the rms speed of hydrogen molecules at 300 K, given that the relative molecular mass of hydrogen is 2.

Solution

(i) mass of hydrogen molecule $= \dfrac{2 \times 10^{-3}}{6.02 \times 10^{23}}$ kg $= 3.332 \times 10^{-27}$ kg

(ii) $c_{rms} = \sqrt{\dfrac{3KT}{m}} = \sqrt{\dfrac{3 \times 1.38 \times 10^{-23} \times 300}{3.332 \times 10^{-27}}} = 1931\ ms^{-1}$

Example 2

A gas contains a mixture of oxygen and hydrogen molecules. Oxygen has a relative molecular mass of 32 and hydrogen has a relative molecular mass of 2. The rms speed of the oxygen molecules in the mixture is 220 ms^{-1}.

Calculate the rms speed of the hydrogen molecules.

Solution

$\tfrac{1}{2}m_{H_2}\langle c^2_{H_2}\rangle = \tfrac{3}{2}kT$ and $\tfrac{1}{2}m_{O_2}\langle c^2_{O_2}\rangle = \tfrac{3}{2}kT$

Since the temperature is the same for both, then:

$$\tfrac{1}{2}m_{H_2}\langle c^2_{H_2}\rangle = \tfrac{1}{2}m_{O_2}\langle c^2_{O_2}\rangle$$

Cancelling the ½ factors and rearranging gives:

$\langle c^2_{H_2}\rangle = \dfrac{m_{O_2}}{m_{H_2}}\langle c^2_{O_2}\rangle$ and using the relative molecular masses gives:

$$\langle c^2_{H_2}\rangle = \frac{32}{2}\langle c^2_{O_2}\rangle = 16\langle c^2_{O_2}\rangle$$

and taking square roots of both sides gives:

$$c^{rms}_{H_2} = 4c^{rms}_{O_2} = 4 \times 220 = 880\ ms^{-1}$$

Answer: 880 ms^{-1}

The Internal Energy of a Gas

The internal energy of a **real** gas is the sum of the potential and kinetic energy of its molecules. All gas molecules in motion possess **translational kinetic energy**. In a monatomic gas such as helium, this is the **only** energy possessed by the atoms. However, molecules consisting of two or more atoms may also possess **rotational** kinetic energy and **vibrational** kinetic energy. Polyatomic molecules may also possess potential energy if the inter-atomic **bonds are compressed or stretched.** In general, the internal energy of a **real** gas is randomly distributed as **kinetic and potential energy.**

However, **ideal** gases are **monatomic** and for ideal gases it is assumed that there are **no forces of attraction between the atoms.** So, **ideal gases possess no potential energy**. The internal energy of the molecules of an ideal gas is therefore **entirely kinetic.**

Exercise 8

Examination Questions

1 (i) Calculate the root mean square speed of nitrogen molecules at 20°C. The mass of a nitrogen molecule is 4.6×10^{-26} kg.

(ii) Sketch a graph to show how the mean square speed $\langle c^2 \rangle$ (not the rms speed) of nitrogen molecules depends on the Celsius temperature, θ.

[CCEA Module 4 June 2004]

2 A molecule of nitrogen uses all its kinetic energy to reach a maximum height of 13.4 km above the surface of the Earth. At ground level the gravitational potential energy of the molecule is taken as zero. By considering the conservation of energy, calculate the temperature of the gas molecule at ground level to enable it to do this. The mass of the nitrogen molecule is 4.65×10^{-26} kg. Assume that the acceleration of free fall is constant over the height involved.

[CCEA Module 4 June 2007]

3. The kinetic theory of gases may be used to derive the equation: $pV = \frac{1}{3}Nm\langle c^2 \rangle$

(i) State in words what the term $\langle c^2 \rangle$ represents.

(ii) In order to obtain a numerical value for the term $\langle c^2 \rangle$ for a given gas it is necessary that the number of gas molecules per unit volume should be very large. Why is this?

(iii) In the kinetic theory it is supposed that the gas molecules make elastic collisions with the walls of the container. Imagine that the collisions were **inelastic**, not elastic. Describe and explain the consequence of this with regard to the temperature of the gas.

(iv) There are no attractive forces between the molecules of an ideal gas. Describe and explain the consequence of this with regard to the two components of internal energy.

(v) Why is the zero of the Kelvin scale sometimes called **absolute** zero?

(vi) If you could observe the molecules of an ideal gas, how would you know if the gas was at absolute zero?

[CCEA Module 4 June 2002 (amended)]

4 (i) Describe an experiment to investigate how the volume of a gas varies with temperature in degrees Celsius while its pressure remains constant. The experiment should commence at about room temperature (20°C). Structure your answer using the following headings:

1. Labelled sketch of the apparatus
2. Experimental procedure
3. Suitable graph to display the readings

(ii) Explain how the result obtained leads to the concept of absolute zero of temperature.

[CCEA Module 4 June 2007]

5 (a) Describe how the pressure that gas molecules exert on the walls of a container changes when the temperature of the gas increases. Explain your answer.

(b) A cylinder has a fixed volume of 1.36×10^{-3} m^3 and contains a gas at a pressure of 1.04×10^5 Pa when the temperature is 15°C.

(i) Calculate the number of gas molecules in the container.

(ii) Calculate the new pressure of the gas when the temperature is increased to 25°C.

(iii) Calculate the increase in kinetic energy of all the gas molecules in the container caused when the temperature is increased to 25°C.

[CCEA Module 4 Specimen Paper published 2007]

Specific Heat Capacity

When heat is supplied to an object the temperature of the object generally rises. The rise in temperature $\Delta\theta$, in K, of the object depends on the following:

Q the quantity of heat energy supplied, in J
m the mass of the object, in kg
the material of which the object is made

These ideas are brought together in the formula below:

$$Q = mc\Delta\theta$$

Definition:

The specific heat capacity, c, of a material is the quantity of heat energy needed to raise the temperature of 1 kg of the material by 1 K.

The units of specific heat capacity are J kg^{-1} K^{-1} (or J kg^{-1} °C^{-1}).

The range of specific heat capacities is very large.

The specific heat capacity of some common substances is given in the table below.

Substance	c / J kg^{-1} K^{-1}
Hydrogen	14 300
Water	4 200
Ethanol	2 400
Ice	2 100
Stainless Steel	510
Copper	385
Mercury	140

The high specific heat capacity of hydrogen makes it suitable as a coolant in large electrical turbines, in spite of the dangers associated with this explosive gas.

In general the specific heat capacity of gases is higher than that of liquids, and the specific heat capacity of liquids is higher than that of solids.

Exercise 9

1 Express the units of specific heat capacity in terms of SI base units.

Electrical Methods of Measuring Specific Heat Capacity

These methods involve supplying a measured quantity of heat to an object of known mass and measuring the temperature increase produced.

Metal Solid

A metal cylinder has two holes in it, one to hold an electrical heater and other to hold a thermometer. A small amount of oil or glycerine in the hole containing the thermometer is used to **improve the thermal contact** between the thermometer and the metal.

The mass of the metal cylinder is measured using a balance and the **initial temperature** of the metal is measured with a thermometer.

The amount of energy can be found using a voltmeter, ammeter and stop clock.

The **power** of the heater
= current × voltage = **I × V**

The stop clock is used to determine the time the heater is switched on.

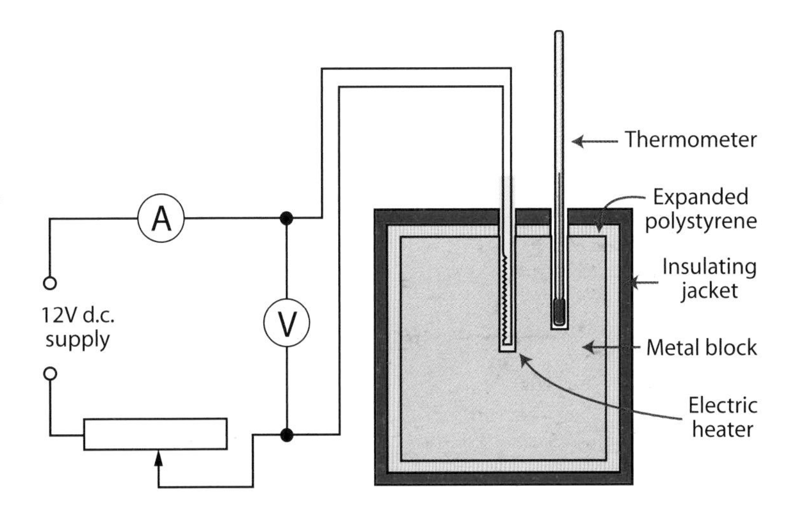

Energy supplied = I × V × t (time in seconds)

The experimentalist must **go on reading the temperature of the metal block for several minutes after the heater is switched off.** This is because the temperature of the heater is higher than that of the block and it takes some time for equilibrium between them to be established. The final temperature is taken as the **highest temperature** reached by the block **after the heater is switched off.**

To reduce heat loss and improve accuracy the metal block is wrapped in an insulator, such as expanded polystyrene, **to reduce heat loss by conduction.** The expanded polystyrene is then covered with **shiny aluminium foil to reduce heat loss by radiation,** and an **outer insulating jacket reduces heat loss by convection.**

Typical results are given below.

Mass of metal cylinder, m, in kg:	0.9
Voltage across heater in V:	12.0
Current in heater in A:	5.0
Power of heater in W:	60
Time heater switched on in s:	300
Heat supplied, Q, in J:	180 000
Initial temperature in °C:	19.5
Final (highest) temperature in °C:	59.5
Temperature rise in °C:	40.0

$$\text{Specific capacity, } c = \frac{Q}{m\Delta\theta} = \frac{18000}{(0.9 \times 40)} = 500 \text{ J kg}^{-1}\,°\text{C}^{-1}$$

Liquids

The method used here is much the same as that employed for solids. However the liquid has to be placed in a container and this will also heat up as energy is supplied to the liquid. To minimise this, a container of small heat capacity should be chosen. **The container,** called a calorimeter, **should also be insulated as with the experiment with the metal block.** In addition, in this experiment **an insulating lid is added** to reduce heat loss due to **convection and evaporation.**

The heat supplied, Q, is used to raise the temperature of the liquid **and the container.**

The relevant equation is:

$$Q = m_L \times c_L \times \Delta\theta + m_c \times c_c \times \Delta\theta$$

where m_L is the mass of liquid
m_c is the mass of the calorimeter
$\Delta\theta$ is is temperature rise in liquid
c_L is the specific heat capacity of the liquid
c_c is the specific heat capacity of the calorimeter

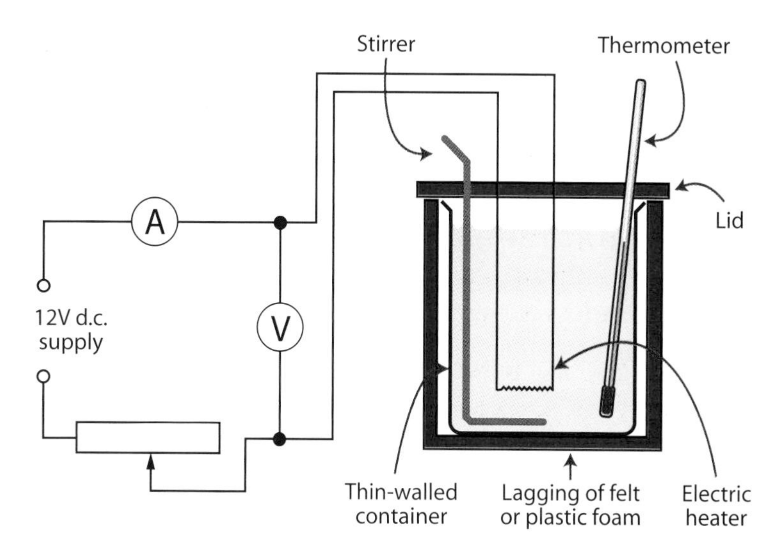

Provided the liquid is stirred well during the experiment the temperature of the liquid and that of the calorimeter will be the same. The temperature **rise** in the liquid, $\Delta\theta$, will therefore be the same as the temperature rise in the calorimeter.

As with the solid the energy supplied can be determined using the ammeter, voltmeter and stop clock method.

It is common in this procedure for the experimental value for specific heat capacity to be larger than the generally accepted value. This is because heat is **always** lost to the environment, resulting in a lower rise in temperature than would otherwise be expected. Since in the calculation of specific heat capacity the rise in temperature is part of the denominator, the calculated value for "c" is larger than the true value.

One way to reduce the error in this experiment is to cool the liquid in a refrigerator to a temperature of around 5°C. When poured into the calorimeter the liquid temperature quickly rises until both calorimeter and liquid are in thermal equilibrium. Heating begins when the liquid and calorimeter are both about 5°C **below** room temperature and continues until the liquid temperature is about 5°C **above** room temperature. During the time when the calorimeter and contents are below room temperature the liquid and the container both absorb heat from the environment. When the calorimeter and contents are above room temperature the liquid and the container both lose heat to the environment. By doing this it is hoped that the heat lost to the environment cancels the heat gained from the environment, and results in a value for the specific heat capacity closer to that which is generally accepted.

Typical results for water are given below. Note the starting and finishing temperatures are respectively below and above room temperature.

Mass of water, m, in kg:	0.41
Voltage across heater in V:	12.0
Current in heater in A:	5.0
Power of heater in W:	60
Time heater switched on in s:	300
Heat supplied, Q, in J:	18 000
Initial temperature in °C:	13.5
Final (highest) temperature in °C:	23.5
Temperature rise $\Delta\theta$ in °C:	10.0
Mass of calorimeter, m_c, in kg:	0.160
SHC of calorimeter, c_c, in J kg^{-1} °C^{-1}:	385

Heat gained by calorimeter $= m_c \times c_c \times \Delta\theta$

$= 0.160 \times 385 \times 10.0 = 616$ J

Heat supplied to water $= 18000 - 616 = 17384$ J

$$\text{Specific heat capacity of water, c} = \frac{\text{Heat supplied to water}}{m\Delta\theta}$$

$$= \frac{17384}{0.41 \times 10} = 4240 \text{ J kg}^{-1} \text{ °C}^{-1}$$

Exercise 10

1 A block of metal having mass 4 kg and temperature 25 °C is heated to a temperature of 80°C. How much heat energy is required to do this? Assume no heat is lost to the surroundings and that the specific heat capacity of the metal is 385 J kg^{-1} K^{-1}.

2 A 12 V electric heater draws 2.0 A for 12 minutes to raise the temperature of a 3.0 kg block of metal. If the initial temperature is 20°C, what would the final temperature be? Assume no heat is lost to the surroundings.

(Specific heat capacity of the metal = 500J kg^{-1} K^{-1}.)

3 A car of mass 1400 kg is travelling at 30 ms^{-1}. The driver applies the brakes which have a total mass of 104 kg. Calculate the increase in the temperature of the brakes.

What assumption have you made? (Specific heat capacity of brake material = 600 J kg^{-1} K^{-1}.)

4 The diesel engine of a ferry using water as a coolant. Every minute 35 kg of water enters the engine at 285 K and leaves again at 375 K. How much heat energy is transferred to the water every minute? The specific heat of water is 4200 J kg^{-1} K^{-1}.

Exercise 11

Examination Questions

1 The mass of a spring is 0.24 g. The spring is made from a metal with a specific heat capacity of 450 J kg^{-1} K^{-1}.

(i) Define specific heat capacity.

(ii) Each time the spring in a ball-point pen is compressed it stores 4 mJ of strain energy. When the spring is released 10% of the strain energy stored in it is converted to heat energy in the spring.

Calculate the number of times the spring must be compressed and released to cause the temperature of the spring to increase by 1 K. Assume all the heat remains in the spring.

[CCEA January 2007 (Amended)]

2 (a) An electric drill is used to make a hole in a piece of copper. During this operation, the drill uses a current of 2.50 A when it is connected to a 230 V supply. The mass of the copper is 0.650 kg. 55% of the electrical energy is converted to heat in the piece of

copper. Calculate the temperature rise in the copper after 15.0 s.
(Specific heat capacity of copper = 380 J kg^{-1} K^{-1}.)

(b) Assume that the figure of 55% for the conversion of electrical energy to heat energy quoted in (a) is correct. State whether the temperature rise in your answer to (a) is greater or less than what occurs in practice. Explain your answer.

[CCEA January 2006]

3 The apparatus below was used to determine the specific heat capacity of milk.

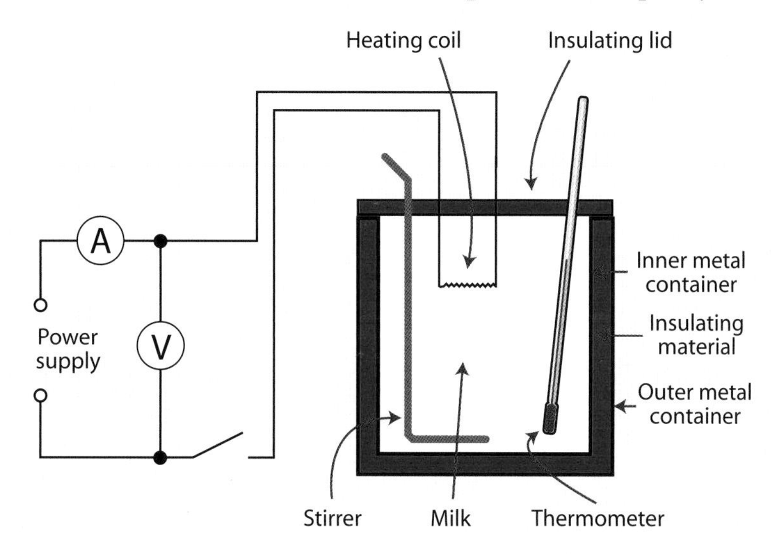

The results below were obtained over a 5.00 minute period for a sample of milk of mass 0.126 kg. This sample of milk completely filled the inner container.

The experimenter closed the switch and simultaneously started the stopwatch, recording the starting temperature, the potential difference and the current.

After 5.00 minutes he opened the switch and recorded the finishing temperature, but did nothing else during this period.

Potential difference	12.4V
Current	3.74A
Starting temperature	19.0 °C
Finishing temperature	43.0 °C

(a) Use these results to calculate the specific heat capacity of the milk. Remember to include the appropriate unit.

(b) Look carefully at the apparatus shown above and describe one experimental technique to improve the accuracy of the result that should be employed during this experiment. Explain why this technique is necessary.

(c) Assume the appropriate technique has been used during this experiment. State the major source of inaccuracy that still exists in this experiment and explain how it may be allowed for in the calculation.

[CCEA January 2008]

4.3 Uniform Circular Motion

Students should be able to:

4.3.1 Demonstrate an understanding of the concept of angular velocity;

4.3.2 Recall and use the equation $v = r\omega$;

4.3.3 Apply the relationship $F = ma = \frac{mv^2}{r}$ to motion in a circle at constant speed;

CD players, satellites, spin-dryers, the hammer thrower in the photograph and fairground rides all use circular motion. What makes an object move in a circle?

Credit: foxypar4 (via Flickr)

Newton's first law tells us that an object continues to move in a **straight** line unless a resultant force acts on it. So to make something move in a circle **we need a force.**

The hammer thrower in the photograph opposite makes the hammer move in a circle using the tension in the wire. In which direction does this force act?

Your own experiences of circular motion may lead you to the wrong answer here. Imagine yourself on the 'chairoplane' ride in the photograph. What force do you feel as you swing round in a circle?

It **feels** as if you are being pushed outwards. People often talk, **wrongly,** about an outwards or "centrifugal" force. Centrifugal forces **do not exist.**

The chair exerts an **inward** force on you. By Newton's Third Law, you exert an equal **outward** force on the chair.

Credit: iStockPhoto

The Basics

One complete circle is equivalent to 360° or 2π radians, as shown in the diagram on the right.

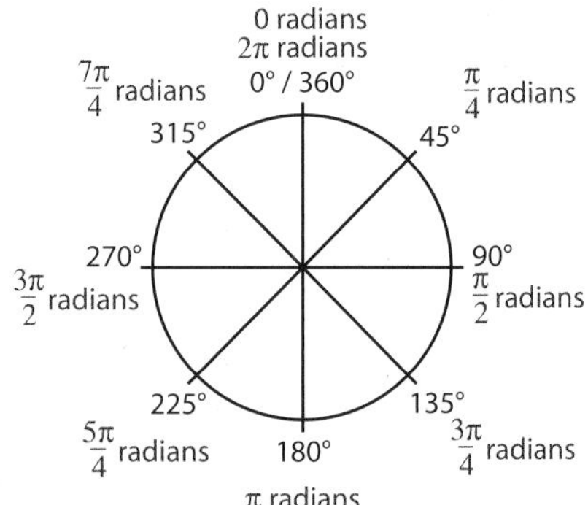

The rate at which a washing machine or a car engine turns can be measured in revolutions per minute (rpm).

A typical spin speed of a washing machine is 1 000 rpm or 2000π radians per second. The maximum rotational speed of a car engine is typically 6500 rpm or 13 000π radians per second. In Physics we use **angular velocity** to measure how quickly something rotates. Angular velocity ω, is the rate of change of angular displacement (the angle turned through in one second).

Radian : the angle subtended at the centre of a circle by an arc of the circle whose length is equal to the radius.

Angular velocity can be measured in degrees per second but the preferred unit is **radians per second.**

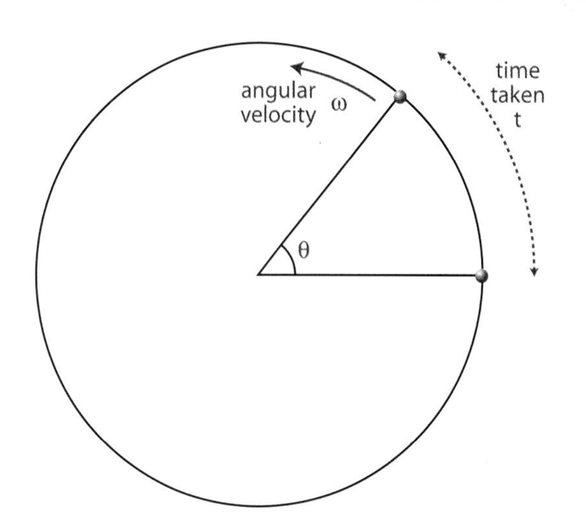

$$\text{Angular velocity} = \frac{\text{Angle turned through}}{\text{Time taken}}$$

$$\omega = \frac{\theta}{t}$$

degrees → radians $\times \frac{2\pi}{360}$

radians → degrees $\times \frac{360}{2\pi}$

Worked Examples

Example 1 What is the angular velocity of the seconds hand of a clock?

Solution In 60 seconds the hand sweeps through 2π radians, so:

$$\omega = \frac{\theta}{t} = \frac{2\pi}{60} = \frac{\pi}{30} \text{ radians per second}$$

Example 2 What is the angular velocity of the Earth as it rotates about its axis?

Solution The Earth spins once on its axis every 24 hours or 86400 seconds

$$\omega = \frac{\theta}{t} = \frac{2\pi}{86400} = 7.27 \times 10^{-5} \text{ radians per second}$$

The Important Equations

In radian measure, the angle is defined as the ratio of the arc length to the radius of the circle, i.e.:

$$\theta = \frac{S}{r}$$

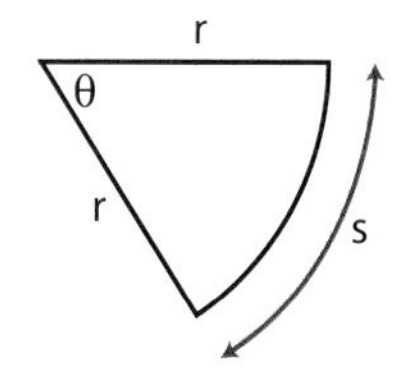

If the arc length s is the same as the radius r then the angle between the two radii is equal to 1 radian.

Suppose a particle moves from P to P' at a constant speed v along a circular arc of length s in a time t. Since the direction in which the particle is moving is continually changing, the particle's velocity must be continually changing also. The particle must therefore be accelerating.

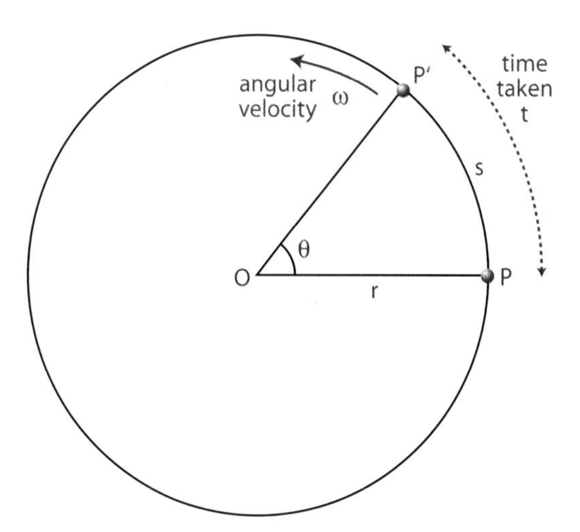

$$\text{Angle} = \frac{\text{Arc length}}{\text{Radius}}$$

$$\theta = \frac{s}{r}$$

The angle θ swept out by the radius vector is called the **angular displacement.**
By the definition of the radian:

(i) $s = r\theta$ where θ is measured in radians and s and r are measured in metres

The **angular velocity** of the particle, ω, is defined as the rate of change of angular displacement. Since the particle is moving at a steady speed, v, the angular velocity, ω, is constant and we can write:

(ii) $\omega = \dfrac{\theta}{t}$ where ω is the angular velocity measured in radians per second

An object moving in a circle with a constant angular velocity has a constant linear or tangential speed. However its linear velocity is not constant because its **direction is changing** from one moment to the next.

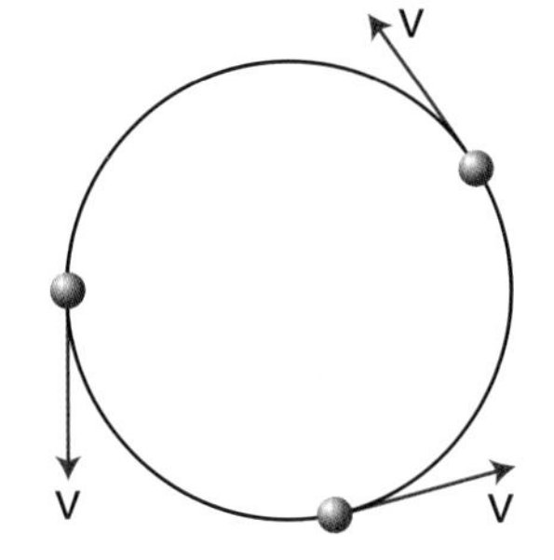

Since speed is the rate of change of distance with time, then combining equations (i) and (ii) above enables us to write:

$$v = \frac{s}{t} = \frac{r\theta}{t} = r\omega$$ where v is the speed in ms^{-1}

The **periodic time** of the motion, T, is defined as the time taken for the particle to travel once round the circle. Since the radius vector sweeps out 360° or 2π radians as the particle moves once round the circle, then:

$$T = \frac{s}{v} = \frac{2\pi r}{r\omega} = \frac{2\pi}{\omega}$$ where T is the periodic time (or simply, the period) in seconds

The **frequency**, f, of the motion is the number of revolutions made per second. Since the particle takes T seconds to make one revolution, we can write:

$$f = \frac{1}{T} = \frac{\omega}{2\pi}$$ where f is the frequency measured in Hertz (Hz)

Worked Examples

Example 1

The seconds hand of a clock is 15.0 cm long. Write down its period and frequency. Calculate the speed of its tip. [Take the tip to be the point at maximum distance from the centre.]

Solution

Period, T = time to make 1 revolution = 60.0 s

$$\text{Frequency, } f = \frac{1}{T} = \frac{1}{60}\text{ Hz} = 0.167\text{ Hz}$$

$$\text{Angular velocity, } \omega = 2\pi f = \frac{2\pi}{60} = 0.105\text{ rad s}^{-1}$$

$$\text{Speed of tip, } v = r\omega = 0.15 \times 0.105 = 0.0157\text{ ms}^{-1}$$

Example 2

An astronaut in training is rotated at the end of a horizontal rotating arm of length 5 m. The arm makes 42 revolutions per minute. Calculate the frequency, period and angular velocity of the rotating arm and speed of the astronaut.

Solution

$$\text{Frequency, } f = \text{Number of revolutions per second} = \frac{42}{60} = 0.70\text{ Hz}$$

$$\text{Period, } T = \frac{1}{f} = \frac{1}{0.7} = 1.43\text{ s}$$

$$\text{Angular velocity, } \omega = 2\pi f = 2\pi \times 0.7 = 1.4\pi = 4.4\text{ rad s}^{-1}$$

$$\text{Speed, } v = r\omega = 5 \times 4.4 = 22\text{ ms}^{-1}$$

Example 3

An electron in a hydrogen atom moves in a circular path of radius 50 pm with a constant speed of 2.2 Mms^{-1}. Calculate its angular velocity and its frequency.

Solution

$$\text{Angular velocity, } \omega = \frac{v}{r} = \frac{2.2 \times 10^{6}}{50 \times 10^{-12}} = 4.4 \times 10^{16}\text{ rad s}^{-1}$$

$$\text{Frequency, } f = \frac{\omega}{2\pi} = 7.0 \times 10^{15}\text{ Hz}$$

Example 4

Planet Venus orbits the Sun in a roughly circular orbit of mean radius of 108 million kilometres every 224 days. Calculate the angular velocity and the speed of Venus.

Solution

$$\omega = \frac{2\pi}{T} = \frac{2\pi}{(224 \times 24 \times 3600)} = 3.24 \times 10^{-7} \text{ rad s}^{-1}$$

$$v = \omega r = 3.24 \times 10^{-7} \times 1.08 \times 10^{11} = 3.51 \times 10^{4} \text{ ms}^{-1}$$

Example 5

The United States Air Force uses a centrifuge with a 5 m radius to train pilots. This centrifuge has period of 3.0 s. Calculate (i) the pilot's angular velocity and (ii) the pilot's linear or tangential speed.

Solution

$$\text{Angular velocity, } \omega = \frac{2\pi}{T} = \frac{2\pi}{3} = 2.094 \text{ rad s}^{-1}$$

$$\text{Speed, } v = r\omega = 5 \times 2.094 = 10.47 \text{ ms}^{-1}$$

Example 6

Calculate the tangential velocity of a point on the Earth's equator. The radius of the Earth is 6400 km and it takes the Earth 1 day to complete one revolution.

Solution

$$v = \frac{2\pi r}{T} = \frac{(2\pi \times 6.4 \times 10^{6})}{(24 \times 60 \times 60)} = 465.4 \text{ ms}^{-1}$$

Centripetal Acceleration

Any particle moving in a circular path at a constant speed must be accelerating because the direction of its motion, and hence its velocity, is constantly changing. The acceleration must have no component parallel to the direction of the velocity as otherwise the particle would get faster or slower. It follows therefore that the acceleration is perpendicular to the velocity at any instant. Thus **the acceleration is always directed towards the centre of the circle.** The word **centripetal** is used to describe such acceleration. Centripetal means "towards the centre".

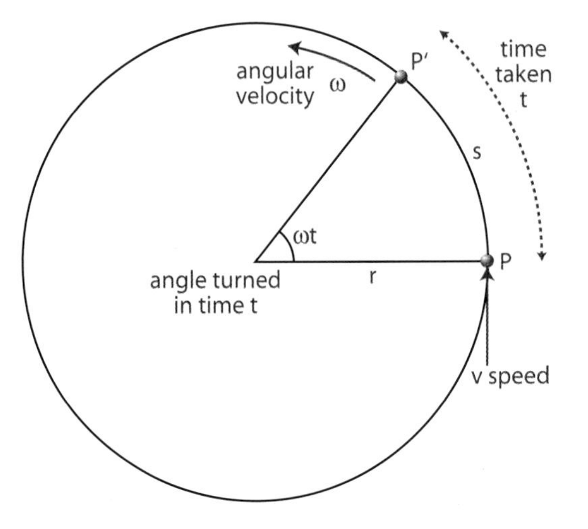

The magnitude of the centripetal acceleration is denoted by the symbol **a** and is given by the equations:

$$a = v\omega = \frac{v^2}{r} = r\omega^2$$

where ω is the angular velocity in rad s^{-1}
v is the tangential speed in ms^{-1}
r is the orbital radius in m

Since the orbiting particle is being accelerated there must also be an accelerating force in accordance with Newton's Second Law. This force, **F**, is given by the equation:

$$F = ma = mv\omega = \frac{mv^2}{r} = mr\omega^2$$

where **m** is the mass of the orbiting particle
and the other symbols have their usual meanings

Since the force and the acceleration are always in the same direction, **the force acts towards the centre of the circle and is known as the centripetal force.**

The CCEA specification does not require candidates to derive these equations for centripetal acceleration and centripetal force. However, recall of the equations and an ability to use them are almost invariably examined.

Worked Examples

Example 1

A disc spins at 45 revolutions per second. A speck of dust on the disc has a mass of 1 μg and is 5 cm from the centre. Calculate the centripetal force on the speck of dust.

Solution

$$\omega = 2\pi f = 2\pi \times 45 = 90\pi \text{ rad s}^{-1}$$

$$F = ma = mr\omega^2 = 1 \times 10^{-6} \times 0.05 \times (90\pi)^2 = 0.004 \text{ N}$$

Example 2

At what angular velocity must a centrifuge spin so that a particle placed 20 cm from the centre experiences a centripetal acceleration equal to the acceleration due to gravity at the Earth's surface?

Solution

$$\omega = \sqrt{\frac{a}{r}} = \sqrt{\frac{9.81}{0.2}} = \sqrt{49.05} = 7 \text{ rad s}^{-1}$$

Causes of the Centripetal Force

The centripetal force is caused by some physical phenomenon. It is important to understand that the **circular motion does not produce the force.** Rather, **the force is needed for circular motion to take place.** Without this force the object would travel in a straight line along the tangent to the curve. There is no outward centrifugal force on the particle. Suppose an object was moving in a circular path at the end of a string. If the string breaks the centripetal force disappears and the object flies off along the tangent to the circle.

The table below identifies the cause of the centripetal force in four different situations.

Physical Situation	Cause of the Centripetal Force
A planet orbiting the Sun	**Gravitational force** between the Sun and the planet
Electrons orbiting the nucleus of an atom	**Electrical force** between the positively charged nucleus and the negatively charged electron
A "conker" being whirled in a circle at the end of a string	The **tension** in the string
A racing car going round a circular track	The **friction force** between the tyres and the track

Worked Examples

Example 1

Calculate the centripetal force on the Earth as it orbits the Sun. You may assume the mass of the Earth is 6×10^{24} kg and that its mean orbital radius is 1.5×10^{11} m. Take 1 year as 3.2×10^{7} seconds.

Solution

$$\omega = \frac{2\pi}{T} = \frac{2\pi}{3.2\times10^{7}} = 1.964\times10^{-7}\ \text{rad s}^{-1}$$

$$F = mr\omega^2 = 6\times10^{24} \times 1.5\times10^{11} \times \left(1.964\times10^{-7}\right)^2 = 3.47\times10^{22}\ \text{N}$$

Example 2

At what minimum speed must a motorcyclist ride over a hump-back bridge of radius 12 m if he just loses contact with the road? Take g as 9.81 ms^{-2}.

Solution

If contact is just lost, reaction = 0 and

centripetal force = weight of motorcycle and machine, so:

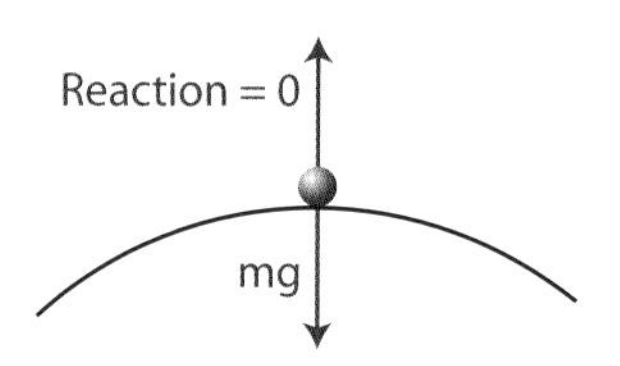

$$\frac{mv^2}{r} = mg \qquad \text{Cancelling m and rearranging gives:}$$

$$v^2 = rg = 12\times9.81 = 117.72$$

$$v = 10.8\ \text{ms}^{-1}$$

3 The breaking force in a length of string is 5 N. What is the maximum number of revolutions per minute which can be made with a conker of mass 60 g at one end of an 80 cm length of this string? Ignore gravitational effects.

Solution

$$F = mr\omega^2 = mr(2\pi f)^2 = 4\pi^2 mrf^2$$

$$f^2 = \frac{F}{(4\pi^2 mr)} = \frac{5}{(4\pi^2 \times 0.06 \times 0.8)} = 2.639$$

Number of revolutions per second, $f = 1.624$

Number of revolutions per minute $= 60f = 97.5$

Motion in a Vertical Circle

The diagram below represents an object of mass m being whirled clockwise at a constant speed v in a vertical circle at the end of a piece of string of length L. When we take the force of gravity into account, the resultant force on the object is **not** constant.

At point A, the tension T_1 in the string is given by:

$$T_1 + mg = \frac{mv^2}{L}, \quad \text{so} \quad T_1 = \frac{mv^2}{L} - mg$$

At point C, the tension T_2 in the string given by:

$$T_2 - mg = \frac{mv^2}{L}, \quad \text{so} \quad T_2 = \frac{mv^2}{L} + mg$$

At points B and D, the tension in the string provides the centripetal force $\frac{mv^2}{L}$. The weight, acting at right angles to the string, does not contribute to the centripetal force.

As the object moves from A to B to C, the tension increases sinusoidally, reaching a maximum at C. As it moves from C to D to A the tension decreases sinusoidally, reaching a minimum at A.

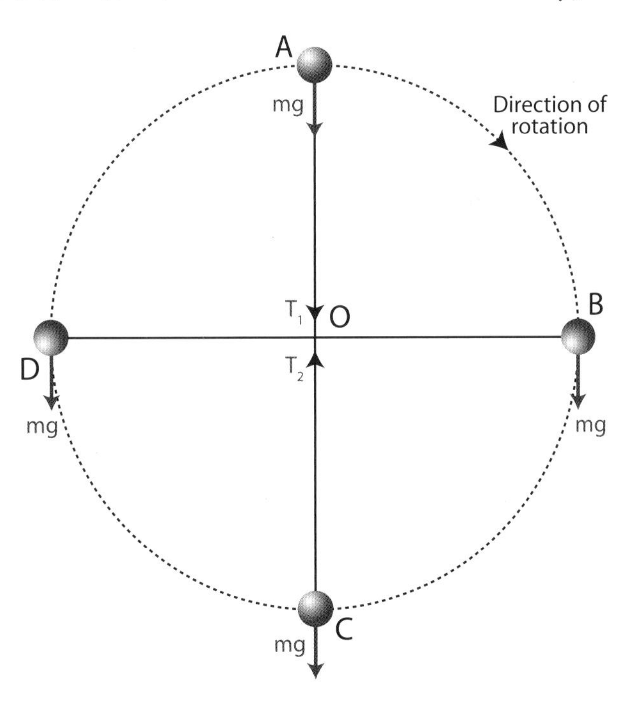

Circling at an Angle

The diagram shows a conical pendulum.

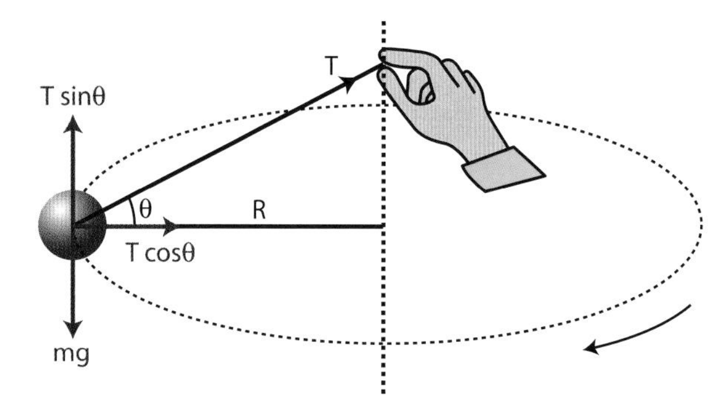

The tension in the string is T and θ is the angle between the string and the horizontal. Then the vertical component tension in the string (T sin θ), balances the weight of the orbiting mass. The horizontal component of the tension provides the centripetal force. So,

$$T\cos\theta = \frac{mv^2}{R} \quad \textbf{and} \quad T\sin\theta = mg \quad \text{where R is the radius of the circle}$$

Dividing one equation by the other gives:

$$\frac{T\sin\theta}{T\cos\theta} = \tan\theta = \frac{mg}{(mv^2/R)} = \frac{gR}{v^2}$$

Exercise 12

Examination Questions

1 (a) A small spherical mass attached to one end of a light string is rotated with constant angular velocity in a horizontal plane as shown below. Explain briefly why it is impossible for the string to be horizontal as the mass rotates in the horizontal plane.

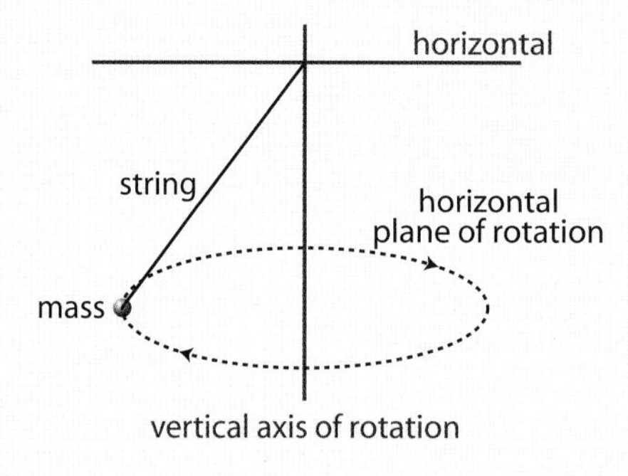

(b) The plane of rotation of the mass is now changed from a horizontal plane to a vertical plane as shown opposite. The mass is again rotated at a constant angular velocity. The string is always taut.

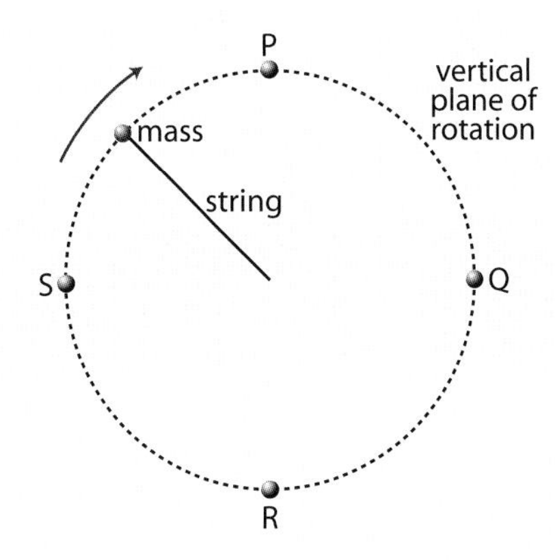

Point P is the highest point in the vertical plane of revolution and Q, R and S are 90 degree intervals around one revolution. Copy the axes below and sketch a graph to show how the tension in the string varies as the spherical mass makes one revolution from P past the points Q, R and S.

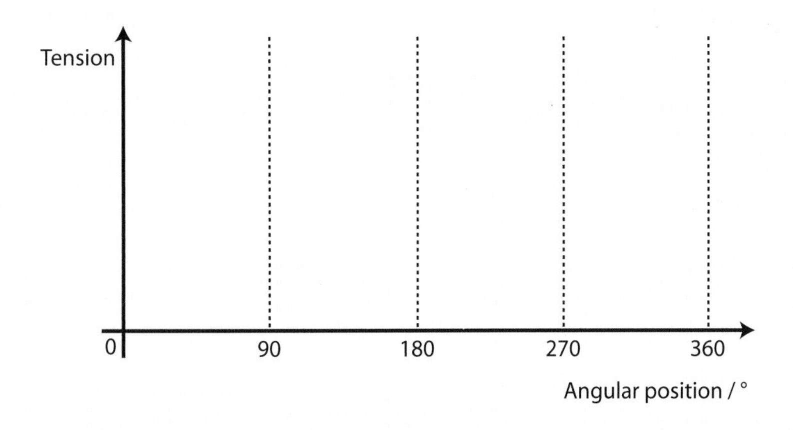

(c) A mass of 135 g, attached to one end of a light string, is rotated in a vertical circular orbit of radius 320 mm at a constant angular velocity of 8.50 rad s^{-1}. Calculate the minimum tension in the string.

[CCEA Module 4 2007]

2 A person is swinging a ball at the end of a string so that it moves with a uniform angular velocity in a horizontal circle.

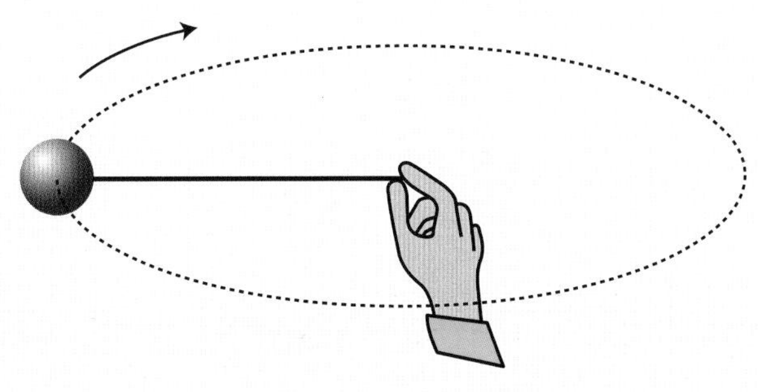

(a) The diagram shows a plan view of the ball moving in its circular path.

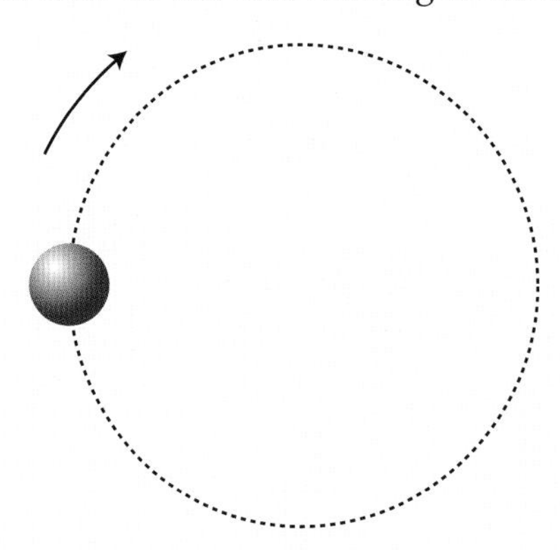

(i) On the diagram mark the path the path the ball would follow if the string were to break when the ball is at the position shown.

(ii) The force acting on the ball as it moves in its circular path with uniform speed is said to be centripetal (towards the centre of the circle). Explain why it **must** be in this direction.

(b) The ball has mass 0.15 kg and moves in a circle of radius 0.60 m. It makes 2.0 revolutions per second.

(i) Assume the ball rotates with the string in the horizontal plane. Calculate the tension T in the string.

(ii) In fact, the weight of the ball makes it impossible for the string to be horizontal. The real situation is sketched below.

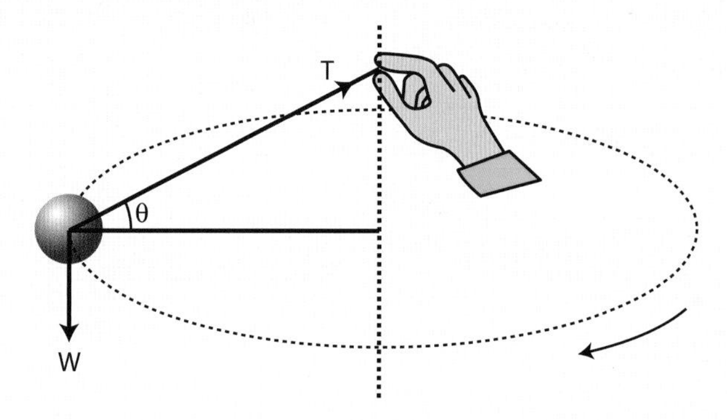

Assume the horizontal component of the tension has the value calculated in (b)(i). Determine the angle θ.

[CCEA Module 4 2006]

3 (a) (i) Define angular velocity.

(ii) Describe the function of a centripetal force.

(iii) State the direction of the centripetal force.

(b) A mass of 2.50 kg rotates in a horizontal circle of radius 1.20 m, as shown in the diagram below.

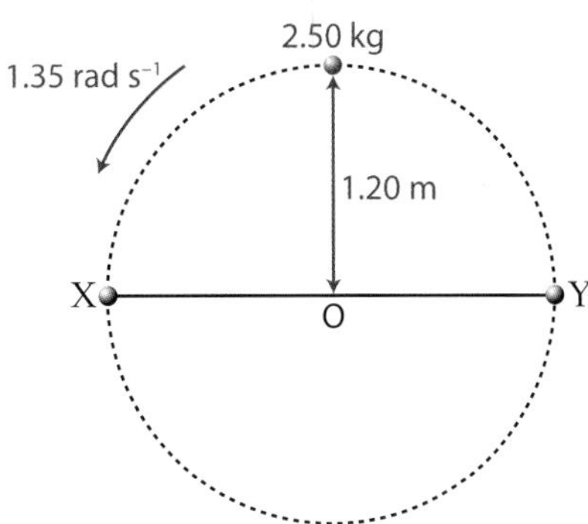

The mass has a uniform angular velocity of 1.35 rad s^{-1}. Points X and Y are at opposite ends of a diameter of the circular path.

(i) Calculate the magnitude of the centripetal force acting on the mass.

(ii) Calculate the magnitude of the change in velocity of the mass as it moves from X to Y.

[CCEA Module 4 January 2006]

4 A pot on a potter's wheel completes 8.0 revolutions in 4.5 seconds.

(a) (i) Calculate the angular velocity of the wheel.

(ii) Calculate the difference between the linear velocity of a point on the wheel 5 cm from the centre and one at 15 cm from the centre.

(b) (i) A lump of clay is placed on the potter's wheel at a distance r from the centre of the wheel. On the grid below, sketch a graph showing how the centripetal acceleration of the clay changes as r increases from zero.

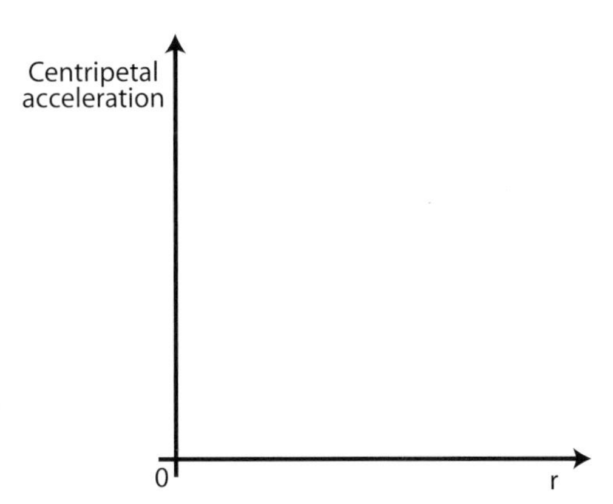

(ii) At what value of r is the centripetal acceleration 19 ms^{-2}?

(c) A lump of clay of mass 0.20 kg is placed 10 cm from the centre of the wheel and the speed of the wheel is increased gradually from rest. The maximum frictional force between the clay and the wheel is 4.2 N. Calculate the angular velocity of the wheel just as the clay starts to slip.

5 A DVD rotating inside a player **must** maintain a constant linear velocity of 3.84 ms^{-1} in order for the information on it to be read correctly. The laser in the DVD player starts at the inside of the disc and moves outwards.

(a) (i) State and explain what must happen to the angular velocity of the disc as the DVD is played.

(ii) Calculate the radius at which the DVD will have an angular frequency of 14.6 revolutions per second when the linear velocity is 3.84 ms^{-1}.

The linear velocity of a CD is 1.3 ms^{-1} compared to the value of 3.84 ms^{-1} for a DVD.

(b) Explain whether dust particles are more likely to remain on a CD or a DVD when the discs are in motion. Assume that the maximum frictional force between a dust particle and DVD is the same as between a dust particle and CD.

[CCEA Module 4 Specimen Paper published 2007]

4.4 Simple Harmonic Motion

Students should be able to:

4.4.1 Define simple harmonic motion using the equation $a = -\omega^2 x$ where $\omega = 2\pi f$;

4.4.2 Perform calculations using $x = A \cos \omega t$;

4.4.3 Demonstrate an understanding of s.h.m. graphs to include measuring velocity from the gradient of a displacement time graph;

4.4.4 Know and be able to use the terms free vibrations, forced vibrations, resonance and damping in this context;

4.4.5 Understand the concepts of light damping, overdamping and critical damping;

4.4.6 Describe mechanical examples of resonance and damping;

Any object that is initially displaced slightly from a stable equilibrium point will oscillate about its equilibrium position. It will, in general, experience a restoring force that is proportional to its displacement from the equilibrium point.

A simple pendulum is a system that behaves in this way. It consists of a mass, called the bob, attached to a length of string. When hanging vertically the resultant force on the bob is zero, the tension in the string is equal and opposite to the weight of the bob. When the pendulum bob is moved a little to one side and released it oscillates backwards and forwards. The resultant force always acts towards the equilibrium position so the bob will vibrate around this point.

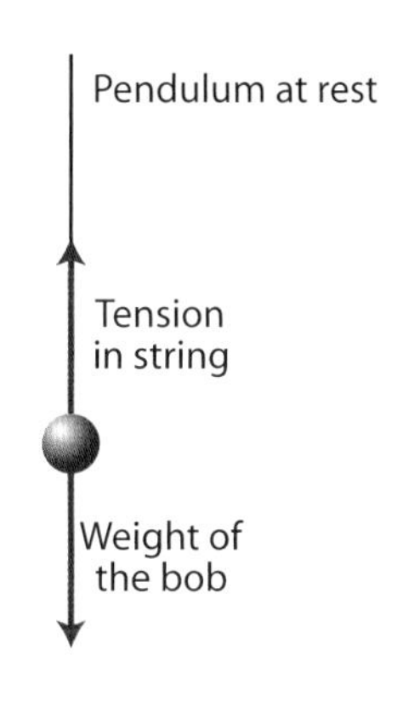

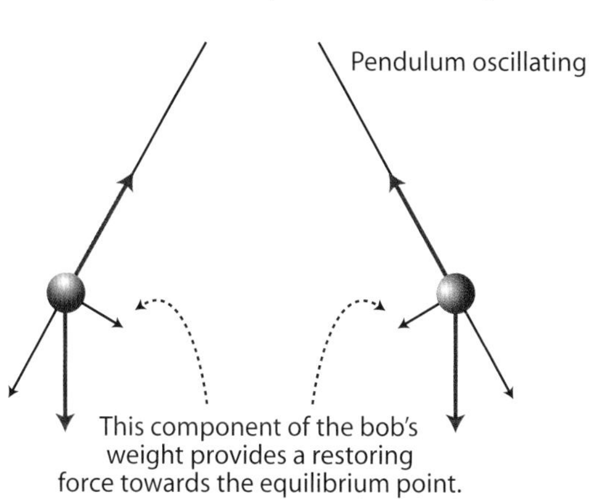

Defining Simple Harmonic Motion (SHM)

A particle moves with simple harmonic motion if

1 its acceleration is proportional to its displacement from a fixed point and

2 the direction of the acceleration is always towards that fixed point.

The defining equation for simple harmonic motion is:

$$a = -\omega^2 x$$

where a is the acceleration in ms^{-2}
x is the displacement in m
ω^2 is a constant in s^{-2}

The minus indicates that the acceleration and displacement are in opposite directions.

Newton's 2nd law tells that a resultant force **F** is needed to produce this acceleration.

The graphs below illustrate these relationships between acceleration, force and displacement.

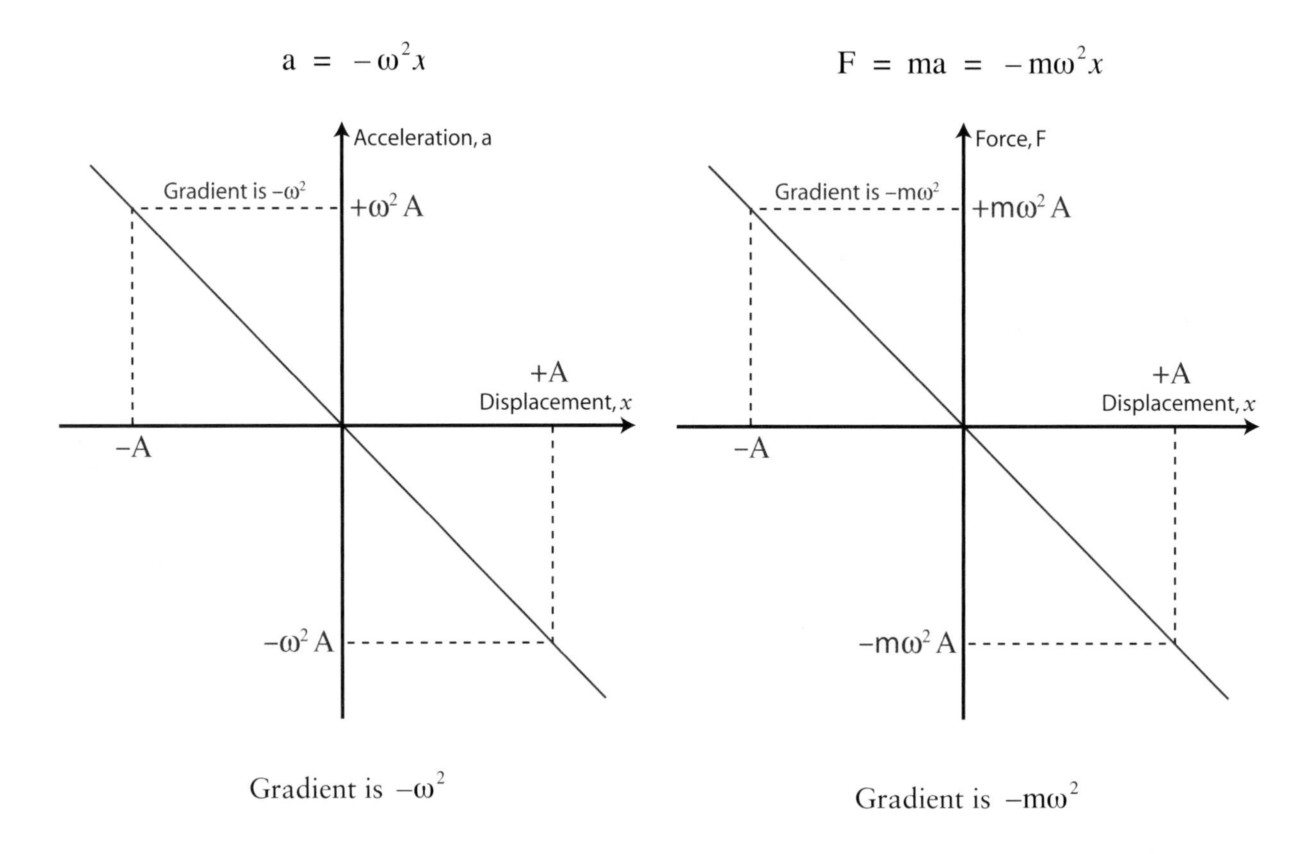

Gradient is $-\omega^2$

Gradient is $-m\omega^2$

The amplitude or maximum displacement of the motion is denoted by the letter A.

In simple harmonic motion the object's displacement from the equilibrium position, velocity and acceleration **all** vary with time. To derive the appropriate equations that describe each of these physical quantities it is necessary to return to circular motion.

Consider a point P moving in circle of radius A with angular velocity ω.

The projection of P onto the diameter XY is the point R. As P moves in a circle starting at Y then to X and finally back to Y, the point R moves along the diameter from Y to X and back to Y.

The point R moves along the diameter with simple harmonic motion.

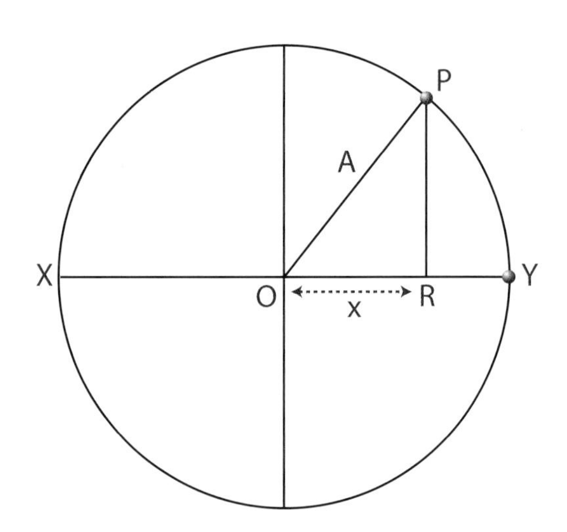

The periodic time T is the time it takes R to complete one oscillation. This is of course the same as it takes P to complete one circle. As with all vibrations, the frequency, f, is the number of vibrations made per second.

$$f = \frac{1}{T} = \frac{2\pi}{\omega}$$

where T is measured in s
ω is measured in rad s^{-1}
f is measured in Hz

The maximum displacement of the point R from the centre of oscillation O is called the amplitude. In this case it equals the radius of the circle OY or OP and is denoted by A. The displacement x of R from O varies with time.

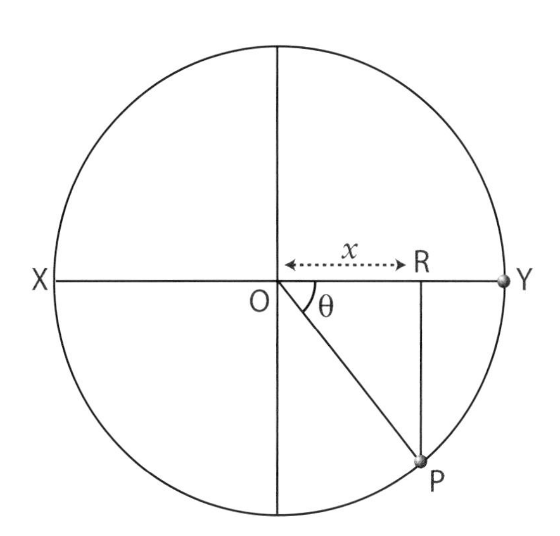

Suppose at time t = 0 R is at Y, i.e. at extremity of the oscillation.

Suppose further at time t, P has moved through an angle $\theta = \omega t$. At this time the displacement of R from O is x.

Then: $\cos\theta = \cos\omega t = \frac{x}{OP}$

But, since $OP = OA =$ amplitude, A, then: $\cos\theta = \frac{x}{A}$

Rearranging gives: $x = A\cos\omega t$

This is the displacement equation for SHM.

Displacement, velocity and acceleration during Simple Harmonic Motion

In the diagram below the object R moves along the line YX with simple harmonic motion. The centre of oscillation is the point O. At time t = 0 the object is at Y and moving to towards O. The graphs overleaf show how the displacement x, (represented by the vector OR), the velocity v and the acceleration a of the object vary with time t. The time to complete one oscillation is the period T.

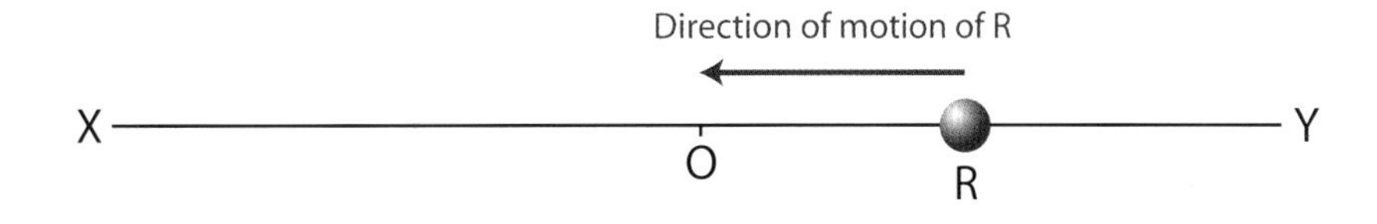

The displacement x varies sinusoidally. The value at any instant is given by

$$x = A\cos\omega t$$

where A is the amplitude of the oscillation

Note that given the displacement-time graph, we can obtain the velocity at any instant by drawing the appropriate tangent at the point and finding the gradient.

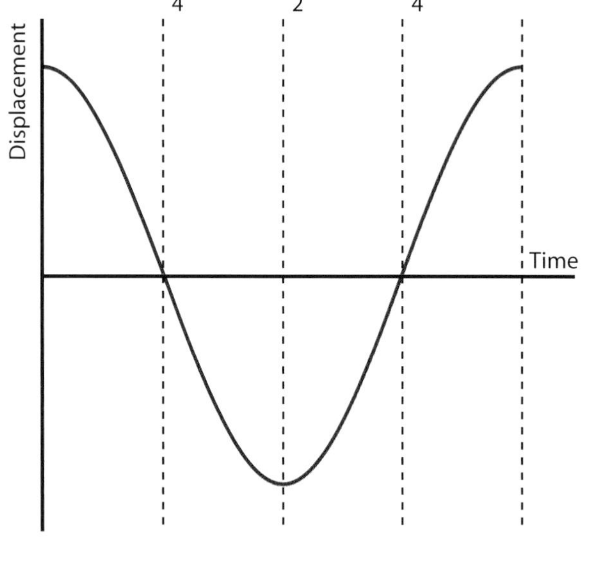

The velocity at any instant is equal to the gradient of the displacement-time graph at that instant. While the general equation for v is not required by the CCEA specification, it turns out to be:

$$v = -\omega A\sin\omega t$$

The velocity has a maximum value of ωA at the instant the object passes through the centre of oscillation.

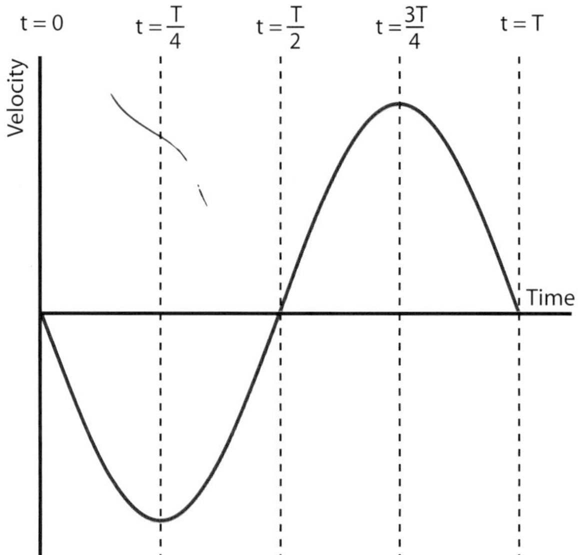

The acceleration at any instant is equal to the gradient of the velocity-time graph at that instant.

$$a = -\omega^2 x = -\omega^2 A\cos\omega t$$

The acceleration has a maximum value of $-\omega^2 A$ at the instant the object reaches the extremities of its oscillation. The acceleration is zero when the object reaches the centre of the oscillation.

The minus sign tells us that the acceleration is always in the opposite direction to the displacement from O.

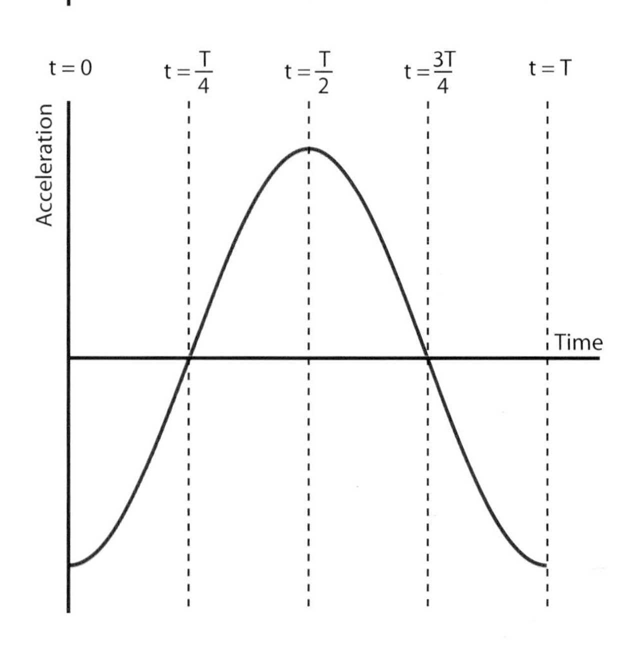

The specification makes it clear that candidates are expected to know how to perform calculations based $x = A\cos\omega t$. This expression for displacement is based on time t = 0 occurring at the point of maximum positive displacement. For the sake of completeness, the equations below show how the displacement and acceleration change with time and also how these equations would change if the time t = 0 occurred at the equilibrium position with the particle moving in the direction of positive displacement. There is no requirement to know the equation showing how velocity changes with time.

	Based on time t = 0 at x = A
Displacement, x	$x = A\cos\omega t$
Acceleration, a	$a = -\omega^2 x = -\omega^2 A\cos\omega t$

Velocity and Displacement

How does the velocity varies with **displacement** from the centre of oscillation?

The equations below show how the displacement and velocity vary with time.

$$x = A\cos\omega t \qquad v = -\omega A\sin\omega t$$

$$x^2 = A^2\cos^2\omega t \qquad v^2 = \omega^2 A^2\sin^2\omega t$$

The relationship between sine and cosine tells us that $\sin^2\theta + \cos^2\theta = 1$. So,

$$\sin^2\omega t + \cos^2\omega t = 1$$

$$\omega^2 A^2\sin^2\omega t + \omega^2 A^2\cos^2\omega t = \omega^2 A^2$$

Substituting for x and v this into the above gives:

$$v^2 + \omega^2 x^2 = \omega^2 A^2$$

$$\text{So: } v^2 = \omega^2 A^2 - \omega^2 x^2 \text{ and hence:}$$

$$v = \pm\omega\sqrt{(A^2 - x^2)}$$

The ± indicates that the velocity can be positive or negative i.e. to left or to the right or up or down, in other words it refers to direction. The equation above is **not** required by the specification, but it is so useful that it is reproduced here for the sake of completeness.

Summary of Equations for Displacement, Velocity and Acceleration

	Displacement	Velocity	Acceleration
Variation with time	$x = A \cos \omega t$	$v = -\omega A \sin \omega t$	$a = -\omega^2 A \cos \omega t$
Maximum value	Amplitude = A	At fixed point, Max velocity = $\pm\omega A$	At extreme displacement, maximum acceleration = $\omega^2 A$
Minimum value	At fixed point, displacement = o	At extreme displacement, minimum velocity = 0	At fixed point, acceleration = o
Variation with displacement		$v = \pm\omega\sqrt{(A^2 - x^2)}$	$a = -\omega^2 x$

Other Equations

Angular velocity (sometimes called angular frequency): $\omega = \frac{2\pi}{T} = 2\pi f$

Worked Examples

Example 1

A steel strip is clamped at one end and made to vibrate at a frequency of **5 Hz** and with an amplitude of **50 mm.**

(a) (i) Calculate the period, T and the angular velocity, ω.

Solution $T = \frac{1}{f} = \frac{1}{5} = 0.2$ seconds

$\omega = 2\pi f = 2\pi \times 5 = 10\pi = 31.4 \text{ rad s}^{-1}$

(ii) Calculate the acceleration at maximum displacement.

Solution $a = \omega^2 x = (10\pi)^2 \times 5 \times 10^{-2} = 5\pi^2 = 49.3 \text{ mm s}^{-2}$

(b) At time t = 0, the displacement is +50 mm. Calculate

(i) the displacement at time t = 0.04 s.

Solution $x = A \cos \omega t = 50 \cos (10\pi \times 0.04) = 15.5$ mm

(ii) the times during the first period at which the distance from the fixed point is 33 mm.

Solution

In the 1st quarter cycle,	In the 2nd quarter cycle,	In the 3rd quarter cycle, the displacement is –33 mm exactly half a period (0.1 s) after the time in the first quarter cycle.	In the 4th quarter cycle, the displacement is –33 mm exactly half a period (0.1 s) after the time in the second quarter cycle.
$x = A \cos \omega t$	$x = A \cos \omega t$	$t = 0.027 + 0.1$	$t = 0.073 + 0.1$
$33 = 50 \cos 10\pi t$	$-33 = 50 \cos 10\pi t$	$t = 0.127$ s	$t = 0.173$ s
$\cos 10\pi t = \frac{33}{50} = 0.660$	$\cos 10\pi t = \frac{33}{50} = -0.660$		
$10\pi t = \cos^{-1} 0.660 = 0.850$	$10\pi t = \cos^{-1} -0.660 = 2.292$		
$t = \frac{0.850}{10\pi} = 0.027$ s	$t = \frac{2.292}{10\pi} = 0.073$ s		

Note that in the calculations above it is essential that the calculator is set to "radian mode" when calculating the cosines and inverse cosines.

Example 2

An object oscillates in a straight line with simple harmonic motion. Its displacement varies with time as shown in the graph below. The displacement x is given by an equation of the form $x = A \cos \omega t$

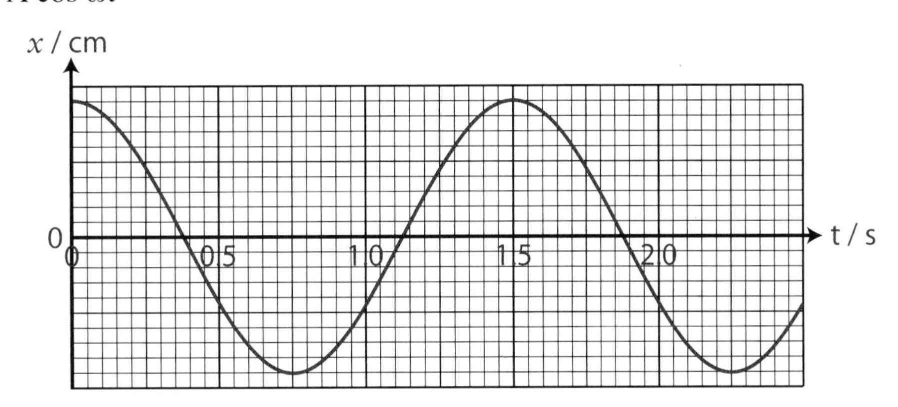

(a) Making use of the graph find the value of ω in this equation.

Solution The period is the time between the two peaks (from 0 to 1.5 s) or the time between the two troughs (from 0.75 to 2.25 s) and is therefore 1.5 s

$$T = \frac{2\pi}{\omega} \quad \text{so: } 1.5 = \frac{2\pi}{\omega}$$

$$\text{Rearranging: } \omega = \frac{2\pi}{1.5} = 4.19 \text{ rad s}^{-1}$$

(b) The amplitude of the oscillation is 1.8 cm. Sketch a graph to show how the velocity of the object varies with time. Include appropriate values of velocity and time on your graph.

Solution

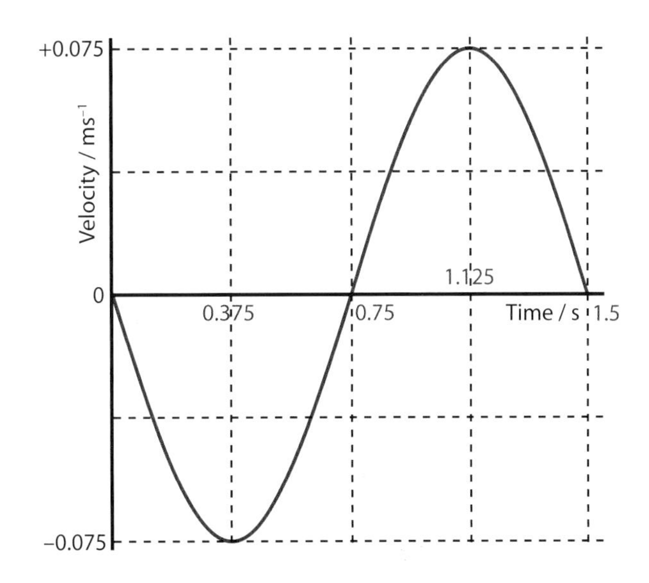

The velocity is zero when the object is at the extremes of its oscillation. This happens at 0, 0.75, 1.5 and 2.25 seconds. The velocity–time graph will cut the time axis at these times.

The maximum velocity = $\pm\omega A = 4.19 \times 0.018 = \pm 0.075 \text{ ms}^{-1}$. The object has its maximum velocity when it is at the centre of the oscillation. This happens mid way between those times when it is at the extremes of its oscillation: 0.375, 1.125 and 1.875 seconds.

The ± indicates the direction. At 0.375 s the gradient of the displacement–time graph is negative so the velocity has a negative value. It could be moving to the left, in which case a positive velocity is taken as movement to the right.

(c) The object has a mass of 100 g. Sketch a graph to show how the kinetic energy of the object varies with time. Include appropriate values of kinetic energy and time on your graph.

Solution

$$\text{Maximum kinetic energy} = \tfrac{1}{2}mv^2_{max}$$

$$= \tfrac{1}{2} \times 0.1 \times 0.075^2 = 2.8 \times 10^{-4} \text{ J}$$

Kinetic energy is a scalar. When the velocity has a negative value the kinetic energy will always have a positive value.

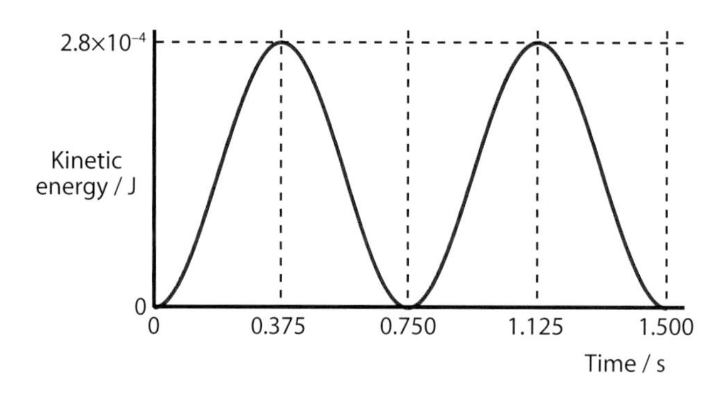

[CCEA June 2002 modified]

Example 3

The motion of a piston in the cylinder of an engine may be assumed to be simple harmonic motion along a horizontal straight line. The amplitude of the oscillation is 52.0 mm. The piston makes 480 complete cycles in 60.0 s.

(a) Calculate the period and angular velocity of the piston.

Solution Period, $T = \dfrac{60}{480} = 0.125 \text{ s}$

Angular velocity, $\omega = \dfrac{2\pi}{T} = \dfrac{2\pi}{0.125} = 16\pi \text{ rad s}^{-1}$

(b)(i) Calculate the maximum acceleration of the piston.

Solution Maximum acceleration $= \omega^2 A$

$= (16\pi)^2 \times 52.0 \text{ mms}^{-2} = 131.38 \text{ ms}^{-2}$

(ii) State where the maximum acceleration occurs in the motion of the piston.

Solution Maximum acceleration occurs when the piston is at maximum displacement ($x = \pm 52.0$ mm) from its equilibrium position.

[CCEA January 2006]

Example 4

A particle moves with simple harmonic motion and has a period of π seconds. At time t = 0 the particle has maximum positive displacement of 50 mm.

(a) Determine the angular velocity ω for this motion and write down expressions for its displacement and acceleration at time t.

Solution

$$\text{Angular velocity, } \omega = \frac{2\pi}{T} = \frac{2\pi}{\pi} = 2 \text{ rad s}^{-1}$$

$$\text{Displacement, } x \text{ (in mm)} = A \cos \omega t = 50 \cos 2t$$

$$\text{Acceleration, a (in mms}^{-2}) = -\omega^2 x = -\omega^2 A \cos \omega t$$

$$= -4 \times 50 \cos \omega t = -200 \cos \omega t$$

(b) At what times during the first period is the particle 25 mm from its equilibrium (central) position?

Solution

Since $x = 50 \cos 2t$, we are required to solve $25 = 50 \cos 2t$ for all $0 \le t \le \pi$

This requires $\cos 2t = 0.5$, and hence $2t = \cos^{-1}(0.5) = 1.0472$ s

Thus the particle first has a displacement of 25 mm at a time of 0.52 s

The particle then passes through the equilibrium position at time $\frac{\pi}{4}$ (0.785 s), and shortly afterwards is displacement is -25 mm. Hence $2t = \cos^{-1}(-0.5) = 2.094$ s. So the time, t, is $2.094 \div 2 = 1.05$ s

The particle then passes through the negative extremum at time $\frac{\pi}{2}$ (1.571 s) and then starts moving back towards its starting position. It has a displacement of -25 mm once again exactly half a period after its displacement was first 25 mm. This occurs at time $0.52 + \frac{\pi}{2} = 2.09$ seconds. The particle passes through the equilibrium position for the second occasion at time of $\frac{3\pi}{4}$ (2.36 s). Finally, it has a displacement of 25 mm once again at time $T - 0.52 = 2.62$ s.

Example 5

To commence oscillation, a pendulum bob is pulled aside a distance of 0.040 m to the right of the equilibrium position and then released. The pendulum oscillates with simple harmonic motion and its period is 1.60 s. Find the displacement of the bob 0.60 s after its release.

[CCEA 1991 Amended]

Solution Since the timing is started from the instant when the bob is at the extremity of the oscillation the appropriate equation for the displacement is $x = A \cos \omega t$

$$T = \frac{2\pi}{\omega} = 1.60 \text{ s}, \text{ so: } \omega = 3.93 \text{ s}^{-1}$$

$$A = 0.040 \text{ m}$$

$$\omega t = 3.93 \times 0.60 = 2.36 \text{ rad}$$

Setting the calculator to radian mode,

$$x = A \cos \omega t = 0.040 \cos 2.36 = -0.028 \text{ m}$$

The minus sign indicates that at this time the pendulum bob is to the left (negative displacement) of the equilibrium position.

Exercise 13

Examination Questions

1 (a) State the characteristics of the acceleration of a body moving in simple harmonic motion.

(b) A body executes simple harmonic motion with a periodic time of 625 ms.
The maximum acceleration of the body is 3.01 ms^{-2}. Show that the amplitude of the oscillation of the body is approximately 0.030 m

(c) Making use of the given data and the quantities calculated, state the equation for the displacement of the body, given that at time t = 0 the body has maximum positive displacement.

[CCEA June 2005 Amended]

2 (a) A body executes simple harmonic motion in a straight line.

(i) Sketch a graph to show how its displacement from it equilibrium position varies with time.

(ii) Sketch a graph to show how the body's velocity varies over the same period of time. What is the phase difference between the velocity and the acceleration?

(iii) Sketch a graph to show how the body's acceleration varies over the same period of time. What is the phase difference between the acceleration and the displacement?

(b) The level of water in a harbour varies periodically in such a way that it may be considered to execute simple harmonic motion with a period of 12.5 hours.
At low tide, the level of the water is 2.40 m above the sea bed.
At high tide it is 5.60 m above the sea bed.

(i) What is the amplitude of oscillation of the water level?

(ii) Calculate the level of water above the sea bed 2.4 hours after low tide.

[CCEA January 2004 Amended]

3 A particle moves along a straight line with simple harmonic motion.

(a) State how the force acting on the particle depends on its displacement from the equilibrium position.

(b) The amplitude of the motion of this particle is A. Write down the displacement from the equilibrium position at which the acceleration of the particle is half its maximum acceleration. Give your answer in terms of A.

[CCEA January 2003]

4 The diagram shows a cube resting on a horizontal, frictionless table. The cube is attached to a spring, the other end of which is fixed to a rigid support. The spring obeys Hooke's law in both compression and extension.

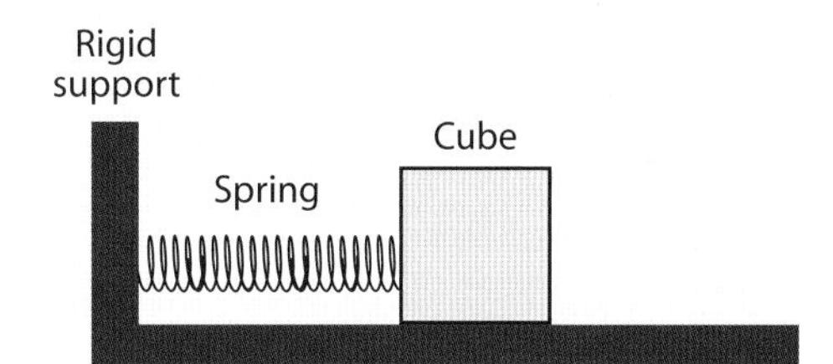

(a) It is found that it takes 25 mJ of work to compress the cube against the spring and compress the spring by 75 mm. Find the force constant of the spring (the constant of proportionality in Hooke's law). Use the fact that the work done in compressing the spring is $\frac{1}{2} kx^2$, where k is the force constant and x is the compression.

(b) When the spring is compressed by 75 mm and the cube released, it undergoes simple harmonic motion with an initial acceleration of 3.0 ms^{-2}.

The period of oscillation, T, is given by $T = 2\pi\sqrt{\frac{m}{k}}$, where m is the mass of the cube. Calculate the mass of the cube.

[CCEA January 2003 Amended]

Exercise 14

Supplementary Questions

1 The bob of a simple pendulum of length l oscillates with a period $T = 2\pi\sqrt{\frac{l}{g}}$. It is possible to draw a linear graph which illustrates the relationship between the frequency of the pendulum and its length l. Draw such a graph labelling the axes with the appropriate quantities.

[CCEA 1991 Amended]

2 (a) Define simple harmonic motion.

(b) A load, m, of mass 20 g is hung from one end of a spring of force constant k (spring constant) 8.8 N m^{-1}. The other end of the spring is clamped. The mass is then set into vertical oscillations.

These oscillations are simple harmonic with a period T given by $T = 2\pi\sqrt{\frac{m}{k}}$. Calculate the frequency of the oscillations.

[CCEA January 2005 Amended]

3 The displacement x of a particle undergoing simple harmonic motion is given by $x = A \cos \omega t$.

(a) State what the quantities A and ω stand for. Give a possible unit for each.

(b) Below is a graph of a cosine function of time. You may assume that the axes are calibrated suitably. Describe how you would use the graph to determine a value for ω for this function.

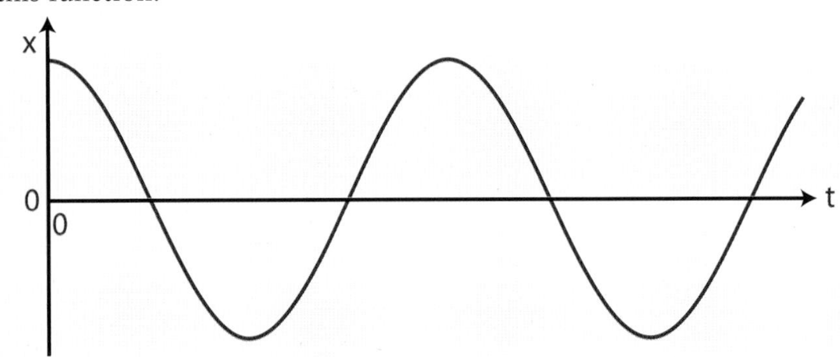

[CCEA June 2004]

Forced Oscillations and Resonance

If a pendulum is set oscillating the amplitude of the oscillation gradually decreases. Eventually it stops oscillating. The swinging pendulum involves a transfer of energy from kinetic to potential and back again to kinetic. This will continue forever if friction is absent. Of course this is not the case and the resistive forces due to the air and friction between the string and the suspension point gradually cause energy to be transferred to heat. Resistive forces that act on an oscillating system are known as **damping** forces. How an oscillating system behaves depends on the size of the damping forces.

A forced oscillation is when any external force which varies with time is used to make an object oscillate. For example pushing a child on a swing is an example or using your hand to make a mass on a spring oscillate by moving your hand up and down.

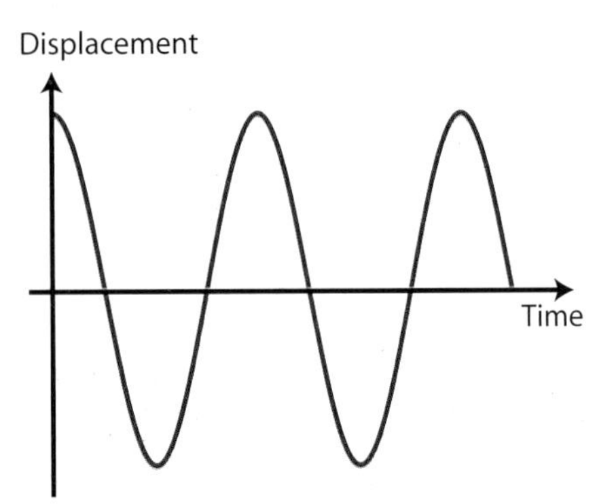

Undamped oscillations are said to be free oscillations (i.e. perfect SHM). The displacement varies periodically as shown. However the amplitude remains constant with time.

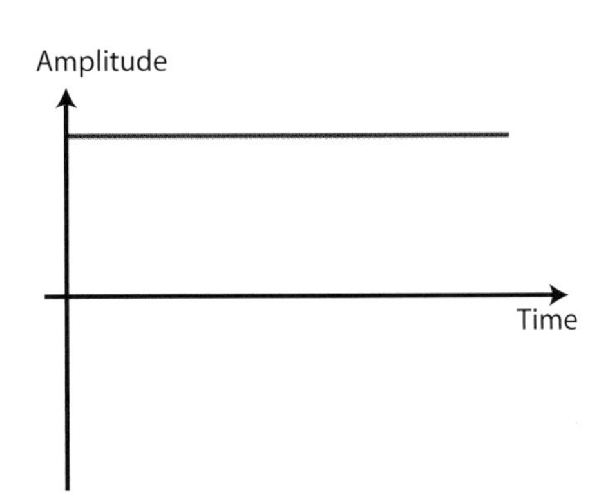

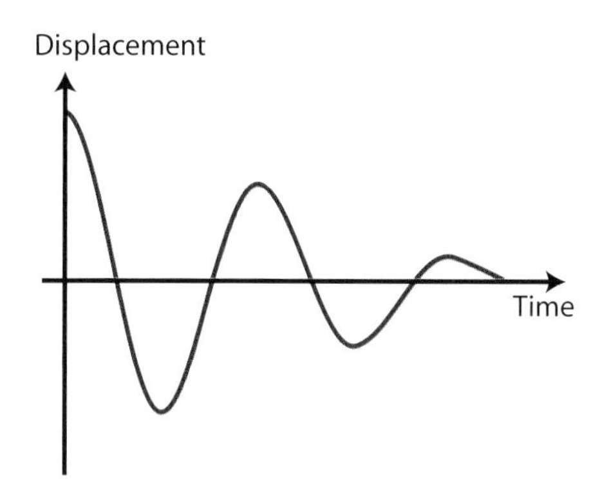

When the system is **lightly damped** the displacement will vary with time as with free oscillations. However the amplitude of the oscillation will gradually get smaller and eventually the oscillations will cease.

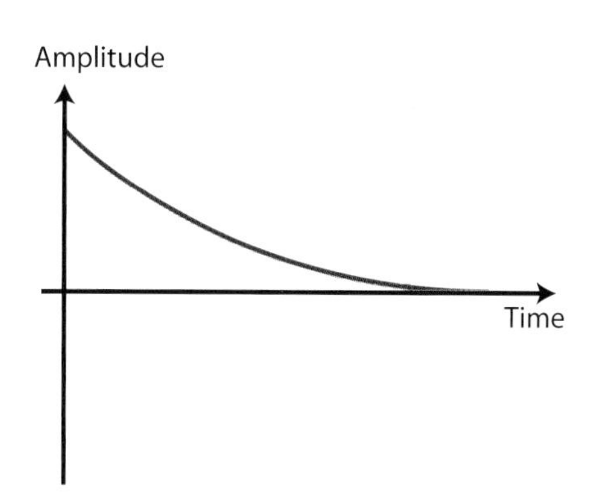

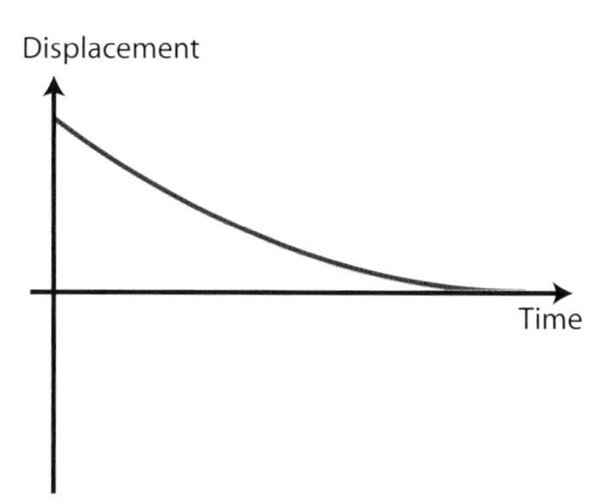

When **heavily damped** no oscillations occur and the system returns very slowly to its equilibrium position.

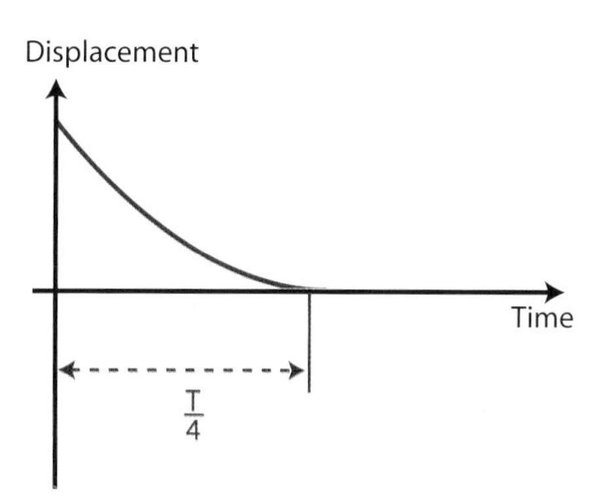

When the damping forces are such that the system can return to its equilibrium position in the shortest possible time then we have what is called **critical damping.**

The shortest possible time for any oscillating system to return to its equilibrium position is ¼ T where T is the periodic time. Electrical meters are critically damped so that the pointer moves to the reading as quickly as possible without any oscillation.

Forced oscillations can be demonstrated using Barton's pendulums. This consists of a number of paper cone pendulums of various lengths. Each cone is loaded with a paper clip inside it. All are suspended from the same string as a 'driver' pendulum which has a heavy bob . The length of this driver pendulum is the same as one of the paper cone pendulums.

The driver pendulum is pulled to one side and released. After a time all the pendulums oscillate with very nearly the same frequency as the driver but with different amplitudes.

The paper cone with the same length as the driver pendulum has the largest amplitude. This is resonance. The length of the pendulum determines its periodic time and so its frequency. Resonance takes place when the frequency of the driving force is the same as the frequency of the oscillating system.

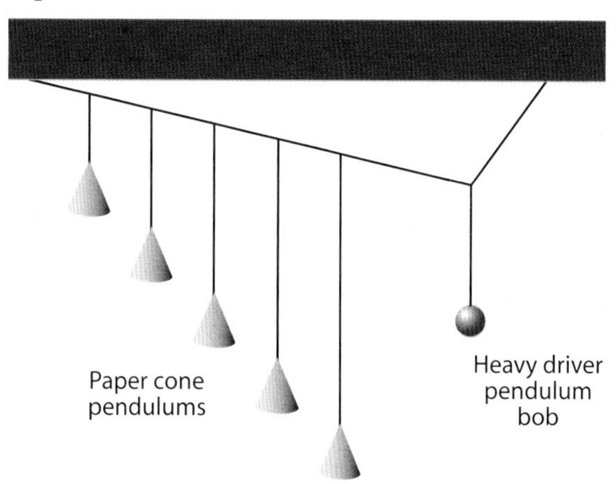

The graph shows how the amplitude of a forced oscillation depends on the frequency of the force causing it to vibrate (driver frequency). The amplitude reaches a maximum when the frequency of the driving force equals that of the natural frequency of the oscillating system.

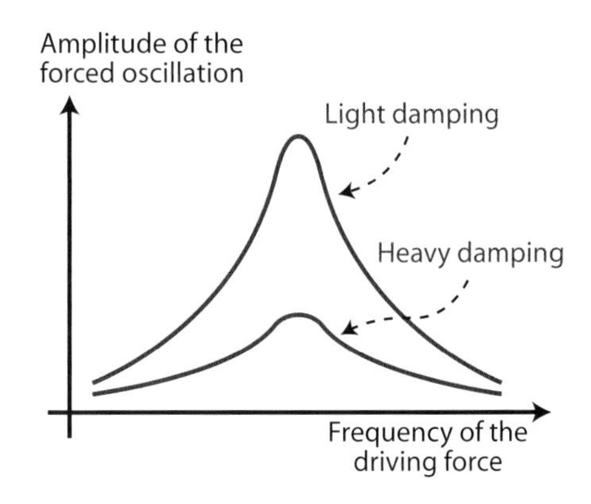

When the damping force is small (light damping) the peak is sharp. However when the frictional or damping forces are greater the peak is broader and in fact the maximum occurs at a slightly lower frequency.

Example of Resonance

Resonance can be a destructive phenomenon in mechanical systems. The photographs show the destruction of a suspension bridge over the Tacoma Narrows in the USA in 1940. The speed of the wind blowing over the bridge on a particular day caused a forced oscillation which matched the natural frequency of the bridge resulting in its destruction.

Wind tunnels are now used to ensure such effects are avoided when designing structures ranging from bridges to the wings of an aircraft.

Musical instruments rely on resonance to make a sound that is more audible. A guitar string has a number of frequencies at which it will naturally vibrate. These natural frequencies are known as the harmonics of the guitar string. Each of these natural frequencies or harmonics is associated with a standing wave pattern. Plucking a string so that it vibrates produces a sound that is barely audible. In a guitar it is the vibrations of the air inside the body of the guitar that produces the louder sound. The vibrations of the string causes the air inside the guitar to oscillate at the same frequency. This is resonance.

A tuning fork when struck produces a sound that is barely heard. However when it is held over an air column of appropriate length a much louder sound is heard, the tuning fork causes the air in the column to vibrate at its natural frequency and so resonance is produced. Musical instruments use this technique to make a louder sound. Many wind instruments consist of an air column enclosed inside of a hollow tube. Though the tube may be more than a metre in length, it is often curved upon itself one or more times in order to conserve space. Blowing into a mouthpiece creates a vibration which in turn causes the air column in the instrument to resonate so a louder sound is produced.

Resonance is used in electrical circuits allowing us to receive radio and television signals. The natural frequency of an electrical circuit can be altered. When it is the same as the frequency of the electromagnetic wave then the magnetic component of that wave acts as a driving force causing the electrons in the circuit to vibrate with a large amplitude and so an electric current of the same frequency as the electromagnetic wave is induced in the radio or television receiver.

The heating of food in a microwave oven is another example of resonance. Microwaves of a specific frequency are generated. This frequency corresponds to the frequency of rotation of the water molecules. This time it is the electric component of the electromagnetic wave that is used. As the microwaves force the water molecules to rotate faster they move around colliding with other water molecules so heating up the water and the food it is in contact with.

Exercise 15

Past Examination Questions

1 Write an account of the variation with time of the amplitudes of objects vibrating with simple harmonic motion. You may illustrate your answer with appropriate sketch graphs.

[CCEA June 2005]

2 A mechanical system which is undergoing forced vibration may show resonance. Explain briefly what is meant by resonance. Illustrate your answer with reference to an example of a mechanical vibrating system, explaining how resonance occurs.

[CCEA January 2005]

3 (a) A mass oscillates along a straight line. Sketch labelled graphs to show the variation of the amplitude with time when

(i) the motion is simple harmonic, i.e. the oscillations are not damped,

(ii) the oscillations are damped.

(b) Systems involving forced vibrations have a characteristic frequency response. Describe one example of such a system.

[CCEA May 2003]

4 (a) A body moves with simple harmonic motion in a straight line. During this motion, the force on the body is proportional to the displacement from the equilibrium position and is in the opposite direction to the displacement.

Below is a graph of the acceleration, a, of the body as a function of its displacement, x, from the equilibrium position.

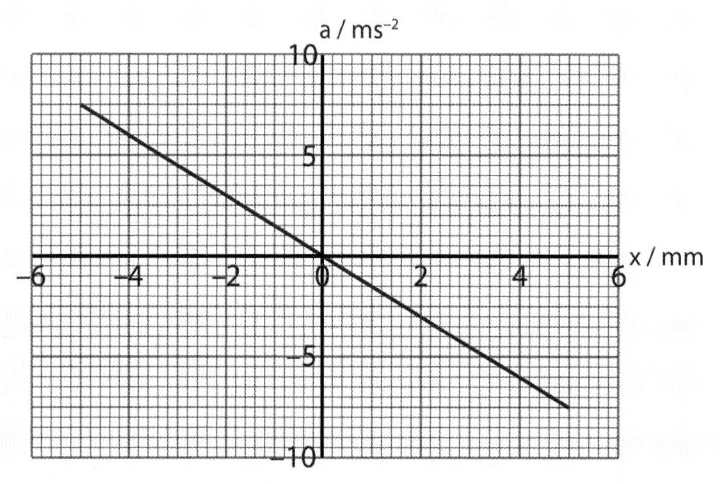

(i) Explain how the graph shows that the force on the body is proportional to the displacement of the body from the equilibrium position, and that the force is in the opposite direction to the displacement.

(ii) From the graph find the amplitude and period of the motion.

(b) Write revision notes, suitable for this examination, on the subject of **Resonance and Damping.** Bullet point notes, illustrated by sketches and/or graphs, will be sufficient.

[CCEA Summer 2006]

4.5 The Nucleus

Students should be able to:

- 4.5.1 Be able to describe evidence for the existence of atomic nuclei, to include alpha particle scattering;
- 4.5.2 Know and interpret the variation of nuclear radius with nucleon number; and
- 4.5.3 Use the equation $r = r_0 A^{1/3}$ to estimate the density of nuclear matter.

The Nucleus

By the end of the nineteenth century the notion that atoms were indivisible particles of matter was beginning to crumble. The study of cathode rays, positive rays and radioactivity had made it clear that atoms contained particles of positive and negative charge. However it was unknown how these charges were arranged inside the atom.

In 1902, Lord Kelvin expressed the opinion that an atom might consist of a sphere of positive charge with negative electrons dotted about inside it. This idea was taken up in 1903 by Sir J. J. Thomson, who developed his "Plum Pudding Model" in which the atom was regarded as a positively charged sphere in which the negatively charged electrons were distributed like currants in bun in sufficient numbers to make the atom as a whole electrically neutral.

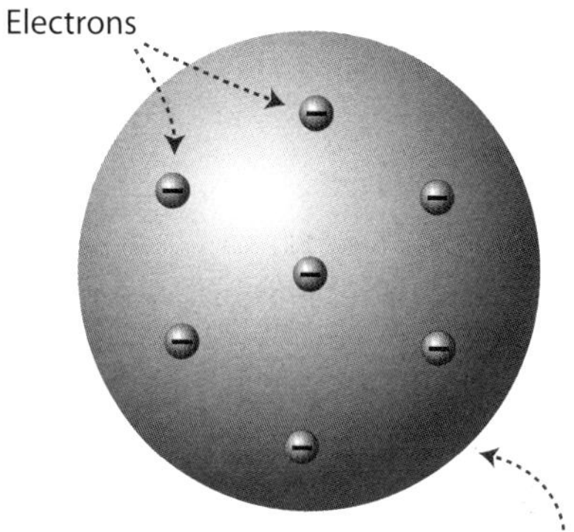

Atomic Structure

In 1906, the New Zealand physicist, Ernest Rutherford at McGill University in Canada noticed that α-particles went easily through mica without making holes in it as a bullet might. This led him to suspect that the α-particles were passing right through the atoms themselves rather than pushing atoms out of the way.

Rutherford also noticed that some of the α-particles were deflected out of their straight-line paths as they went through the mica, and he thought that this was caused by electric repulsion between the positively charged part of the mica atoms and the positive α-particles. Shortly afterwards Rutherford left Canada to become Professor of Physics at Manchester where, with post-graduate students, Hans Geiger and Ernest Marsden, he carried out a series of experiments on the scattering of α-particles by thin metal films.

The most celebrated of these experiments was with thin gold foil. A source of α-particles was contained in an evacuated chamber. The α-particles were incident on a thin gold foil whose plane was perpendicular to their direction of motion.

The alpha particles were detected by the flashes of light (scintillations) they produced when they hit a glass screen coated with zinc sulphide. The experiment had to be carried out in a vacuum to prevent collisions between alpha particles and gas atoms deflecting the alpha particles.

Rutherford found that:

- most of the alpha particles were undeflected;
- some were scattered by appreciable angles;
- about 1 in 8000 was 'back-scattered' through a very large angle indeed.

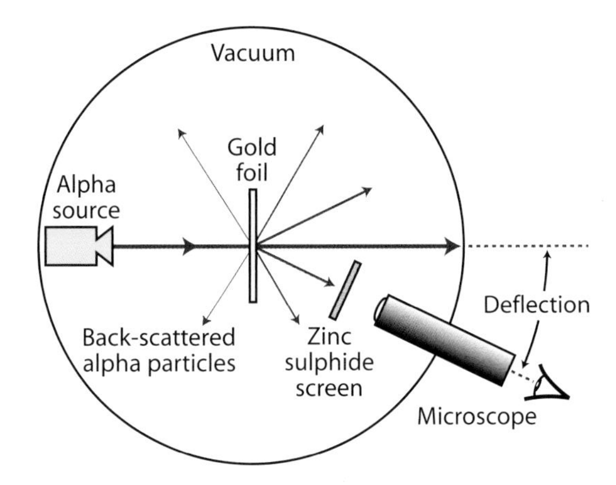

Diagram of the Rutherford alpha-particle scattering apparatus

The back-scattering was so unexpected that Rutherford later wrote:

> "It was quite the most incredible event that ever happened to me in my life. It was almost as incredible as if you fired a 15-inch shell at a piece of tissue paper and it came back and hit you."

Rutherford correctly deduced:

- The majority of the alpha particles passed straight through the metal foil because they did not come close enough to any repulsive positive charge at all.
- All the positive charge and most of the mass of an atom formed an exceptionally small, dense core or nucleus.
- The negative charge consisted of a "cloud of electrons" surrounding the positive nucleus.
- Only when a positive alpha particle approached sufficiently close to the nucleus, was it repelled strongly enough to rebound at high angles.
- The small size of the nucleus explained the small number of alpha particles that were repelled in this way.
- Most of an atom is empty space.

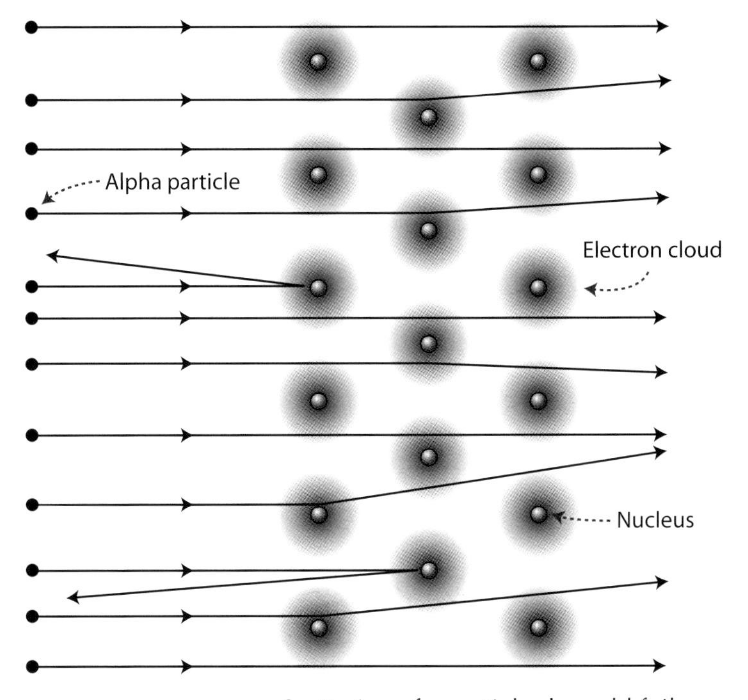

Scattering of a-particles by gold foil.

The Size of a Nucleus

It was a challenging exercise for physicists to get some notion of the size of a nucleus by experiment. This is because the nucleus does not have a sharp edge; rather we should describe the edge of a nucleus as being "fuzzy". Nevertheless, physicists obtained an early idea of the size of a nucleus by firing α-particles at it. The distance of closest approach is an approximate upper limit of the size of a nucleus. Early experiments suggested that the nucleus might be considered as a sphere of radius $\approx 10^{-15}$ m.

Suppose we assume that the volume of a nucleon (proton or neutron) in any nucleus is about the same. Then the volume of the nucleus is directly proportional to the total number of nucleons contained within it (that is, the **mass number** A). If we also suppose the nucleus to be spherical, then the radius of the nucleus is directly proportional to $A^{1/3}$. This allows us to write:

$$r = r_0 A^{1/3}$$

where r is the radius of a given nucleus
A is the mass number (number of nucleons within the nucleus)
r_0 is the constant of proportionality

What is the significance of r_0? Readers will observe that it is the constant of proportionality in the equation above. However, it is also the radius of the nucleus for which A = 1, that is, it is the radius of the proton. The value of r_0 is found by experiment to be **about** 1.2×10^{-15} m or 1.2 femtometres (sometimes written 1.2 fm or 1.2 fermi). However, it must be emphasised that the value of r_0 is not a well-defined constant and different values are obtained using different experimental techniques.

Worked Example

Uranium–235 has 92 protons and 143 neutrons in its nucleus. Use the equation $r = r_0 A^{1/3}$ to find the radius of this nucleus. Take r_0 as 1.2 fm.

Solution

The mass number A is 235, so

$$r = r_0 A^{1/3} = 1.2 \times 235^{1/3} \text{ fm} = 1.2 \times 6.17101 \text{ fm} \approx 7.4 \text{ fm}$$

Nuclear Density

We can show that the density of nuclear matter is constant, that is, all nuclei have the same density.

Volume, V, of a sphere of radius r: $\dfrac{4\pi r^3}{3}$

Mass, M, of spherical nucleus of radius r: $M = \rho \dfrac{4\pi r^3}{3}$ where ρ is the density of nuclear matter

$$M = \rho \frac{4\pi r_0^3 A}{3} \quad \text{since } r = r_0 A^{1/3}$$

If we write the mass M of a nucleus of mass number A as Am, where m = mass of a nucleon, then

$$Am = \rho \frac{4\pi r_0^3 A}{3}$$

and cancelling A on both sides leaves us with

$$m = \rho \frac{4\pi r_0^3}{3}$$

and rearranging gives

Density of nuclear matter = $\rho = \dfrac{3m}{4\pi r_0^3}$ which is constant since m and r_o are constant.

Using $r_o = 1.2 \times 10^{-15}$ metres and m, the mass of the proton, as 1.66×10^{-27} kg, we can calculate the density of nuclear matter to be

$$\frac{3 \times 1.66 \times 10^{-27}}{4\pi \times (1.2 \times 10^{-15})^3} = 2.3 \times 10^{17} \text{ kg m}^{-3}$$

which is an amazing **230 000 000 000 000 times greater than water!**

The huge density of nuclear matter in comparison with everyday matter is a reflection of the fact that in ordinary matter there is a great deal of empty space between the nucleus and the orbiting electrons. There is no empty space between the particles inside the nucleus!

Exercise 16

Examination Questions

1 Your Data and Formulae Sheet gives the following equation for the radius of a nucleus.

$r = r_0 A^{1/3}$ **Equation 1.1**

(a) (i) In Equation 1.1, what do the symbols r, r_o and A represent?

(ii) Which of the values listed below is the best approximation to r_o ?

1 μm ☐ 1 nm ☐ 1 pm ☐ 1 fm ☐

(b) (i) Use your value of r_o and Equation 1.1 to calculate the density of nuclear matter.

(ii) Comment on the value of the density of nuclear matter you have calculated in comparison with the density of everyday matter. Suggest a reason for any difference between these values.

[CCEA June 2007]

2 (i) The volume of a particular nucleus may be taken to be about 4.6×10^{-43} m³. The nucleus may be treated as a sphere. Estimate the radius of the nucleus.

(ii) Your Formulae Sheet gives the equation: $r = r_0 A^{1/3}$ for the nuclear radius r. Use your answer to (a) (i) to find the nucleon number *A* for this nucleus. Take r_o to be 1.2 fm.

[CCEA June 2006]

3 Your Data and Formulae Sheet gives the equation

$r = r_0 A^{1/3}$ **Equation 1.1**

for the radius r of a nucleus with nucleon number A.

A graph of the variation of r with A is shown opposite.

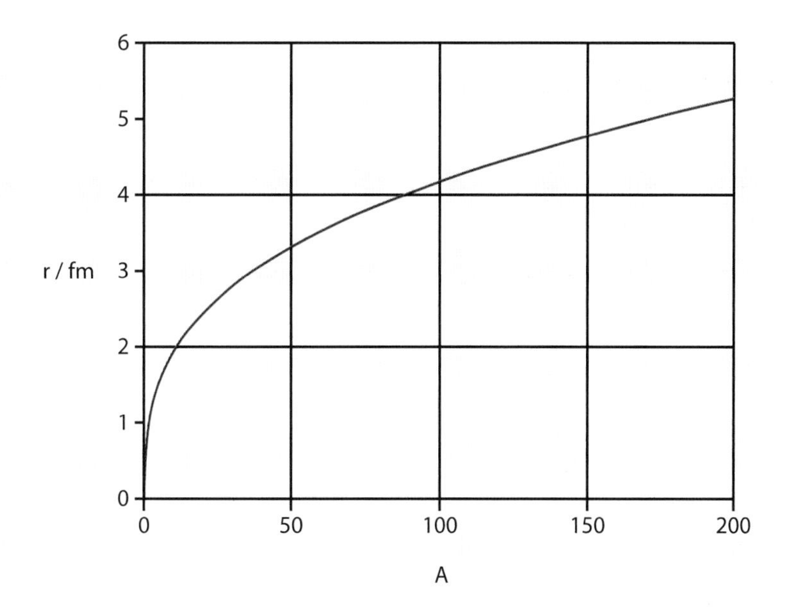

(i) Use the graph above to find the radius of the nucleus represented by the symbol $^{150}_{60}Nd$ This nucleus has a nucleon number of 150.

(ii) Calculate the density of the $^{150}_{60}Nd$ nucleus.

(iii) A student is told to plot a **linear** graph to display the relationship between r and A. What functions should he plot on the vertical and horizontal axes of the graph?

[CCEA June 2003 (amended)]

4 To investigate the structure of the atom, Rutherford, Geiger and Marsden directed a beam of alpha particles at a very thin gold foil. They found that the particles were scattered at angles varying from 0° to 180° with respect to the direction of the incident beam. Practically all the particles had a zero angle of scattering, that is they went straight through the foil, but a very few came straight back, that is the scattering angle was 180°.

(i) State the conclusion made from the fact that practically all the alpha particles had a zero angle of scattering.

(ii) State two conclusions made from the fact that very few alpha particles had a scattering angle of 180°.

(iii) Rutherford's experiment was carried out in an evacuated tube. Explain why the vacuum is essential.

[CCEA June 2006]

5 (i) Describe the structure of an atom of aluminium–27 ($^{27}_{13}Al$) in terms of protons, neutrons and electrons.

(ii) Use the equation $r = r_0A^{1/3}$ to find the radius of the aluminium–27 nucleus. Take r_o as 1.2 fm and give your answer in metres.

[CCEA June 2004]

4.6 Nuclear Decay

Students should be able to:

4.6.1 Understand how the nature of alpha-particles, beta particles and gamma-radiation determines their penetration and range;

4.6.2 Calculate changes to nucleon number and proton number as a result of emissions;

4.6.3 Appreciate the random nature of radioactive decay;

4.6.4 Model with constant probability of decay, leading to exponential decay;

4.6.5 Use the equations $A = \lambda N$ and $A = A_0 e^{-\lambda t}$, where *A* is the activity;

4.6.6 Define half-life;

4.6.7 Use $t_{\frac{1}{2}} = \dfrac{0.693}{\lambda}$;

4.6.8 Describe an experiment to measure half-life;

The Nucleus

Every atom has a central positively charged nucleus with a diameters of around 10^{-15} m.

Atomic diameters are around 10^{-10} m, so the atom is typically 100 000 times bigger than its nucleus.

Over 99.9% of the mass of an atom is in its nucleus.

Atomic nuclei are totally unaffected by chemical reactions.

Nuclei contain protons and neutrons. These are collectively known as **nucleons.** Electrons are not normally found in the nucleus. The properties of nucleons are compared with those of the electron in the table below.

	Electron	Proton	Neutron
Relative Mass*	$^1/_{1840}$	1	1
Actual Mass	9.109×10^{-31} kg	1.673×10^{-27} kg or 1836 m_e	1.675×10^{-27} kg or 1839 m_e
Relative Charge*	–1	+1	0
Charge	-1.60×10^{-19} C	$+1.60 \times 10^{-19}$ C	Zero

* relative to the proton m_e is the mass of the electron

Nucleons are held together by one of the four fundamental forces of nature - the strong interaction.

This nuclear force acts over very short distances and is much stronger than the electric force of repulsion that exists between protons within the nucleus.

Some Definitions

Atomic Number - Z: this is the number of protons in the nucleus.

Mass Number - A: this is the total number of nucleons in the nucleus.

Isotopes: these are nuclei with the same number of protons but differing numbers of neutrons.

A nucleus is described using these two numbers and the **chemical symbol of the element:**

$$^{A}_{Z}X$$

So, for example, the isotope of uranium having 238 nucleons in its nucleus, of which 92 are protons, is given the symbol:

$$^{238}_{92}U$$

Hydrogen has three stable isotopes, hydrogen (1p), deuterium (1p,1n) and tritium (1p, 2n).

$$^{1}_{1}H \quad ^{2}_{1}H \quad ^{3}_{1}H$$

Some elements have unstable isotopes whose nuclei disintegrate randomly and spontaneously. This effect known as radioactivity.

Radioactivity

This was first noticed by Becquerel in 1896 when he observed that some photographic plates which had been stored close to a uranium compound had become fogged. In his honour, the name Becquerel (Bq) is given to the unit of activity.

1 Bq is 1 disintegration per second

Types of Radiation

Alpha (α) Radiation

- Alpha radiation is made up of a stream of alpha particles emitted from large nuclei.
- An alpha particle is a helium nucleus with two protons and two neutrons, and so has a mass number of 4. The symbol for an α-particle is $^{4}_{2}\alpha$ or $^{4}_{2}He$
- Alpha particles are positively charged and so will be deflected in a magnetic field.
- Alpha particles have poor powers of penetration and can only travel through about 4 centimetres of air. They can easily be stopped by a sheet of paper.
- Since α particles move relatively slowly (about 6% of the speed of light) and have a high momentum they **interact with matter producing intense ionisation** – a typical α-particle can produce about **100 000 ion-pairs per cm of air** through which it passes.
- Alpha decay is described by the equation below.

Decaying parent nucleus		Daughter nucleus remains		α-particle emitted
$^{A}_{Z}X$	$\longrightarrow$	$^{A-4}_{Z-2}Y$	+	$^{4}_{2}He$

Note that the number of nucleons (Mass number) is conserved (A to A – 4 + 4) and the charge (Atomic number) is also conserved (Z to Z – 2 + 2).

Beta (β) Radiation

- Beta radiation is emitted from nuclei where the number of neutrons is much larger than the number of protons.
- A beta particle is a very fast electron (up to 98% of the speed of light) and thus it has relative atomic mass of about 1/1840. The symbol for a β-particle is therefore $^{0}_{-1}\beta$ or $^{0}_{-1}e$
- β-particles are emitted by nuclei with too many neutrons to be stable. One of the neutrons changes into a proton and an electron. The proton remains inside the nucleus but the electron is immediately emitted from the nucleus as a β-particle.
- As beta particles are negatively charged, they will be deflected in a magnetic field. This deflection will be greater than that of alpha particles, as the beta particles have much smaller mass to charge ratio.
- Beta particles move much faster than alpha particles and so they interact less with matter than β-particles and have a greater penetrating power.
- Beta particles can travel several metres in air, but are stopped by 5 mm thick aluminium foil.
- Beta radiation has an ionising power between that of alpha and gamma radiation.
- The decay equation for β emission is shown below.

Decaying parent nucleus		Daughter nucleus remains		β-particle emitted
$^{A}_{Z}X$	$\longrightarrow$	$^{A}_{Z+1}Y$	+	$^{0}_{-1}He$

Note that the total number of nucleons (mass number) does **not** change, but the atomic number (Z) of the daughter nucleus (Z + 1) is greater than that of the parent (Z) by 1.

Gamma (γ) Radiation

- Unlike the other types of radiation, gamma radiation does not consist of particles. γ-rays are short wavelength, high energy electromagnetic radiation emitted from unstable nuclei.
- The wavelength of γ-rays is characteristic of the nuclide that emits it. The wavelengths of γ-rays are typically in the region 10^{-10} to 10^{-12} m. (X-rays can also have wavelengths in this range, but the method of production is what differentiates X-rays from γ-rays.)
- Like alpha and beta radiation, gamma radiation comes from a disintegrating unstable nucleus.
- As there are no particles, gamma radiation has no mass.
- As there are no charged particles, a magnetic field has no effect on gamma radiation.
- Gamma radiation has great penetrating power, travelling several metres in air.
- A thick block of lead or concrete is used to greatly reduce the effects of gamma radiation, but

cannot stop it completely. A lead block about 5 cm thick will absorb around 90% of the γ-rays

- Gamma radiation has the weakest ionising power as it interacts least with matter.
- The decay equation for this is:

Nuclide in excited state		Nucleus remains		γ-radiation emitted
${}^{A}_{Z}X^{*}$	$\longrightarrow$	${}^{A}_{Z}X$	+	γ

Exercise 17

1 A radioactive decay series can be represented on a graph of mass number, A, (y–axis) against atomic number, Z. Part of a table for such a series is given below:

Element (symbol)	Atomic number	Mass number	Decays by emitting	Leaving element
U	92	238	α	Th
Th	90	234	β	Pa
Pa	91	234	β	
	92	234	α	
	90	230		Ra
Ra	88	226		Rn
Rn	86			Po
Po		218	α	Pb
Pb				Bi
Bi	83			Po

(a) In what ways do mass number and atomic number change in:
(i) α–decay and (ii) β–decay?

(b) Copy and complete the table above.

(c) Plot the points on a graph of mass number (y–axis) against atomic number (x–axis) to show the decay of each element. Join the points with arrows to show the decay.

(d) Explain why the emission of a gamma ray cannot be shown on such a graph.

(e) Identify two pairs of isotopes using the table.

2 Copy and complete the following decay equations:

Alpha decay of uranium–238 $\quad {}_{92}U \rightarrow {}^{234}Th + {}_{2}He$

Beta decay of carbon–14 $\quad {}^{14}C \rightarrow {}_{7}N + \beta$

The Law of Radioactive Decay

Radioactive nuclei disintegrate **spontaneously and randomly**. This means that we can neither tell **which** particular nuclei in a given sample are going to decay nor can we tell **when** they are going to decay. However, if we have sufficiently large numbers of nuclei in our sample, the random decay of individual nuclei averages out in such a way as to be governed by empirical laws. The rate of disintegration cannot be speeded up or slowed down by any known means (temperature, pressure, particle size, chemical reactions etc).

The number of unstable nuclei (and hence the activity) decreases exponentially.

$$N = N_0 e^{-\lambda t}$$

where N_0 is number of unstable nuclei present at time = 0

N is number of unstable nuclei present after a time = t

λ is decay constant

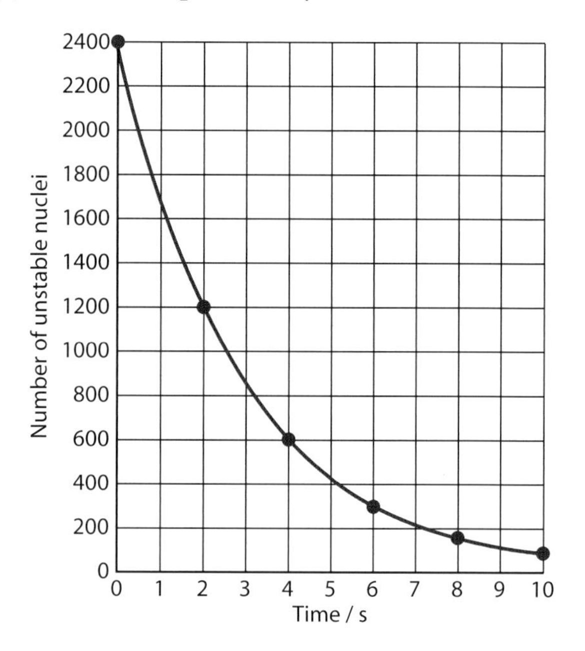

The rate of decay of a particular species (nuclide) is directly proportional to the number of unstable nuclei of that nuclide present.

Therefore if there are **N** unstable nuclei present at time **t**:

Rate of Disintegration with time = $-\lambda N$

λ is a constant of proportionality called the decay constant. It is measured in units of s^{-1}.

The minus sign is present because as t increases N decreases.

The rate of disintegration of unstable nuclei is called **the activity** of the source. Activity is measured in disintegrations per second or becquerel. (Bq).

Activity, A = $-\lambda N$

Note that the **activity also decreases exponentially** with time since it is directly proportional to the number of unstable nuclei present.

$$A = A_0 e^{-\lambda t}$$

where A_o is the initial activity at time t = o

A is the activity at a time t = t

λ is the decay constant

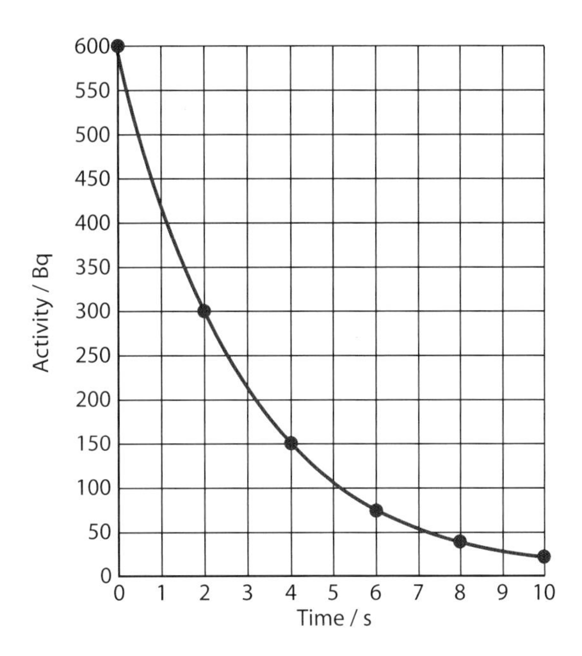

Half–Life ($T_{1/2}$)

The half-life of a radioactive nuclide is the time taken for half those radioactive nuclei present to disintegrate.

$$N = N_0 e^{-\lambda t}$$

After a one half life has passed only half the unstable nuclei remain.

$$\frac{N_0}{2} = N_0 e^{-\lambda T_{1/2}}$$

Dividing across by No

$$\frac{1}{2} = e^{-\lambda T_{1/2}}$$

Taking natural logs of both sides

$$\log_e 1 - \log_e 2 = -\lambda T_{1/2}$$

And since $\log_e 1 = 0$, on re-arranging we get

$$\log_e 2 = 0.693 = \lambda T_{1/2}$$

$\log_e$ is more frequently written as ln, so we arrive at

$$\ln 2 = 0.693 = \lambda T_{1/2}$$

$$T_{1/2} = \frac{0.693}{\lambda}$$

The reader will note that the following is an alternative definition for half-life.

The half-life of a radioactive material is the time taken for the activity of that material to fall to half of its original value.

Worked Examples

Example 1

1 A sample of radioactive phosphorous has an activity of 1000 Bq. Seven days later the activity was found to have decreased by 29%.

(a) What is the decay constant for this isotope of phosphorous? Give your answer in days^{-1}.

(b) Calculate the half-life of the isotope of phosphorous in days.

(c) Calculate the number of atoms of phosphorous in the sample.

Solution

(a) $A = A_0 e^{-\lambda t}$, so taking logs to base e of both sides gives

$$\ln A = \ln A_0 - \lambda t$$

$$\lambda = \frac{(\ln A_0 - \ln A)}{t}$$

$$= \frac{(\ln 1000 - \ln 710)}{7} = 0.0489 \text{ days}^{-1}$$

(b) $T_{1/2} = \dfrac{0.693}{\lambda} = \dfrac{0.693}{0.0489} = 14.2$ days

(c) Since activity is in Bq (disintegrations per second), λ must be converted to units of s^{-1}.

$$N = \frac{A}{\lambda} = \frac{1000}{0.0489 \div (24 \times 3600)} = 1.77 \times 10^9 \text{ atoms}$$

Example 2

(a) A small volume of a solution containing a radioactive isotope has an activity of 1.2×10^4 disintegrations per minute. This solution is injected into the blood stream of a patient. After 30 hours a 1 cm^3 sample of the blood is found to have an activity of 0.5 disintegrations per minute. Estimate the volume of blood in the patient. The half life of the isotope is 15 hours.

(b) Another radioisotope of the same activity and emitting the same type of radiation of the same energy but having a half-life of 2.6 years is available. Give one advantage and one disadvantage of using this for estimating the volume of blood in the patient.

Solution

(a) Total initial activity = 1.2×10^4 disintegrations per minute

Since $T_{1/2} = 15$ hours, then total final activity after 30 hours (2 half-lives) is

$\frac{1}{4}(1.2 \times 10^4) = 3000$ disintegrations per minute

Activity in 1 cm^3 sample of the blood is 0.5 disintegrations per minute

So total volume of blood = 3000 / 0.5 = 6000 cm^3 = 6 litres

(b) Advantage: very long half life means that the activity of the sample is very low, so it is likely to cause least radiation damage to patient.

Disadvantage: time required to get same activity from 1 cm^3 blood is two half-lives – but 5.2 years is totally unacceptable time to medical staff and patient. Moreover, exceptionally sensitive equipment would be required to measure the change in activity over a 30 hour period.

Example 3

A certain isotope used in medical physics has a half-life of 3.0 minutes. It is taken from a locked cupboard and prepared for injecting into a patient's vein. The time taken to make up the required solution is exactly 2.0 minutes and at that time the isotope has an activity of 500 Bq. Calculate the activity at the time when the isotope was removed from the cupboard.

Solution

$$\lambda = \frac{0.693}{T_{1/2}} = \frac{0.693}{3} = 0.231 \text{ minutes}^{-1}$$

We must find the activity at time t = – 2.0 minutes:

$$A = A_0 e^{T_{1/2}} = 500 \times e^{-0.231 \times (-2.0)} = 500 \times e^{0.462} = 500 \times 1.5872$$

$$= 794 \text{ Bq}$$

Note: The same result is reached whether the calculation uses minutes or seconds.

Measuring the Half-life of a Radioactive Substance

There are two common methods used in schools and colleges to measure the half-life of a radioactive substance. One involves the use of an ionisation chamber and a source of radon gas (^{220}Rn). This isotope of radon emits α–particles. The other method involves the use of a Geiger-Müller tube, a scaler or ratemeter and a source of protactinium (^{234}Pa). This isotope of protactinium emits β–particles.

Finding the Half-Life of Radon–220 with an Ionisation Chamber

An ionisation chamber consists of an aluminium can about the size of one used for baked beans. A metal rod (the negative electrode) can be mounted centrally within the chamber. The metal can is insulated from its mounting but the whole chamber screws directly onto the input of the DC amplifier so that the connection between its central negative electrode and the amplifier is as short as possible.

When in operation, a voltage of about 9 V is maintained between the central electrode and the can itself. Suppose now there is a source of α–particles within the chamber. These α–particles cause considerable ionisation of the air inside the chamber. The electron-ion pairs so formed will move apart under the influence of the electric field between the central electrode and the metal can. When these particles strike the appropriate electrode, they cause a tiny electric current, called the ionisation current, to flow. This small ionisation current is fed into a DC amplifier which amplifies it about 100 million times until it is large enough to work a moving-coil milliammeter.

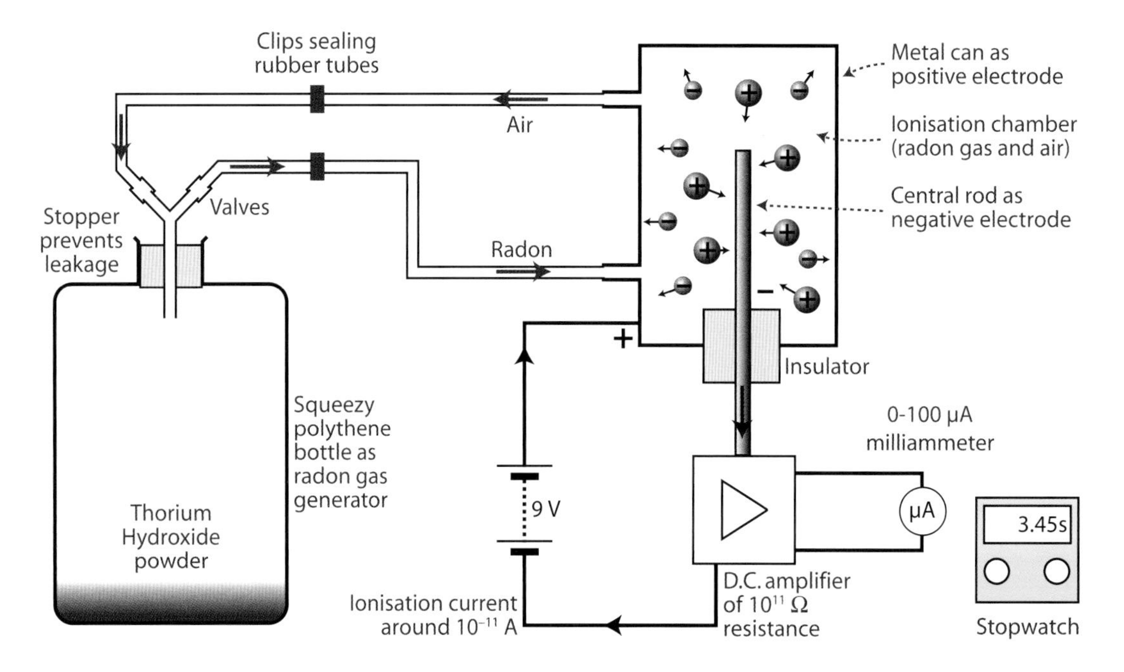

Apparatus used to find the half-life of radon-220 using an ionisation chamber and DC amplifier.

Adjusting the DC amplifier

- Set the input selector switch to the set-zero position so that the input of its amplifier is short-circuited and the input is zero.
- Adjust the set-zero control to give a zero output current on the milliammeter.
- Select the highest resistance input, usually 1011 Ω.

Filling the ionisation chamber with radon gas

- Connect the polythene bottle radon generator to the ionisation chamber with the two thin rubber tubes.
- Release both clips on the tubes.
- Squeeze the bottle two or three times until the reading on the milliammeter just goes past full-scale deflection.
- Refit both clips on the tubes.

Taking readings

- Start the clock as the milliammeter current reading falls to full-scale deflection.
- Read and record the current every 10 seconds for about three minutes.

As the number of radon gas atoms in the chamber gets smaller and smaller they emit fewer and fewer α–particles. These produce less and less ions and so a smaller ionisation current flows: **the ionisation current is directly proportional to the number of radon atoms remaining and hence to the activity of the gas within the chamber.** The time taken for the ionisation current to fall to half its initial value is therefore the half-life of radon.

Treatment of the Results to Find the Half-life

There are **two** possible approaches. One is to plot a graph of the current readings against time and draw a **smooth curve** through the points. Then read from the graph three time intervals during which the current had fallen by 50%. These three times represent the half-life of radon–220 and should all be around 52 seconds. Take the mean of these three values as the half-life of radon–220.

Another approach is to plot a linear graph. Since the ionisation current, I, is directly proportional to the activity of the gas within the chamber, then

$$I = I_0 e^{-\lambda t}$$

where I_o is the current at time t = 0 and λ is the decay constant.

Taking natural logs of both sides gives

$$\ln I = \ln I_0 - \lambda t$$

Comparing this last equation with that for a straight line

$$y = c + mx$$

we see that a graph of ln I (y–axis) against time (x–axis) is a straight line of gradient –λ and y–axis intercept $\ln I_o$

We therefore plot a graph of ln I (y–axis) against time (x–axis), draw the straight line of best fit and determine its gradient (–λ). We then find the half-life by calculating the value of 0.693 / λ.

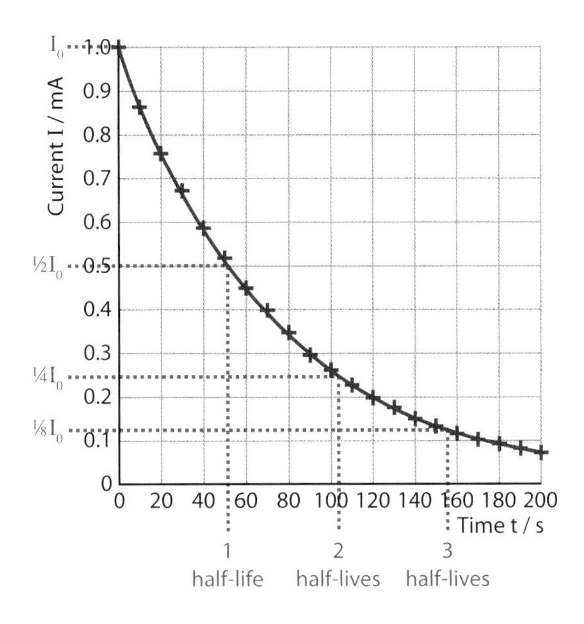

Smooth curve method to find half-life

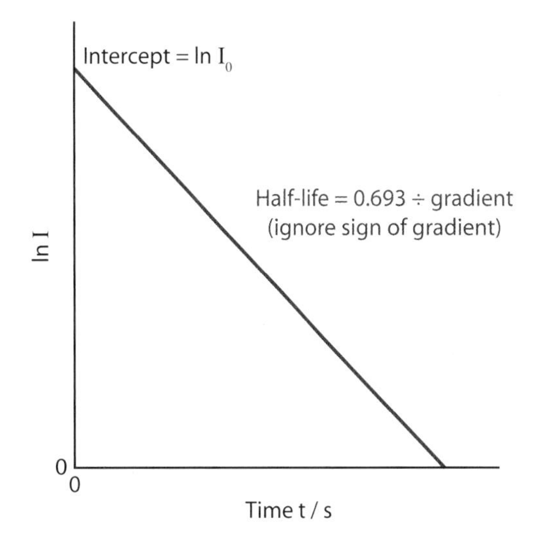

Linear graph method to find half-life

Measuring the Half-life of Protactinium–234 using a Scaler (Counter)

This experiment involves the use of a Geiger-Müller (GM) tube and a counter to measure the activity of a sample of protactinium. When alpha, beta or gamma radiation enters the GM tube, it causes some of the argon gas inside to ionise and give an electrical discharge. This discharge is detected and counted by the counter. If the counter is connected to its internal speaker, you can hear the click when radiation enters the tube.

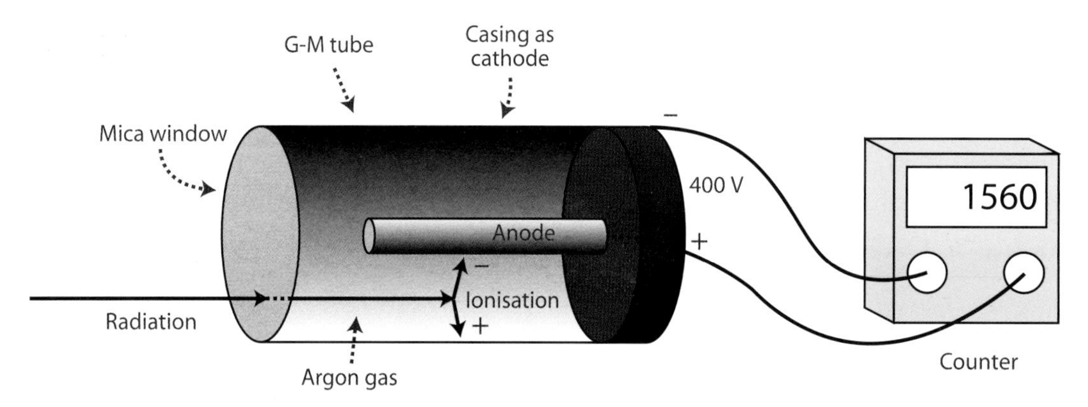

Section through a GM tube

However, even in the absence of all known sources of radioactivity, the GM tube and counter still detects radiation. This is known as **background radiation** and it comes from the Sun, cosmic rays from space, hospital nuclear physics departments, nuclear power stations, granite rocks and so on.

Before we use a GM tube to carry out any quantitative work on radiation we must first measure the background radiation if we are to correct for it in our experiment.

To Measure the Background Radiation

First remove known sources of radiation from the laboratory, then set the GM counter to zero. Switch on the counter and start a stopwatch. After 30 minutes read the count on the counter. Divide the count by 30 to obtain the background count rate in counts per minute. A typical figure is around 15 counts per minute.

This background count must always be subtracted from any other count when measuring the activity from a specific source.

Setting up the Protactinium Source

- Protactinium–234 is one of the decay products of uranium–238 and any compound of uranium–238 will have within it traces of protactinium. These traces may be conveniently extracted from it by chemical means.
- The protactinium decays by β–emission into another long-lived isotope of uranium (^{234}U) which is itself α–emitting. The very long half-life indicates low radioactivity, and in any case, not enough to interfere with this experiment. Moreover, the α–particles which are emitted will not penetrate the polythene bottle containing the protactinium.

- The β–activity at any instant of the extracted solution can therefore be used as a measure of the quantity of protactinium still present in it.

A simple practical arrangement is that shown opposite

- A thin-walled polythene bottle is filled with equal volumes of an acid solution of uranyl nitrate and pentyl ethanoate.
- When the liquids are shaken up together the organic ethanoate removes most of the protactinium present. The solutions are not miscible and the protactinium remains in the upper layer when the liquids have once more separated.
- The β–activity of the protactinium is observed with a G-M tube and ratemeter, and the count-rate is recorded at 10 second intervals.
- Allowance is then made for the background count of the G-M tube. If say the measured rate with the GM-tube and ratemeter is 32 counts/minute and the background rate is 15 counts/minute, then the corrected count rate is 32 – 15 = 17 counts/minute.

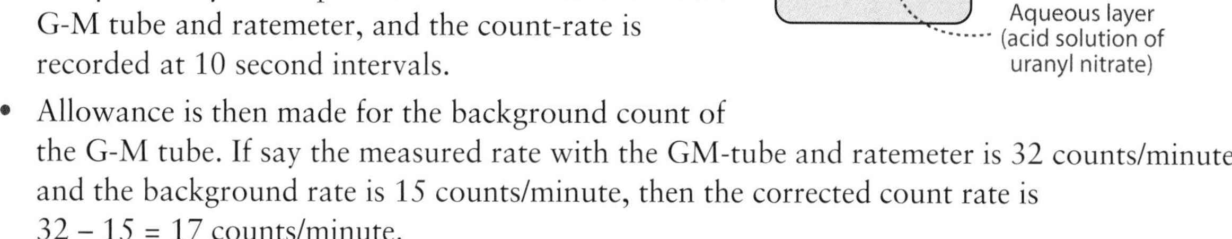

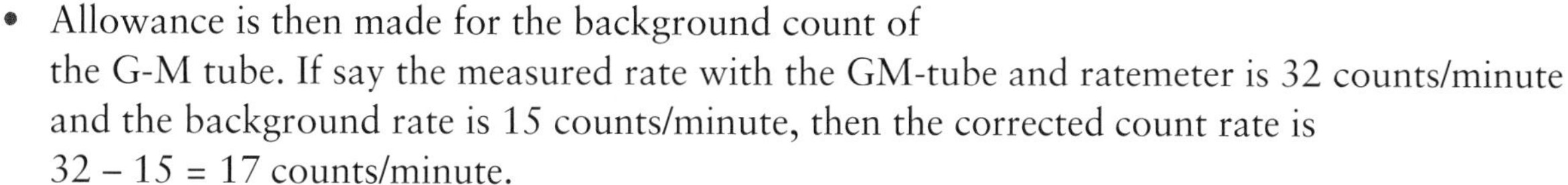

Treatment of the Results to Find the Half-life of Protactinium

The corrected count rate of the protactinium is taken as a measure of its activity, A. By the Law of Radioactive Decay,

$$A = A_0 e^{-\lambda t}$$

where A_o is the activity at time t = 0 and λ is the decay constant.

Taking natural logs of both sides gives

$$\ln A = \ln A_0 - \lambda t$$

Comparing this last equation with that for a straight line

$$y = c + mx$$

we see that a graph of ln A (y–axis) against time (x–axis) is a straight line of gradient –λ and y–axis intercept ln A_o

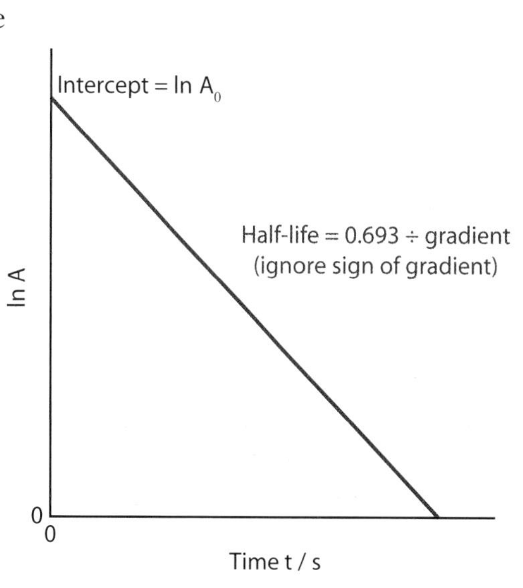

We therefore plot a graph of ln A (y–axis) against time (x–axis), draw the straight line of best fit and determine its gradient (–λ). We then find the half-life by calculating the value of 0.693 / λ.

The generally accepted value for the half-life of protactinium–234 is 68 seconds.

Safety Notes:

- Teachers must satisfy themselves before carrying out these demonstrations that they are fully compliant with the various regulations governing the use of ionising radiations in schools.
- It is essential that the coupling between the can and the DC amplifier is absolutely air-tight and that the rubber tubing connecting radon gas generator to the ionisation chamber is neither worn nor cracked. While α–particles themselves have a short range in air, the radon gas which emits them presents a serious hazard if ingested and every precaution must be taken to prevent radioactive radon getting into the atmosphere.
- At the end of a practical session do not immediately dismantle the ionisation chamber. Leave it for half an hour until the remaining radioactivity has decayed to a very low level.

Exercise 18

Examination Questions

1 (a) Complete the following equations for nuclear decay by inserting the appropriate nucleon number or proton number in the boxes provided.

(i) $^{\square}_{6}\text{C} \rightarrow {}^{14}_{\square}\text{N} + {}^{0}_{-1}\text{e}$

(ii) $^{238}_{\square}\text{U} \rightarrow {}^{\square}_{90}\text{Th} + {}^{4}_{2}\text{He}$

(b) To monitor a patient's thyroid gland, the patient is injected with a radioactive tracer containing 5.0×10^{14} atoms of Iodine–131. The half-life of Iodine–131 is 8.0 days.

(i) Calculate the number of undecayed nuclei of Iodine–131 remaining after 29.0 days.

(ii) Calculate the activity of the injected Iodine–131 after 29.0 days. Give your answer in Bq.

[CCEA June 2006]

2 (a) Your Data and Formulae Sheet gives the following relationship for half-life:

$$T_{1/2} = \frac{0.693}{\lambda} \quad \textbf{Equation 5.1}$$

(i) Define **half-life.**

(ii) Use the equation $A = A_0 e^{-\lambda t}$ to show how **Equation 5.1** arises.

(b) The half-life of $^{238}_{92}\text{U}$ is 4.5×10^{9} years. A sample contains 3.0×10^{21} Uranium–238 nuclei. Calculate the activity of the sample due to the decay of the Uranium–238.
[1 year = 3.2×10^{7} seconds.]

[CCEA June 2003]

4.7 Nuclear Energy

Students should be able to:

4.7.1 Appreciate the equivalence of mass and energy;

4.7.2 Recall the equation $E = mc^2$ and understand that it applies to all energy changes;

4.7.3 Use $E = \Delta mc^2$ in nuclear calculations;

4.7.4 Know how the binding energy per nucleon varies with mass number;

4.7.5 Describe the principles of fission and fusion with reference to the binding energy per nucleon curve

Equivalence Of Mass And Energy

In 1905 Albert Einstein published a remarkable paper which astonished the scientific community. It was called "The Special Theory of Relativity", and it dealt with the speed of light for observers moving with a constant velocity relative to each other. The predictions made by Einstein's Special Theory have now all been shown to be true. However, for the purpose of this chapter, the important assertion made by Einstein was that **there is an equivalence to a mass, m and energy, E** given by the equation $E = mc^2$, where c is the speed of light in a vacuum.

What does the notion that mass and energy are equivalent really mean? It tells us, for instance, that when 1 kg of uranium in a nuclear power station undergoes fission and releases 8×10^{13} J of energy, there is a corresponding reduction in mass of $E \div c^2$ or $8\times10^{13} \div 9\times10^{16}$ which is about 9×10^{-4} kg. Of course, this is a tiny reduction in mass, but that mass reduction is measurable and the measurements confirm Einstein's ideas. Equally, under certain circumstances, very high energy electromagnetic waves, called cosmic rays, can vanish, leaving behind two particles, an electron and a positron. And the remarkable thing is that the reduction in the energy due to the annihilation of the cosmic ray is exactly the same as the equivalent mass of the positron and the electron, according to Einstein's equation.

Chemical reactions release relatively little energy compared with nuclear reactions. Burning 1 kg of oil, for example, releases about 5×10^7 J of energy. According to Einstein's equation this results in a reduction in mass of about 5.5×10^{-10} kg. It is now firmly established that **wherever** energy is released there will **always** be a reduction in mass. Raising the temperature of a beaker of water, striking a football so that it moves faster, causing an electric current to make a bulb shine all cause a mass reduction – but, of course, the energy involved is so tiny that the mass reduction **is extremely small.**

The Electron-Volt (eV) and the Unified Atomic Mass Unit (u)

The joule and the kilogram are much too large to be useful when dealing with atomic and nuclear processes. A much more appropriate unit for energy is the electron-volt (eV). The electron volt is the kinetic energy possessed by an electron accelerated from rest through a voltage of one volt.

For our purposes it is sufficient to know that:

$$1\text{ eV} = 1.6 \times 10^{-19}\text{ J}$$

$$1\text{ MeV} = 1\text{ million electron-volts} = 1.6 \times 10^{-13}\text{ J}$$

The masses of nucleons, electrons, nuclei and atoms are usually given in **unified atomic mass units (u)**.

$$1\text{u} = \tfrac{1}{12}\text{ of the mass of the carbon-12 atom}$$

$$1\text{u} = 1.66 \times 10^{-27}\text{ kg}$$

It is sometimes helpful to remember that $1\text{u} = 1.49 \times 10^{-10}\text{ J} = 931\text{ MeV}$

Nuclear Binding Energy

If you place 100 g on a top-pan balance and then add another 100 g you would expect the combined mass to be 200 g and you would be right. It will therefore be a surprise to learn that the mass of a nucleus is **always less** than the sum of the masses of its constituent nucleons. This difference in mass is called the **mass defect** of the nucleus.

Mass defect = Total mass of the nucleons – Mass of the nucleus

This reduction in mass arises due to the combining of the nucleons to form the nucleus. When the nucleons are combined to form a nucleus a tiny portion of their mass is converted to energy. This energy is called the **binding energy** of the nucleus. The binding energy is the amount of energy that has to be **supplied** to separate the nucleons completely i.e. to an infinite distance apart.

Recall from Einstein's special theory of relativity that a change in mass **Δm** is equivalent to an amount of energy **E**, where

$$E = \Delta mc^2$$

The binding energy of a nucleus is therefore

Binding Energy (J) = mass defect (kg) × c^2

Binding energies are usually quoted in millions of electron volts (MeV).

To Calculate the Binding Energy of Helium

The He nucleus contains 2 protons and 2 neutrons.

mass of protons $= 2 \times 1.0078\text{ u} = 2.0156\text{ u}$ and

mass of neutrons $= 2 \times 1.0087\text{ u} = 2.0174\text{ u}$ so:

mass of the constituent nucleons $= 4.0330\text{ u}$ but:

mass of the nucleus $= 4.0026\text{ u}$ so:

Mass defect $= 4.0330 - 4.0026 = 0.0304\text{ u}$

To find binding energy we must first convert the mass defect to kg:

$$m = 0.0304 \times 1.66 \times 10^{-27} \text{ kg}$$

$$\text{Binding energy} = mc^2 = 0.0304 \times 1.66 \times 10^{-27} \times \left(3 \times 10^8\right)^2$$

$$= 4.54 \times 10^{-12} \text{ J}$$

Convert this to MeV:

$$\text{Binding energy} = 4.54 \times 10^{-12} \div 1.6 \times 10^{-13} = 28.39 \text{ MeV}$$

A useful measure of the stability of a nucleus is the binding energy per nucleon.

Average BE per nucleon = Binding energy/ Number of nucleons

For He the average binding energy per nucleon is therefore $28.39 \div 4 = 7.1$ MeV

The average binding energy per nucleon varies with nucleon number as shown below.

(A graph of average binding energy per nucleon against atomic number has a similar shape.)

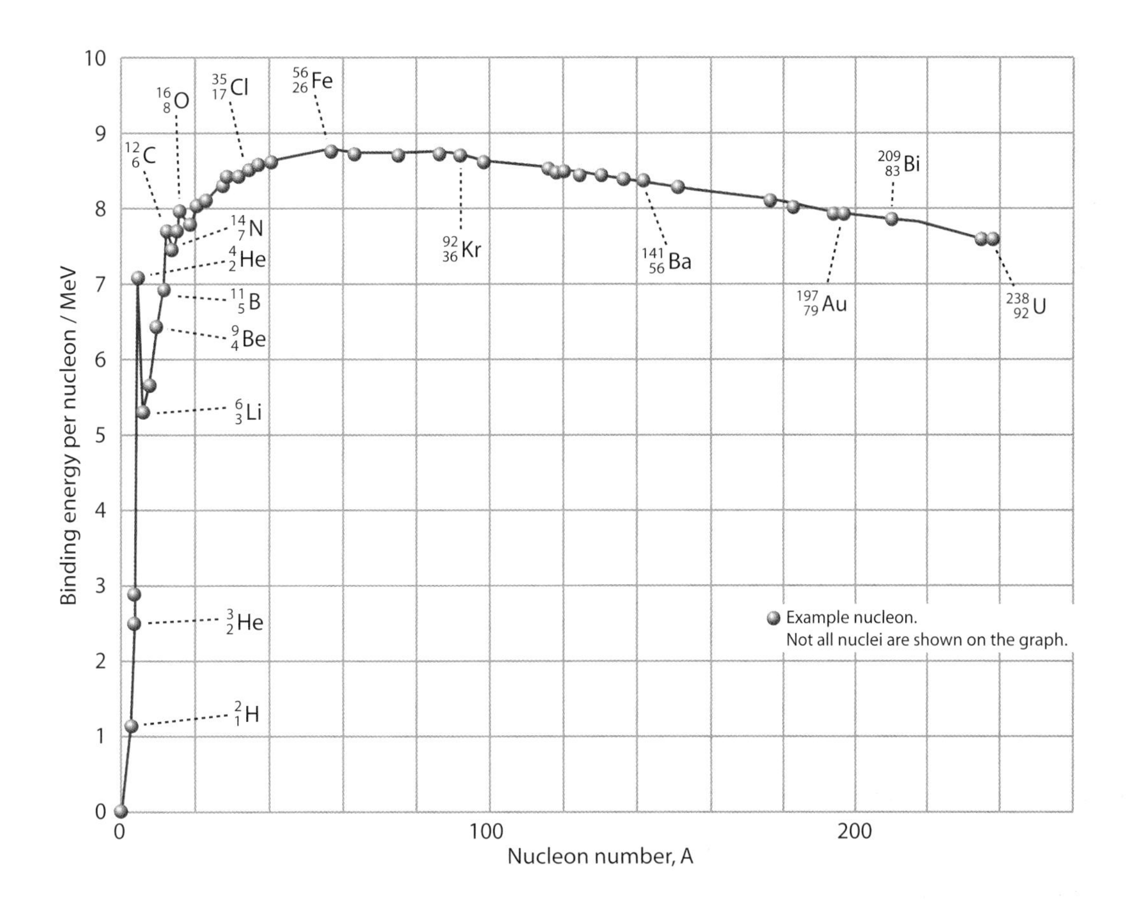

Exercise 19

1 Using the following data, calculate

(a) the binding energy of $^{16}_{8}O$

(b) the average binding energy per nucleon in $^{16}_{8}O$

Give your answers in MeV

Mass of $^{16}_{8}O$ = 15.9905 u   Mass of neutron = 1.0087 u   Mass of proton = 1.0078 u

2 Consider the following decay in which an isotope of thorium decays to radium by α–particle emission:

$$^{228}_{90}Th \rightarrow ^{224}_{88}Ra + ^{4}_{2}\alpha$$

Data:	
Mass of Th nucleus = 227.97929 u	Mass of Ra nucleus = 223.97189 u
Mass of α particle = 4.00151 u	1 u = 1.66×10^{-27} kg

(a) Use the information given in the data box to calculate the mass difference in unified atomic mass units (u) between the LHS and the RHS of the equation.

(b) Express this mass difference in kg

(c) Use Einstein's equation to convert this mass difference into (i) J and (ii) MeV

(d) In what form does this energy appear?

3 An isotope of Aluminium decays by beta emission to Silicon.

$$^{29}_{13}Al \rightarrow ^{29}_{14}Si + ^{0}_{-1}\beta$$

Data:
Mass of Al = 28.97330 u
Mass of Si = 28.96880 u
Mass of beta particle = 0.000549 u

Use the information given in the data box above to calculate the energy released by this decay.

Nuclear Fission

In nuclear fission a massive nucleus is deeply divided and breaks up into two less massive nuclei.

The average binding energy of these fission fragments is higher than that of the heavy nucleus. In the fission process, due to the increase in the total binding energy, some of the mass of the heavy nucleus is converted to kinetic energy of the fission fragments.

Uranium 235 undergoes fission when it absorbs a neutron.

One possible fission reaction is:

$${}^{235}_{92}U + {}^{1}_{0}n \rightarrow {}^{144}_{56}Ba + {}^{90}_{36}Kr + 2{}^{1}_{0}n + \text{energy}$$

Given the masses shown below, we can calculate how much energy is released by this fission reaction.

$$\text{Mass of } {}^{235}_{92}U = 235.12 \text{ u}$$
$$\text{Mass of } {}^{144}_{56}Ba = 143.92 \text{ u}$$
$$\text{Mass of } {}^{90}_{36}Kr = 89.94 \text{ u}$$
$$\text{Mass of neutron} = 1.0087 \text{ u}$$

$$\text{Mass on LHS} = (235.12 + 1.0087) = 236.1287 \text{ u}$$
$$\text{Mass on RHS} = (143.92 + 89.94 + 2 \times 1.0087) = 235.8774 \text{ u}$$
$$\text{Mass reduction} = 236.1287 - 235.8774 = 0.2513 \text{ u}$$
$$= 0.2513 \times 1.66 \times 10^{-27} \text{ kg}$$
$$= 4.17158 \times 10^{-28} \text{ kg}$$

$$\text{Energy released} = mc^2 = 4.17158 \times 10^{-28} \times (3 \times 10^8)^2$$
$$= 3.754422 \times 10^{-11} \text{ J}$$
$$= 3.754422 \times 10^{-11} \div 1.6 \times 10^{-13} \text{ MeV} = 234.7 \text{ MeV}$$

Nuclear Fusion

The binding energy curve for nuclei is shown opposite. Fusion is the joining of lighter nuclei to produce a heavier and more stable nucleus. The fusion process results in the release of energy since the average binding energy of these fusion product is higher than that of the lighter nuclei which join together. In the process some of the mass of the lighter nuclei is converted to kinetic energy of the fusion product that is, the mass of the heavier nucleus is less than the total masses of the two light nuclei that fuse.

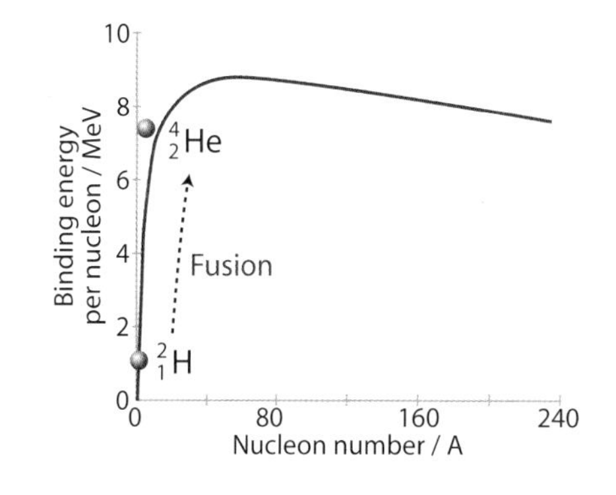

The reaction cannot take place at room temperature because of the repulsive electric force between the positively charged nuclei. It only occurs when the speed of the colliding nuclei is great enough for the nuclei to overcome this repulsive force and they can come close enough for the attractive, but very short-range, strong nuclear force to cause fusion to occur. The hydrogen nuclei can therefore fuse only when the prevailing temperature is high enough (about 15 million degrees Celsius).

Below is the fusion reaction for two deuterium nuclei. When they fuse they produce a nucleus of helium and Q represents the quantity of energy released in the reaction.

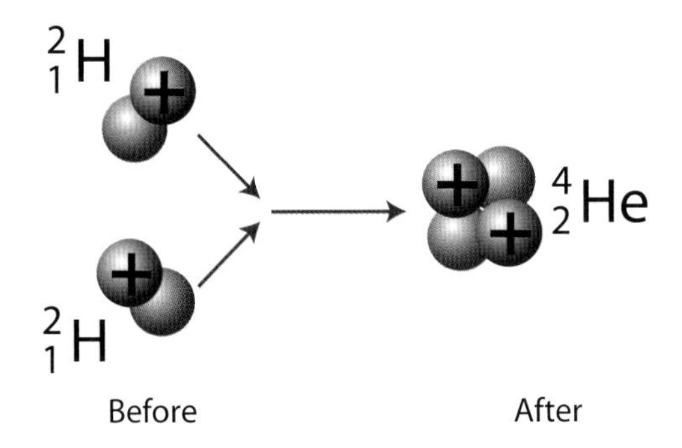

$$^{2}_{1}H + ^{2}_{1}H \rightarrow ^{4}_{2}He + Q \text{ (energy)}$$

We can use Einstein's equation to calculate the numerical value of Q. When we insert the masses we get

$$2.014102 \text{ u} + 2.014102 \text{ u} = 4.002604 \text{ u} + Q$$
$$4.028204 \text{ u} - 4.002604 \text{ u} = Q$$
$$Q = 0.0256 \text{ u}$$

To convert Q to joules we first covert to kg:

$$Q = 0.0256 \text{ u} = 0.0256 \times 1.66 \times 10^{-27} = 4.2496 \times 10^{-29} \text{ kg}$$

Then we use Einstein's mass-energy equivalence formula:

$$E = mc^2$$
$$Q = 4.2496 \times 10^{-29} \times (3 \times 10^{8})^2$$
$$Q = 3.82464 \times 10^{-12} \text{ J}$$

To express this in MeV, divide the energy in joules by 1.6×10^{-13}

$$Q = 3.82464 \times 10^{-12} \div 1.6 \times 10^{-13} \text{ MeV}$$

giving an energy release of **23.9 MeV**.

Exercise 20

1 Use the data below to calculate the energy released when a uranium–236 nucleus undergoes fission according to:

$$^{236}_{92}U \rightarrow ^{146}_{57}La + ^{87}_{35}Br + 3^{1}_{0}n$$

Data: Binding energy per nucleon of
uranium–236 = 7.59 MeV   lanthanum–146 = 8.41 MeV   bromine–87 = 8.59 MeV

2 A nuclear submarine has an average power requirement of 500 kW. It obtains this from the fission of uranium–235. The fuel is enriched uranium containing 3% uranium–235 and 97% uranium 238 by mass. The energy released in each fission is 3×10^{-11} J

(i) Calculate the number of uranium–235 atoms in 1 kg fuel.

(ii) Estimate how long 1 kg of this fuel would last.

3 The mass of the isotope $^{7}_{3}Li$ is 7.018 u. Find the binding energy of the $^{7}_{3}Li$ nucleus given that the mass of the proton is 1.008u and the mass of the neutron is 1.009 u. Take 1 u ≡ 931 MeV.

Exercise 21

Examination Questions

1 The fusion reaction between a lithium nucleus and a deuterium nucleus is represented by the equation:

$$^{6}_{3}Li + ^{2}_{1}H \rightarrow 2^{4}_{2}He$$

The energy released in this reaction is 3.59×10^{-12} J

(a)(i) Convert this energy to MeV.

(ii) Calculate the energy released per nucleon of fuel used. Give your answer in MeV.

(iii) The energy released in the fission of one uranium–235 nucleus is approximately 200 MeV. Comment on your answer to (a)(ii) in comparison with this figure.

(b) State two other advantages of fusion over fission.

[CCEA June 2007]

2 One example of a fusion reaction involving hydrogen is:

$$^{2}_{1}H + ^{3}_{1}H \rightarrow ^{4}_{2}He + ^{1}_{0}n$$

Use the data below to calculate the energy released in this reaction.

Give your answer in J. Show your working clearly.

Nuclear masses:		
	$^{2}_{1}H$	2.014102 u
	$^{3}_{1}H$	3.016030 u
	$^{4}_{2}He$	4.002604 u
	$^{1}_{0}n$	1.008665 u

[CCEA June 2006]

3 The reaction between a lithium nucleus and a deuterium nucleus is represented by the equation:

$$^{6}_{3}Li + ^{2}_{1}H \rightarrow 2\ ^{4}_{2}He$$

The relevant atomic masses are:

$^{6}_{3}Li$	6.0151 u
$^{2}_{1}H$	2.0141 u
$^{4}_{2}He$	4.0026 u

Calculate the energy released in this reaction. Give your answer in J.

[CCEA June 2003]

4 One example of a fusion reaction involving hydrogen is:

$$^{2}_{1}H + ^{2}_{1}H \rightarrow ^{3}_{2}He + ^{1}_{0}n$$

Using the data below, calculate the energy released in this reaction.
Give your answer in MeV.

Nuclear masses:	$^{2}_{1}H$	2.014102 u
	$^{3}_{2}He$	3.016030 u
neutron mass:	m_n	1.008665 u

[CCEA June 2002]

4.8 Nuclear Fission

Students should be able to:

4.8.1 Describe a fission reactor in terms of chain reaction, critical size, moderators, control rods, cooling system and reactor shielding.

In the previous chapter we looked at the principles of nuclear fission. In this chapter we will examine in some detail how physicists bring about the conditions in which controlled nuclear fission can occur, so as to generate the heat needed to produce steam for electricity production. In that chapter we saw that one possible fission reaction process was:

$$^{235}_{92}U + ^{1}_{0}n \rightarrow ^{144}_{56}Ba + ^{90}_{36}Kr + 2\,^{1}_{0}n + \text{energy}$$

But this is only one of several reactions which can occur. Another is:

$$^{235}_{92}U + ^{1}_{0}n \rightarrow ^{141}_{56}Ba + ^{92}_{36}Kr + 3\,^{1}_{0}n + \text{energy}$$

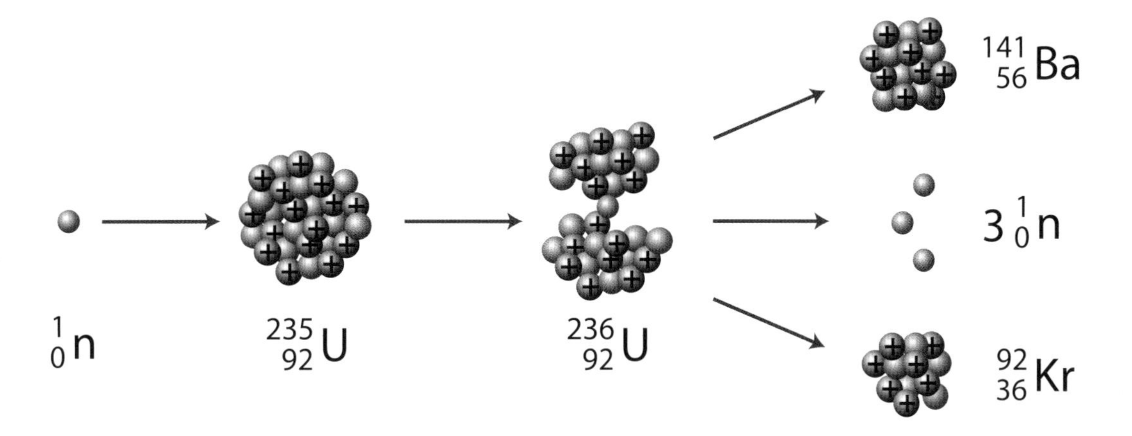

What are the common features of these processes? Fission of uranium:

- Always releases huge* quantities of energy, about 80% of which is carried away by the kinetic energy of the two major fission fragments
- The fission fragments are often radioactive and their subsequent decay accounts for about 10% of the total energy released.
- There are always extremely penetrating and highly dangerous gamma rays produced along with the fission fragments. These gamma rays and the kinetic energy of the sub-atomic particles produced account for the remaining energy released.
- There are always further neutrons produced, on average around 2.5 per fission. These additional neutrons can go on to produce further fission events, yielding more neutrons which produce even more fission events and so on. In this way fission of uranium has the potential to produce a chain reaction.

*Burning 1 atom of carbon (coal) typically releases 5 eV; the fission of 1 uranium nucleus releases more than 200 million eV.

An uncontrolled fission chain reaction in uranium is brought about in an atomic bomb. The focus of this chapter is how to bring about a **controlled chain reaction in a fission reactor**.

There are four common types of nuclear reactor: the Magnox type, the Advanced Gas-cooled Reactor (AGR), the Pressurised Water Reactor (PWR) and the Fast Reactor. Britain's oldest nuclear reactors were of the Magnox type, so-called because the natural uranium fuel was clad in a tube made of magnesium alloy. AGRs use circulating gas (almost invariably carbon dioxide) as the coolant, while PWRs use water under such high pressure that even at a temperature of over 200°C it is still not boiling. Fast reactors use plutonium rather than uranium as the fuel and the coolant is circulating liquid sodium. Fast reactors tend to be used in nuclear-powered submarines. Because the AGR is now the most common type of reactor used in Britain, our future discussion will be limited to that reactor type.

At the moment around 15% of the world's electricity comes from nuclear reactors and that figure is likely to rise as more and more developed countries see nuclear power as a way of meeting their electricity needs while making progress towards their commitment to reduce greenhouse gas emissions.

Fuel for Nuclear Reactors

Natural uranium consists of about 99.3% of uranium–238 and 0.7% uranium–235. Uranium–238 is fissile, but only with very fast neutrons. Conversely, uranium–235 is fissile only with slow neutrons (called thermal neutrons). The neutrons emitted in the fission of uranium–235 are too slow to cause fission in uranium–238, but **need to be slowed down to cause further fission in uranium–235. The neutrons are slowed down by the use of a material called a moderator**, of which the most common are graphite, water (H_2O) or heavy water (deuterium oxide or D_2O). Readers may recall an examination question in the **Momentum** section of this book (page 11) where candidates were asked to prove that graphite is a better moderator than a heavy element such as lead.

To make the uranium within the reactor more likely to undergo fission with slow neutrons it needs to be enriched. This involves adding uranium–235 to natural uranium so increasing the proportion of uranium–235 to uranium–238 within the fuel. Enrichment for the production of civil nuclear power raises the proportion of fissile uranium–235 from about 0.7% in the natural ore to about 3%. Enrichment for military purposes requires a much greater proportion of the uranium to be fissile uranium–235. Since plutonium–239 is also fissile with slow neutrons, this material is sometimes added to natural uranium during fuel reprocessing.

The AGR is an example of a **graphite moderated reactor**. In this reactor the enriched uranium is in long, sealed tubes, called **fuel pins** which are arranged inside a block of graphite. The neutrons released by the fission of uranium–235 collide with the atoms of the graphite moderator and are slowed down to a speed where they are more likely to cause further fission in uranium–235 than be absorbed by the uranium–238.

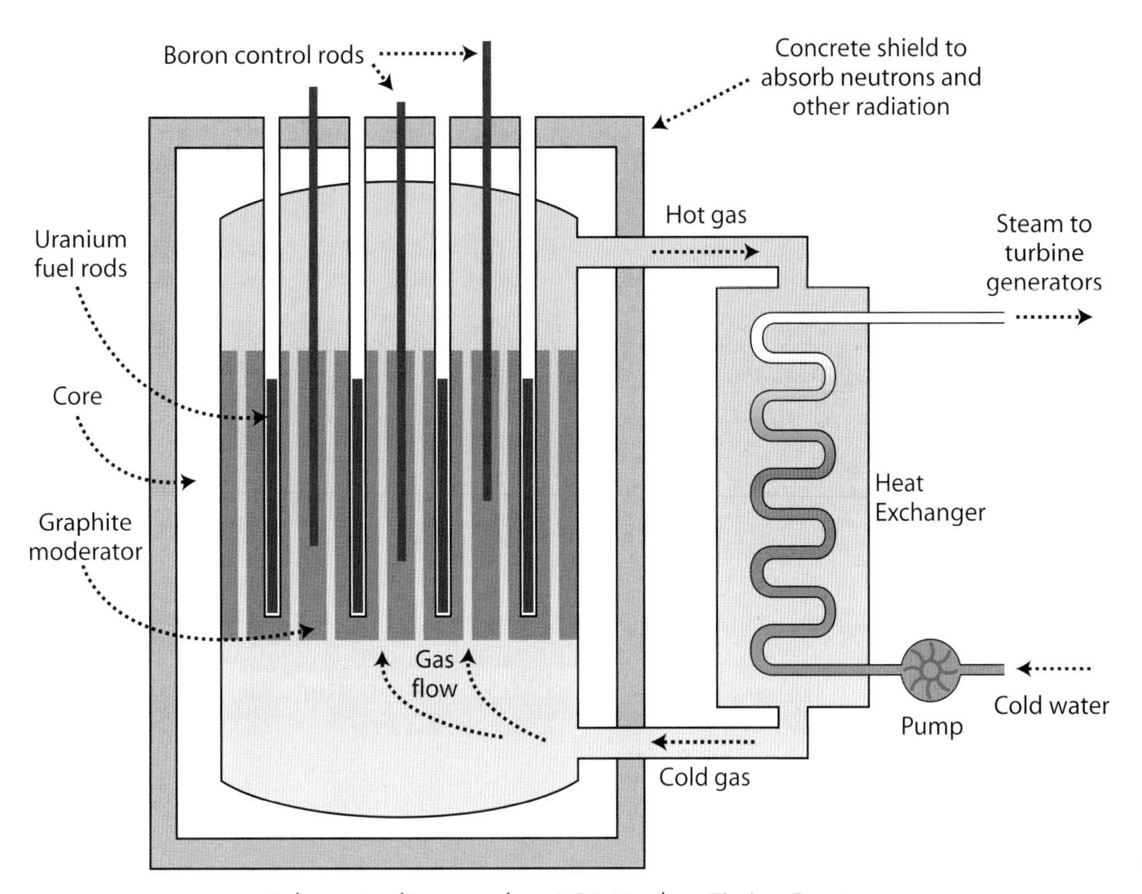

Schematic diagram of an AGR Nuclear Fission Reactor

Wesley Johnston

Hunterston-B (on left), in Scotland, containing a pair of AGR reactors

Critical Size

When a fission event occurs, there are 3 possible fates for the fission neutrons produced. They might escape from the uranium fuel rod where they were formed without causing a further fission; they might be absorbed by a neighbouring nucleus, again without causing fission or they might cause further fission. For the reaction to be sustained, at least one of the neutrons released by each fission event must go on to produce a further fission. The bigger the size of the uranium fuel assembly, the more likely it is that a fission neutron will go on to produce another fission event. We can define the **critical size** of the fuel assembly as that which is just capable of sustaining a chain reaction within it. The critical size for uranium–235 is about that of a small football. Below the critical size, too many of the neutrons which might have induced further fission escape and the chain reaction dies away. A typical nuclear power station has a fuel assembly which is around 5% above the critical size.

Control Rods

If, on average, much more than one of the neutrons produced by fission went on to cause further fission, then the reaction would quickly go out of control. Boron has a high affinity to absorb neutrons. So, **boron-coated steel rods, called control rods, are used to capture excessive neutrons** before they can cause fission. When the control rods are lowered into the reactor the number of available neutrons is decreased and fewer fission events can occur.

Electricity Production

The heat energy produced by the fission reaction is removed by passing a coolant through the reactor. This coolant then passes its energy to the water flowing through a **heat exchanger**, producing steam to drive **the turbines** which turn the **electricity-producing generators**. However, that part of a nuclear power station associated with the turbines and generators is exactly the same as would be seen in a conventional power station burning fossil fuels.

Reactor Shielding

Around every civil reactor is a very thick concrete shield to prevent potentially dangerous radiation, particularly very penetrating gamma waves and neutrons, from reaching the operating personnel and the wider community.

Exercise 22

1 In the context of a thermonuclear reactor explain what is meant by the terms

(i) chain reaction, (ii) critical size, (iii) the moderator, (iv) the control rods, (v) the cooling system and (vi) reactor shielding.

2 What materials are used for (i) the fuel (ii) the moderator (iii) the coolant and (iv) the control rods in a nuclear fission reactor of the AGR type?

3 A reactor core contains 2000 fuel rods containing enriched uranium. Each rod has a mass of 14 kg and 3% of its mass is uranium–235. At full capacity the reactor can produce 2400 MJ of heat energy every second. Fission of 1 kg of uranium–235 produces 1×10^{14} J of heat energy.

(a) How much uranium–235 is there in the core of this reactor?

(b) How much heat energy can the reactor produce in total?

(c) How long would the reactor's nuclear fuel last if it operated at full capacity?

Take 1 year as 3×10^{7} seconds

4.9 Nuclear Fusion

Students should be able to:

- 4.9.1 Understand the conditions required for nuclear fusion;
- 4.9.2 Estimate the temperature required for fusion;
- 4.9.3 Describe the following methods of plasma confinement: gravitational, inertial and magnetic;
- 4.9.4 Appreciate the difficulties of achieving fusion on a practical terrestrial scale;
- 4.9.5 Describe the JET fusion reactor; and
- 4.9.6 State the D-T reaction and appreciate why this is most suitable for terrestrial fusion.

Fusion in the Stars

Almost all of the energy we receive on Earth comes from the Sun as a result of thermonuclear fusion. All the elements which form the basis of our material world as well as the material in all living things on the Earth were formed by fusion in stars.

The Sun consists mainly of hydrogen and helium. At the core of the Sun the temperature is many millions of kelvins resulting in a constant fusion of hydrogen nuclei. The reactions can be summarised by:

$$^{1}_{1}\mathrm{H} + ^{1}_{1}\mathrm{H} + ^{1}_{1}\mathrm{H} + ^{1}_{1}\mathrm{H} \rightarrow ^{4}_{2}\mathrm{He} + \text{other products} + \text{energy}$$

The story of the Universe is one of fusion. Following the Big Bang some 15 000 million years ago (15×10^{9} years) hydrogen and helium formed from the particles available. The hydrogen and helium nuclei had thermal motion that was so great that this prevented them fusing to form heavier nuclei. As the resulting gas cooled down regions of it contracted to form stars and in the stars more helium and heavier nuclei were formed.

The final phases of the evolution of a star depend on its mass. When less massive stars, such as our Sun, run out of hydrogen to fuse, they begin to fuse helium nuclei to form heavier elements. This releases more energy and the star expands into a red giant swallowing up most of the inner planets in the process. More massive stars blow up as supernovae leaving behind a dense neutron star (or a black hole) made up entirely of nuclei (no electrons). The density of such material is enormous (about 1×10^{17} kg m^{-3}).

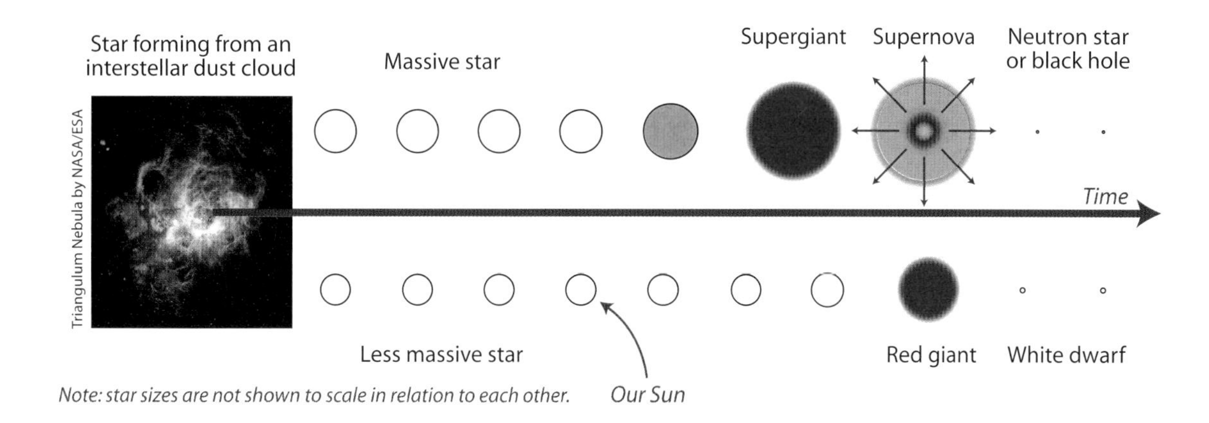

Note: star sizes are not shown to scale in relation to each other.

Conditions Required for Nuclear Fusion

In nuclear fusion two light particles say two protons are brought close together so that they fuse to form a more massive particle.

Protons are positively charged as they come closer they repel each other. The closer they get the greater the repulsive force. How close do they have to be? Certainly closer than the diameter of the nucleus (about 1×10^{-15} m)

Estimate of The Temperature Required for Fusion

If the protons are projected towards each other the repulsion between their positive charges causes their kinetic energy to decrease and their potential energy to increase. The energy needed to bring a pair of protons sufficiently close to bring about fusion is *about* 110 keV. We can use this information to calculate the temperature to which the proton assembly must be heated.

We recall that the temperature of a gas is a measure of the mean kinetic energy of the gas molecules and that

$$\tfrac{1}{2}m\langle c^2\rangle = \text{mean kinetic energy} = \tfrac{3}{2}kT$$

$$\text{where } k = \text{Boltzmann's constant}$$

$$\text{and } T = \text{temperature in K}$$

Suppose we treat the collection of protons as a gas. What temperature is needed to give the protons an average kinetic energy of about 110 keV? (1 keV $=1.6\times10^{-16}$ J)?

$$\text{Mean kinetic energy} = \tfrac{3}{2}kT\text{, so,}$$

$$T = \frac{2 \times \text{mean kinetic energy}}{3 \times k}$$

$$= \frac{2 \times 110 \times 1.6\times10^{-16}}{3 \times 1.38\times10^{-23}} = 850\times10^{6}\ \text{K}$$

Plasma Confinement

To bring protons sufficiently close to bring about fusion on a practical terrestrial scale requires a temperature of 850 million kelvins. How can we achieve such high temperatures? At such temperatures what is matter like? Above 6000K **all matter is in a gaseous state**, so there is no possibility of using a conventional reaction vessel – its walls would simply vaporise! At the sort of temperatures required for fusion, the material is in the form of a **plasma**. In a plasma **all** the electrons have broken free of the atoms, and the gas becomes a mixture of electrons and ions. Once a plasma is created the problem is how to contain it for long enough for fusion to take place. Stars are made of plasma, so how is confinement achieved on the Sun?

In the case of a star, gravitational forces provide the plasma confinement. The inward gravitational pull balances the outward forces created by pressure of the plasma and radiation pressure (photons impacting on the particles of the gas plasma and exerting an outward pressure).

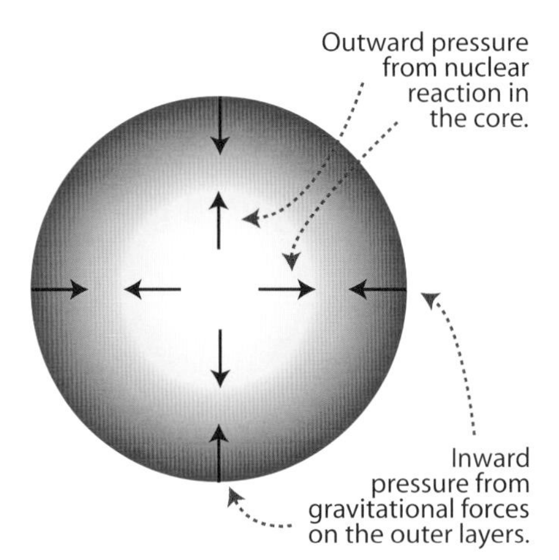

Clearly, **gravitational confinement would not work on the Earth** since we require an enormous mass of material to provide gravitational forces strong enough to balance the forces tending to dissipate the plasma.

Another method of confinement involves **using intense ion or laser beams** directed at **a solid fuel pellet** (such as lithium hydride). **The beams provide the energy to heat the material to the temperature required for fusion.** This is a technique that many physicists are currently researching.

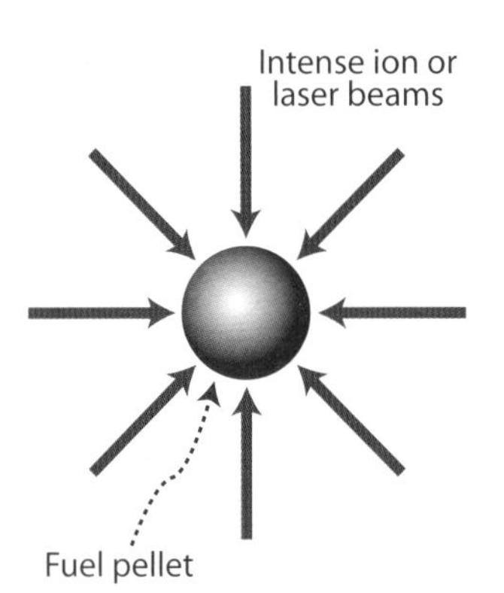

It is called **inertial confinement**. The idea is to produce fusion for long enough to extract the energy before the plasma escapes – rather like pulling a table-cloth away from a table before the tea-cups fall over!

The fusion seen in this situation is of the type:

$${}^{6}_{3}\text{Li} + {}^{2}_{1}\text{H} \rightarrow 2\,{}^{4}_{2}\text{He}$$

So far we have looked at

- **gravitational confinement** and
- **inertial confinement**

But the third technique, **magnetic confinement**, has the promise of being more successful.

Magnetic confinement uses magnetic fields to hold the plasma in what is sometimes called a **magnetic bottle**. The particles in a plasma are charged. If such particles are moving in a magnetic field they experience a force.

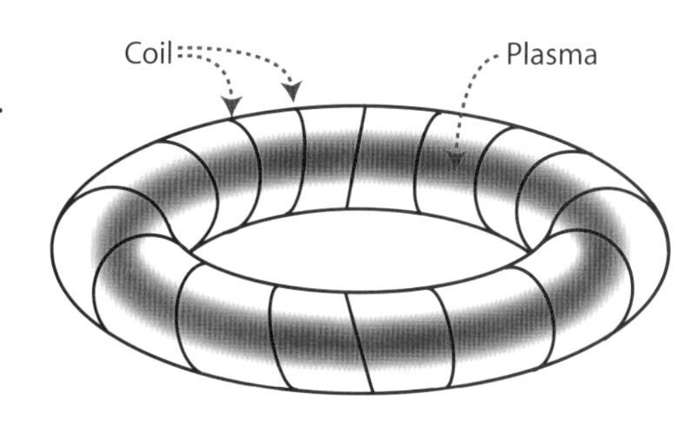

The magnetic field is produced by electric current flowing in a coil wound into a shape known as toroid (doughnut). The magnetic field produced is circular within the highly evacuated toroidal chamber.

Charged particles moving parallel or anti-parallel to the magnetic field do not experience a force and move in a straight line with constant speed. Particles moving at right angles to the magnetic field experience a force that makes them move in a circle.

However, many plasma particles have velocities making some small angle with the magnetic field. We can resolve their velocity into directions parallel and perpendicular to the magnetic field. These particles move in a spiral around the magnetic field lines, as explained below.

Velocity of positive ions, v

Direction of magnetic field, B

Firstly resolve the ion velocity, v, parallel and perpendicular to the direction of the magnetic field.

This component of the velocity, **perpendicular to the magnetic field**, v_{perp}, causes the positive ions to move in a **circular path** around the field line. Recall that the force on a moving charge in a magnetic field is given by:

v_{perp}

Direction of magnetic field, B

$$F = Bqv_{perp}$$

where B = magnetic field strength (flux density)

q = charge and v_{perp} = velocity

The particle moves in a circle of radius r where:

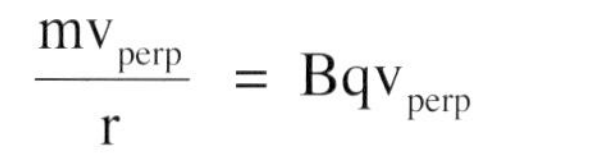

$$\frac{mv_{perp}}{r} = Bqv_{perp}$$

There is no force parallel to the direction of B, so the component of velocity parallel to the magnetic field, causes the positive ions to move in the same direction as the field. The two components of velocity together cause the plasma ions to spiral around the magnetic field lines.

How is the Plasma Heated?

A plasma is a good conductor of electricity. The plasma acts like the secondary coil of a large transformer. The current induced in this circulating plasma secondary heats the particles.

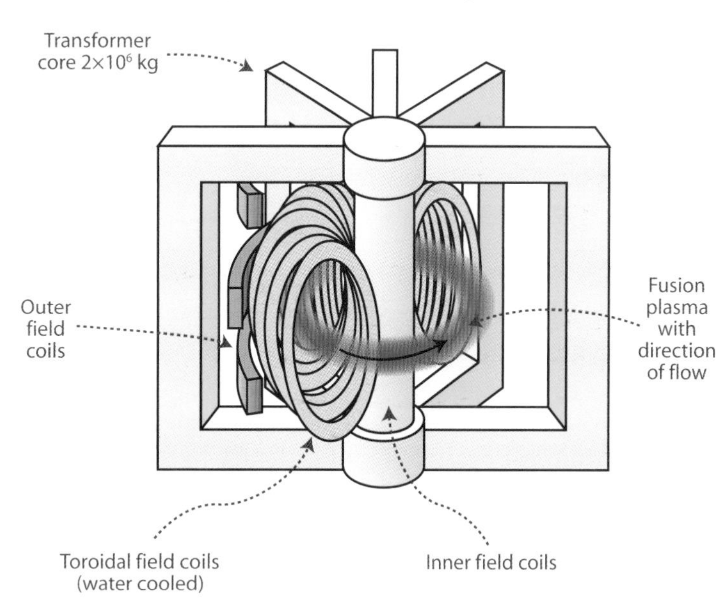

The magnetic confinement system shown opposite is of the Joint European Torus (JET) which is based at the Culham Laboratory near Oxford in England. It is called a **tokamak** from the Russian acronym for a "toroidal chamber with a magnetic field". The

Culham tokamak has achieved temperatures of around **1×10^7 K**. The current flowing in the toroid is of the order of **several million amperes!**

In the simplified diagram of a tokamak shown opposite, the containment magnetic field is produced by toroidal, current-carrying coils. These generate enormous quantities of wasted heat, so the coils must be water-cooled. Current is induced in the plasma to heat it to the temperatures required for fusion.

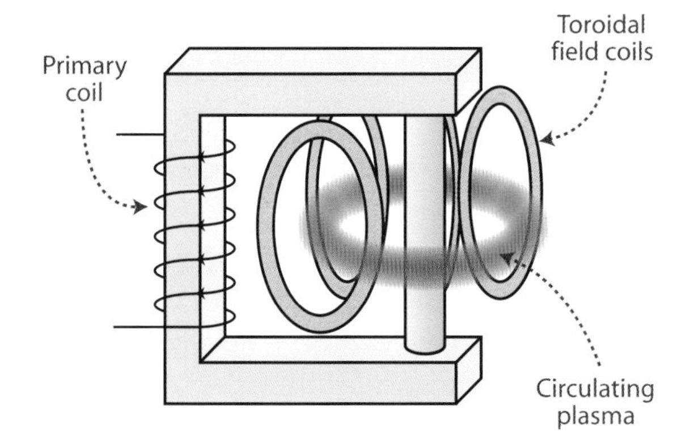

The initial design for the original tokamak is due to two Russian physicists one of whom was Andrei Sakharov, the human rights campaigner who achieved a Nobel Prize for Peace in 1975.

There are many ongoing problems for researchers in this field. One is to keep the plasma well-contained and at sufficiently high temperature for long enough to obtain adequate numbers of fusion reactions. Another is to ensure that the energy produced by fusion exceeds that supplied by the operators to maintain the reactor at its enormously high temperature.

The Deuterium-Tritium (or D-T) Reaction

The process with greatest promise at the moment is known as the **deuterium-tritium (or D-T) reaction:**

$$^{2}_{1}H + ^{3}_{1}H \rightarrow ^{4}_{2}He + ^{1}_{0}n + 17.6\text{ MeV}$$

Suppose that surrounding the reactor is a blanket of lithium. Lithium can absorb the fusion neutron and then fission according to the reaction:

$$^{7}_{3}Li + ^{1}_{0}n \rightarrow ^{4}_{2}He + ^{3}_{1}H + ^{1}_{0}n$$

The neutrons released in both reactions above can go on to sustain a chain reaction in lithium, thus converting the lithium into tritium fuel for the D-T process.

The heat from the kinetic energy given to the helium nuclei would then heat water to give steam, which then would be used to drive turbine-generators.

The D-T process is attractive because:

- There is a ready supply of fuel (deuterium from seawater and tritium from lithium)
- The process is well understood
- There are only limited hazardous waste products (neutron irradiated materials)
- There is a considerable energy yield per kg of fuel used

Advantages of Fusion

- Readily available and virtually inexhaustible supplies of fuel (seawater contains 1 atom of deuterium for every 5000 of ordinary hydrogen and this can be extracted by electrolysis)
- None of the toxic and highly radioactive waste associated with fission. However there are waste products in the form of materials that have been irradiated with neutrons but these pose

much less of a problem than the very long half-life fission fragments associated with fission of uranium

- Greater yield of energy per kilogram of fuel consumed from a hydrogen fusion reactor than from a thermonuclear fission reactor. It has been estimated that a 2000 MW fusion power station would require only 500 grams of deuterium and 1.8 kg of lithium per day.
- Fusion reactors are inherently "fail safe" as fuel is continuously fed into them – stop this and the reaction ceases. In fission reactors all the fuel is place in the reactor to start with as there has to be a least a critical mass before the fission chain reactor can proceed.

Exercise 23

Examination Questions

1 (a) Explain why it is difficult to achieve fusion reactions on a practical terrestrial scale

(b) Describe briefly the main features of the JET fusion reactor.

[CCEA June 2002]

2 Temperatures as high as 8.0×10^8 K are needed for thermonuclear fusion to occur. At this temperature matter exists as a plasma. Estimate the kinetic energy of a particle in a plasma at a temperature of 8.0×10^8 K. Express your answer to two significant figures.

[CCEA June 2006]

3 (a) Describe what a plasma is.

(b) Briefly outline the basic principles of plasma confinement by inertial confinement.

[CCEA June 2007]

4 (a) Write down an equation for any **one** of the fusion reactions which is believed to occur on the Sun.

(b) State two advantages that fusion reactors would have over current fission reactors, if they could be made to work successfully.

(c) Name ant two methods of plasma confinement.

(d) To overcome proton-proton repulsion, nuclei need energies of the order of 110 keV. Use this value to estimate the temperature needed for a proton to have this amount of thermal energy. Show your working clearly.

[CCEA Summer 2004]

Unit 5 (A2 2)

Fields and their applications

5.1 Force Fields

Students should be able to:

5.1.1 Explain the concept of a field of force

Fields of Force

The idea of a force field is familiar to us all. Bring a magnet close enough to a steel ball and we see the ball move towards the magnet's pole. Drop an object from a tall building and it accelerates to the surface below due to the gravitational force on it. These are both illustrations of force fields. Indeed, one of the early experiments on magnets was the use of iron filings to show the shape of the field of force.

When physicists use the term "a field of force", they mean something more specific than the layperson does. For the physicist, a field of force is a region of space within which objects with a particular property experience a force. The table below illustrates the properties involved for three common force fields.

Field	Property Involved
Gravitational	Mass
Electrical	Charge
Magnetic	Moving charge

Note that the property and the field must match. An uncharged mass will experience the gravitational force, but not the electrical force. A stationary particle with both mass and charge will experience both gravitational and electrical force, but not the magnetic force.

This enables us to make the following definitions:

A gravitational field is a region of space within which a mass will experience a force.

An electrical field is a region of space within which a charge will experience a force.

A magnetic field is a region of space within which a moving charge will experience a force.

We now have a mechanism to explain the force between, say, electric charges. Around every charge there exists an electric field.

In the diagram opposite, charge Q_2 is in the field of Q_1 so Q_2 experiences a force.

But charge Q_1 is in the field of Q_2 so Q_1 also experiences a force.

Similar arguments can be put forward to explain the existence of gravitational and magnetic forces.

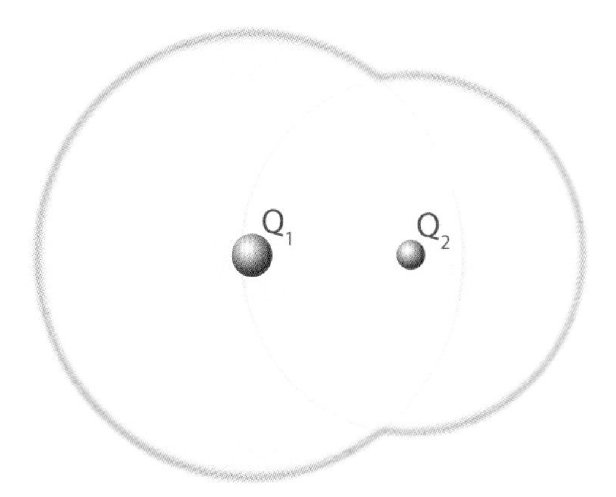

Around every charge is an electric field.

5.2 Gravitational Fields

Students should be able to:

5.2.1 Define gravitational field strength;

5.2.2 Recall and use the equation $g = \frac{F}{m}$;

5.2.3 State Newton's law of universal gravitation;

5.2.4 Recall and use the equation for the gravitational force between point masses, $F = G\frac{m_1m_2}{r^2}$

5.2.5 Recall and apply the equation for gravitational field strength, $g = G\frac{m}{r^2}$ and use this equation to calculate the Earth's mass;

5.2.6 Apply knowledge of circular motion to planetary and satellite motion;

5.2.7 Show that the mathematical form of Kepler's third law (t^2 proportional to r^3) is consistent with the law of universal gravitation;

5.2.8 State the period of a geostationary satellite;

The Strength of a Field of Force

We have already indicated that a force field will exert a force on an object with the specific property associated with that field. However, it is very convenient to describe the strength of the field at any point as the force on **a unit** of that property. Thus, for example, **the gravitational field strength at a point is equal to the force which would be produced on a test mass of 1 kg at that point.**

Similarly, **the electrical field strength at a point is equal to the force which would be produced on a test charge of +1 C at that point.**

These definitions make it clear that field strength is a vector – it has both magnitude and direction. For the moment we will restrict the discussion to gravitational and electrical fields. Magnetic fields will be dealt with later.

The table below illustrates the relationships between the relevant property and the field strength for gravitational and electrical fields.

Field	Symbol for Field Strength	Field Strength Equation	Unit for Field Strength
Gravitational	g	g = force per unit mass = F/m	N kg^{-1}
Electrical	E	E = force per unit charge = F/q	N C^{-1}

Gravitation

Following on the work of Tycho Brahe and Johannes Kepler on the movement of the planets around the Sun, Isaac Newton suggested in 1687 what today is known as Newton's Law of Universal Gravitation. It states the following:

Between every two point masses there exists an attractive gravitational force which is directly proportional to the mass of each and inversely proportional to the square of their separation.

Newton's Law can readily be expressed as an equation:

$$F = G\frac{m_1m_2}{r^2}$$

where m_1 and m_2 are the respective point masses, in kg
r is the distance in metres between them
G is a constant known as the universal gravitational constant

Note that when dealing with real objects, it is sufficient to imagine the entire mass to be concentrated at the centre of mass and to consider that point to be the point mass.

The generally accepted value of G is 6.67×10^{-11} N m^2 kg^{-2}. The small value of G means that gravitational forces between objects can usually be ignored unless at least one of them has a particularly large mass. Certainly gravitation is **not** involved in inter-atomic and inter-molecular forces.

There are two other interesting facts about gravitational forces. The first is that they have **infinite range,** so they can operate over the vast distances which exist between galaxies. The second is that they cannot be shielded. It is well known for example that there can be no magnetic field inside a hollow sphere made of soft iron. Similarly there can be no electric field inside a conducting metal box. However, there is **no known way to shield from a gravitational field.**

Exercise 24

1 By considering the mathematical form of Newton's Law of Universal Gravitation, show that G has derived units N m^2 kg^{-2} and hence determine the SI base units in which G can be measured.

2 In a hydrogen atom an electron of mass 9.11×10^{-31} kg orbits a proton of mass 1.66×10^{-27} kg at a radius of 50 pm. The speed of the electron is 2.2 Mms^{-1}. Calculate:
(a) the centripetal force on the electron
(b) the gravitational force between the proton and the electron.
Comment on the results of these calculations.

3 Calculate the gravitational field strength on a planet where a mass of 5 kg has a weight of 180 N.

4 The gravitational force between two point masses is 36 N when they are 2 m apart. Calculate the gravitational force between the same masses when they are:
(a) 1 m apart
(b) 3 m apart

5 An object weighing 14 N on Earth has a weight of 35 N on planet Xenon. Calculate the gravitational field strength on Xenon. Take g_{earth} = 9.8 N kg^{-1}.

Gravitational Field Lines

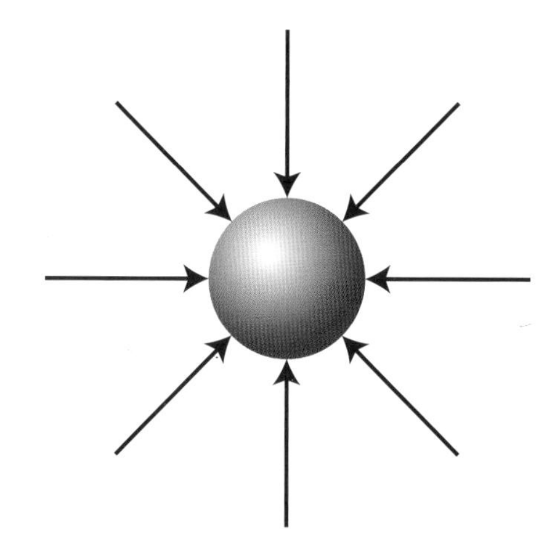

Gravitational Field Lines around a Uniform Sphere (or Planet)

The direction of a gravitational field line at a point shows the direction of the gravitational force on a mass of 1 kg at that point. Clearly this means that field lines can never cross or touch each other – to do so would give rise to an ambiguity as to the direction of the field at the crossing point. If the field line is curved, then the direction of the field is along the tangent at the point in question.

For a uniform sphere the gravitational field pattern is described as **radially inwards,** because all field lines appear to converge at the centre of mass of the sphere.

Notice that as one moves away from the surface of the sphere the field lines get further apart. At the same time the strength of the field decreases. This is generally so – **the closer the field lines, the stronger the gravitational field.**

For the observer on Earth the field lines strike the surface at right angles. The radius of the Earth is so large that the field lines appear as parallel and equally spaced. Such a field is, of course, a **uniform field.**

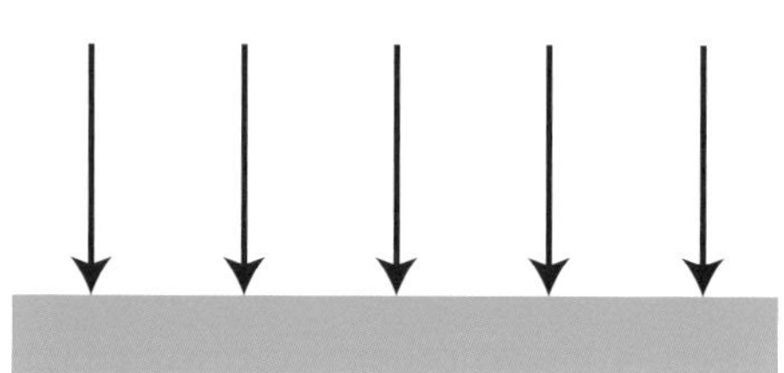

Uniform field at the Earth's Surface

Variation of Gravitational Field Strength with Height above the Earth's Surface

Consider the gravitational force on a mass of 1 kg as it moves away from the surface of the Earth. Then, by Newton's Law, and remembering that the force on 1 kg is the gravitational field strength, g,

$$g = F_{\text{on 1kg}} = \frac{(G\,M_E \times 1)}{r^2} \quad \text{and hence}$$

$$g = G\frac{M_E}{r^2}$$

where r is the distance from the centre of the Earth

$r \geq R_E$ is the radius of the Earth

From this we see that **g varies inversely as r^2** and that **g has a maximum value on the surface of the Earth.** At the surface the field strength, g_o is given by

$$g_0 = G\frac{M_E}{R_E^2}$$

and the mean value of g_o on the Earth's surface is 9.81 N kg^{-1}. The zero in g_o refers to zero height above the surface.

Students should observe that the graph of gravitational field strength, g, against distance from the centre of the Earth is a curve which touches neither the vertical nor the horizontal axis.

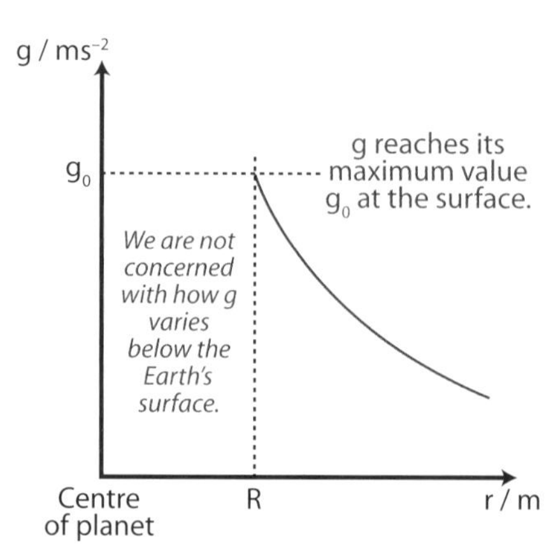

Distance from the planet's centre

The mean field strength at the equator is 9.78 N kg^{-1} while at the poles it is 9.83 N kg^{-1}. This is mainly because the equatorial radius is slightly greater than the polar radius.

The Mass of the Earth

The equation for g_o on the previous page provides an in direct way to estimate the mass of the Earth. The average value of g over the surface of the Earth is generally taken as 9.81 ms^{-2}. The radius of the planet can be found from observations of the fixed stars from points of known separation on the Earth's surface and is generally taken as around 6400 km.

Since:

$$g_0 = G\frac{M_E}{R_E^2}$$

$$M_E = \frac{g_0 R_E^2}{G} = \frac{9.81 \times (6.4 \times 10^6)^2}{6.67 \times 10^{-11}}$$

So, the mass of the Earth, $M_E = 6.0 \times 10^{24}$ kg

Worked Example

Locating a Neutral Point

The Moon has a mass of 7.3×10^{22} kg and orbits the Earth at a mean distance of 384 000 km. The mass of the Earth is 6.0×10^{24} kg.

(a) Taking both the Moon and the Earth as point masses, locate the point, P, on the line joining their centres, where the gravitational field strength is zero. Such a point is called a **neutral point.**

(b) Suggest why it is requires more fuel to take a spacecraft from the surface of the Earth to the Moon than it does to take the same spacecraft from the Moon to the Earth.

Solution

(a) Suppose the distance from P to the Earth is x, the distance from P to the Moon is $3.84 \times 10^8 - x$

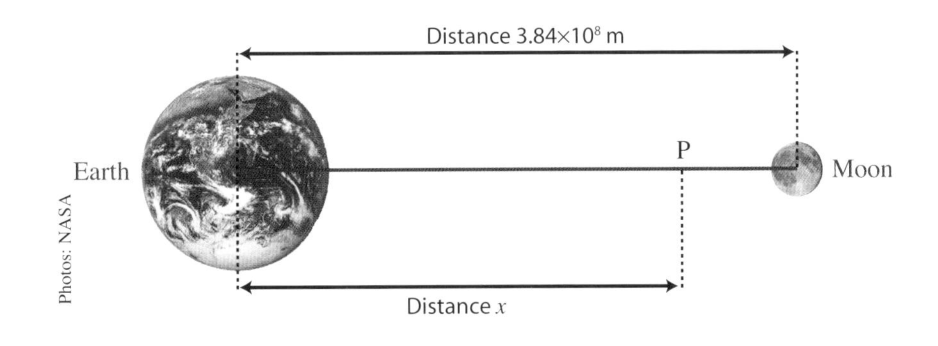

At the neutral point,
Field strength due to Earth = Field strength due to Moon

$$\frac{GM_{earth}}{x^2} = \frac{GM_{moon}}{(3.84 \times 10^8 - x)^2}$$

Cancelling G, substituting for the masses of the Earth and the Moon, and taking the square root of both sides,

$$\frac{\sqrt{6.0 \times 10^{24}}}{x} = \frac{\sqrt{7.3 \times 10^{23}}}{(3.84 \times 10^8 - x)}$$

which simplifies to

$$\left\{\sqrt{7.3 \times 10^{23}} + \sqrt{6.0 \times 10^{24}}\right\}x = 3.84 \times 10^8 \times \sqrt{6.0 \times 10^{24}}$$

$$x = \frac{9.41 \times 10^{20}}{2.72 \times 10^{12}} = 3.46 \times 10^8 \text{ m}$$

so the neutral point is approx. **3.46×10^8 m from the Earth** and **3.8×10^7 m from the Moon.**

(b) Travelling from Earth to the Moon requires more work to be done that when travelling to Earth from the Moon. This is because work is done against the greater gravitational force of the Earth for a much longer distance. In both cases, once you reach the neutral point, the gravitational force of the target pulls you towards your destination. Since the neutral point is much further from the Earth than the Moon, it takes more fuel to travel from the Earth to the Moon than from the Moon to the Earth.

Satellites Orbiting the Earth

(a) The General Case

Consider a satellite of mass m orbiting the Earth in the equatorial plane and at a distance r from the Earth's centre. Assuming a circular orbit, we can use the fact that the gravitational attraction between the satellite and the Earth provides the centripetal force and write:

$$mr\omega^2 = \frac{GM_Em}{r^2}$$

where ω = the angular velocity of the satellite
m = its mass

Cancelling m on each side and rearranging gives

$\omega^2 = \frac{GM_E}{r^3}$ and since $\omega = \frac{2\pi}{T}$ we can simplify further to give

$$T^2 = \frac{4\pi^2 r^3}{GM_E} \quad \text{or} \quad T = 2\pi\sqrt{\frac{r^3}{GM_E}}$$

(b) Just Above the Surface

To a good approximation, a satellite orbiting the Earth at a height of 200 km or less has an orbital radius equal to the radius of the planet itself, R_E. Substituting R_E for r in the previous equation gives:

$$T = 2\pi\sqrt{\frac{R_E^3}{GM_E}}$$

Substituting the numerical values of G, M_E and R_E, we obtain a period of about 85 minutes.

(c) Geostationary Satellites

A geostationary satellite is one which orbits **in the equatorial plane** with **an orbital period of 24 hours.** This means that **its angular velocity is the same as that of the Earth itself on its polar axis. The direction of the satellite's motion is in the same sense as that of the Earth spinning on its axis.** To an observer on the Earth's surface such a satellite appears stationary. We can therefore use the equation from (a) and write:

$$T = 2\pi\sqrt{\frac{r^3}{GM_E}} = 24 \times 3600 \text{ seconds}$$

and rearranging gives

$$r^3 = \frac{GM_E \cdot (24 \times 3600)^2}{4\pi^2}$$

Substituting the numerical values of G and M_E we obtain

$$r^3 = \frac{6.67 \times 10^{-11} \times 6.0 \times 10^{24} \times (24 \times 3600)^2}{4\pi^2} = 7.567 \times 10^{22} \text{ m}^3$$

Taking the cube root gives $r = 4.23\times10^7$ m and since the radius of the Earth itself is 6.4×10^6 m, the orbital height is $4.23\times10^7 - 6.4\times10^6 =$ **3.59×10^7 m**

The orbital height of a geostationary satellite is therefore fixed.

The **velocity** of a geostationary satellite **is also fixed**, since its period is 24 hours and its orbital radius is fixed at 4.23×10^7 m.

Orbital velocity,

$$v = \frac{2\pi R}{T} = \frac{2\pi \times 4.23 \times 10^7}{24 \times 3600} = 3076 \text{ ms}^{-1}$$

Kepler's Third Law

Johannes Kepler has three laws relating to ... ne of them, the third law, states that the square of the ... Sun is directly proportional to the cube of their n... ally, the law can be written:

$$T^2 \propto r^3 \quad \text{or} \quad \frac{T^2}{r^3} = \text{a constant}$$

Showing Kepler's Third Law is Consistent

The CCEA specification requires students ... Law is consistent with Newton's Law of Gravitat... he mathematics also allows us to estimate the ...

Consider a planet of mass m moving abou... ose the angular velocity of the planet is ω and the ...

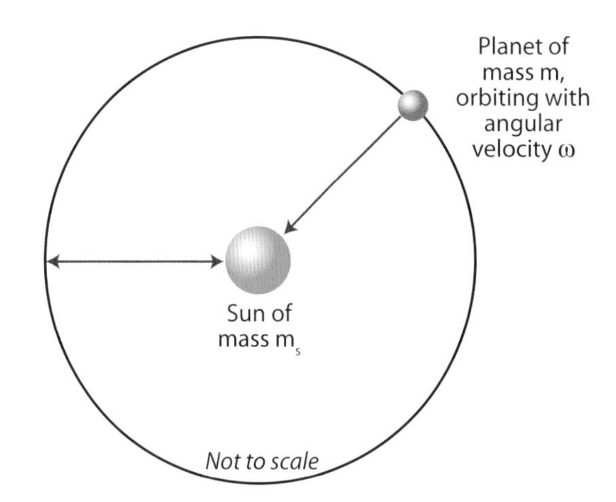

Then, since gravitation provides the centripetal force between the planet and the Sun,

$$F = mr\omega^2 = G\frac{m_s m}{r^2}$$

Dividing by m and making the substitution ω = 2π/T gives

$$r\left(\frac{2\pi}{T}\right)^2 = G\frac{m_s}{r^2}$$

Removing the brackets gives

$$\frac{4\pi^2 r}{T^2} = G\frac{m_s}{r^2}$$

Rearranging to make T^2 the subject gives

$$T^2 = \frac{4\pi^2 r^3}{Gm_s} \quad \text{or} \quad \frac{T^2}{r^3} = \frac{4\pi^2}{Gm_s}$$

G, m_s and π are constants regardless of the planet being considered.

Hence, $\frac{T^2}{r^3}$ is constant and hence Newton's Law of Gravitation and Kepler's Third Law are consistent with each other.

Mass of the Sun

The above shows that $T^2 = \frac{4\pi^2 r^3}{Gm_s}$ or $m_s = \frac{4\pi^2 r^3}{GT^2}$

If now we consider the Earth orbiting at a mean separation of 150 million km from the Sun and with a period of 365.25 days, we can estimate the Sun's mass from

$$m_s = \frac{4\pi^2 r^3}{GT^2} = \frac{4\pi^2 \times \left(1.5 \times 10^{11}\right)^3}{6.67 \times 10^{-11} \times \left(365.25 \times 24 \times 3600\right)^2} = 2.0 \times 10^{30} \text{ kg}$$

which is about 330 000 times greater than the mass of the Earth.

Exercise 25

Examination Questions

1 A lunar landing craft of mass 17 500 kg moves in a circular orbit of radius 2.00×10^6 m around the moon. The gravitational field strength due the moon at this location is 1.23 Nkg^{-1}. Calculate the momentum of the lunar landing craft at this location and state its direction.

[CCEA Summer 2007]

2 (a) Communications satellites are usually placed in geostationary orbits. The radius of such an orbit has a definite value.

(i) State **two** other features of a geostationary orbit.

(ii) The mass of the Earth is 6.0×10^{24} kg. Calculate the radius of a geostationary orbit.

(b) A daily influx of meteorites and meteor dust is well known to scientists, but the total mass added daily to the Earth's surface is difficult to estimate. Some scientists estimate that 7.9×10^7 kg is added per year. Calculate how many years it would take for the gravitational field strength at the surface of the Earth to increase by 0.1%. Assume that the increase in mass continues at a steady rate and that the radius of the Earth is unaltered.

[Radius of Earth = 6.38×10^6 m; gravitational field strength at Earth's surface = 9.81 Nkg^{-1}.]

[CCEA Winter 2007]

3 (a)(i) State, in words, the law of gravitational force between two point masses.

(ii) The laws of force for gravitational and electric fields have similar mathematical forms. However, they differ in some important ways. State one of these differences.

(b) Gravitational force has an important role in explaining the orbits of planets and satellites. What is this role?

(c)(i) Kepler's third law of planetary motion states that, for the simplified case of circular orbits, the square of the period of rotation of the planet in its orbit about the Sun is proportional to the cube of the radius of the orbit. Show that this result is consistent with the law of gravitational force.

(ii) The radius of the Earth's orbit about the Sun is 1.50×10^{11} m. Calculate the mass of the Sun.

[CCEA Summer 2006]

5.3 Electric Fields

Students should be able to:

5.3.1 Define electric field strength;

5.3.2 Recall and use the equation $E = \dfrac{F}{q}$;

5.3.3 State Coulomb's law for the force between point charges;

5.3.4 Recall and use the equation for the force between two point charges,

$$F = \frac{q_1 q_2}{4\pi\varepsilon_0 r^2} = k\frac{q_1 q_2}{r^2} \text{ where } k = \frac{1}{4\pi\varepsilon_0}$$

5.3. … n and determine its SI base units;

5.3. … c field strength due to a point charge,

5.3. … d, the field strength is constant, and recall and

5.3. … gravitational and electric fields;

Electric …

In a celebr… Charles Coulomb, set forth the law of electric charges wh…

Between every two point charges there exists an electrical force which is directly proportional to the charge of each and is inversely proportional to the square of their separation.

Coulomb's Law can be expressed mathematically. If point charges Q_1, and Q_2 are a distance r apart, and F is the force on each, then according to Coulomb's law:

$$F \propto \frac{Q_1 Q_2}{r^2}$$

With a suitable constant, we arrive at the equation:

$$F = \frac{kQ_1 Q_2}{r^2}$$

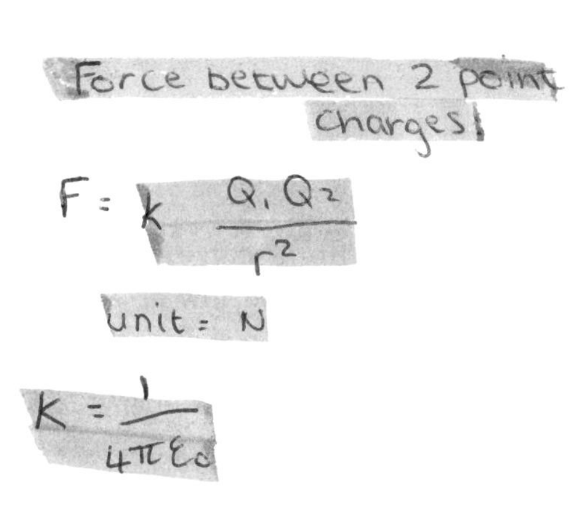

Unlike the force between point masses, the force between … repulsive. The force is attractive when the charges have di… When the charges have the same sign, the force is repulsiv…

The value of the constant, k, depends on the nature of the … point charges. This constant can be found experimentally. … is 8.99×10^9 Nm^2 C^{-2}. However, it is more convenient to us… **permittivity** of the material.

For charges in a vacuum, Coulomb's Law is usually written:

$$F = \frac{Q_1Q_2}{4\pi\varepsilon_0 r^2}$$

where ε_0 is known as the **permittivity of free space** (vacuum) and has the experimental value of $8.85\times10^{-12}\ C^2 N^{-1} m^{-2}$.

It may appear at first sight that the 4π complicates the above equation, but it is a necessary consequence of the SI system of units and indeed, it brings significant simplifications later.

Worked Examples

Example 1

Show from Coulomb's Law that the permittivity of free space, ε_0, has units $C^2 N^{-1} m^{-2}$. Given that the unit of capacitance, the farad (F) is equivalent to the coulomb per volt show that ε_0 has units of Fm^{-1}.

Solution

Coulomb's Law can be written:

$$F = \frac{Q_1Q_2}{4\pi\varepsilon_0 r^2}$$

Rearranging gives:

$$\varepsilon_0 = \frac{Q_1Q_2}{4\pi F r^2}$$

Since charge Q is measured in C, force F is measured in N, distance r is measured in m and the 4π term has no units, then:

Unit for $\varepsilon_0 = C^2 N^{-1} m^{-2}$

Unit for $\varepsilon_0 = C^2 N^{-1} m^{-2} = C^2 (N^{-1} m^{-1}) m^{-1} = C^2 (N m)^{-1} m^{-1} = C^2 J^{-1} m^{-1} = C (JC^{-1})^{-1} m^{-1}$

But the volt, V, is defined as the JC^{-1}, so:

Unit for $\varepsilon_0 = CV^{-1}m^{-1} = Fm^{-1}$

Example 2

Express ε_0 in terms of its SI base units.

Solution

Unit for $\varepsilon_0 = Fm^{-1} = C\ V^{-1}\ m^{-1} = A\ s\ (JC^{-1})^{-1} m^{-1} = A\ s\ (kg\ m^2 s^{-2})^{-1}\ C\ m^{-1}$

$= A\ s\ kg^{-1}\ m^{-2}\ s^2\ (A\ s)\ m^{-1} = \mathbf{A^2 s^4 kg^{-1} m^{-3}}$

Alternatively, unit for ε_0 $= C^2 N^{-1} m^{-2} = (A\ s)^2\ (kg\ ms^{-2})^{-1}\ m^{-2} = (A\ s)^2\ kg^{-1}m^{-1}s^2\ m^{-2}$

$= \mathbf{A^2 s^4 kg^{-1} m^{-3}}$

Electric Field Strength, E

It is useful to recall our earlier definition of the **strength of an electric field, E, at a point as the force which would be produced on a test charge of + 1 C at that point.**

This definition allows us to write:

$$E = \frac{F}{q} \quad \text{or} \quad F = Eq$$

The unit for electric field strength is therefore the NC^{-1}.

For an isolated point charge, Q, therefore, we obtain:

$$E = \frac{Q}{4\pi\varepsilon_0 r^2}$$

Note that electric field strength, E, is a **vector.** Its direction is that of the force on a positive (+) charge. Physicists say that **the electric field around an isolated point charge decreases as $1/r^2$.**

Like the gravitational field around a point mass, the electrical field around a charged sphere is also radial. However, there is an importance difference. Around an isolated positive charge the field lines are radially **outwards,** while around an isolated negative charge the field lines are radially **inwards.** This is to satisfy the requirement that **the direction of the field is the same as that of the force on a positive test charge.**

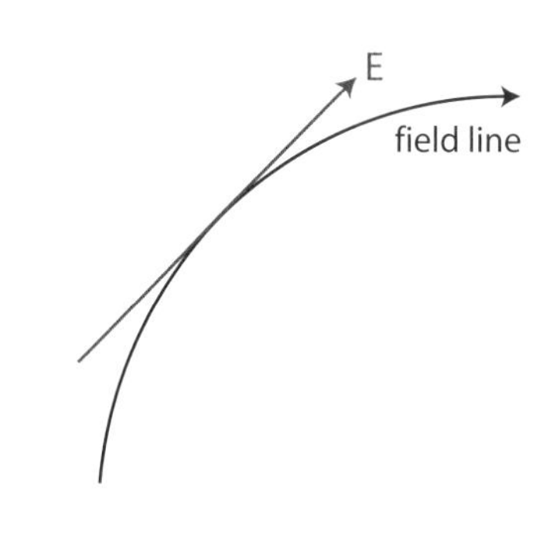

If the field line is curved as in the diagram on the right, then the tangent to the curve at any point gives the direction of the electric field at that point.

The diagrams below show some common field patterns.

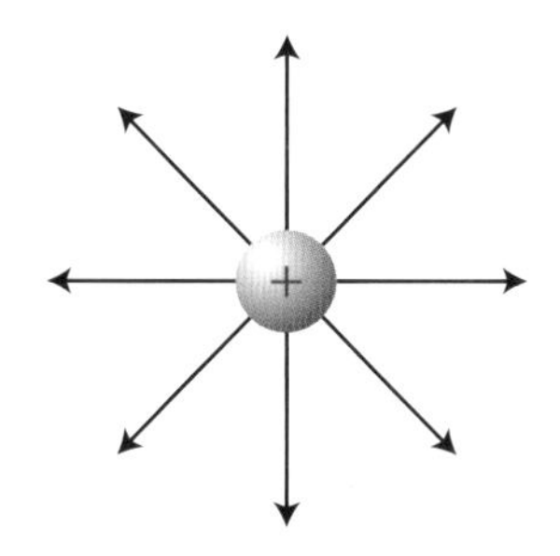

The radially outwards field pattern around a **positively** charged, isolated conductor.

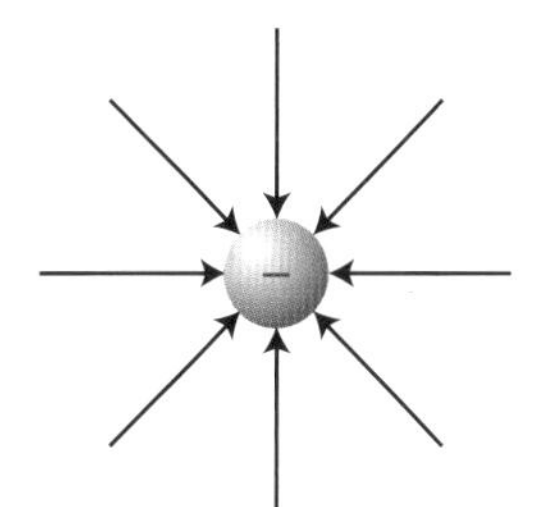

The radially inwards field pattern around a **negatively** charged, isolated conductor.

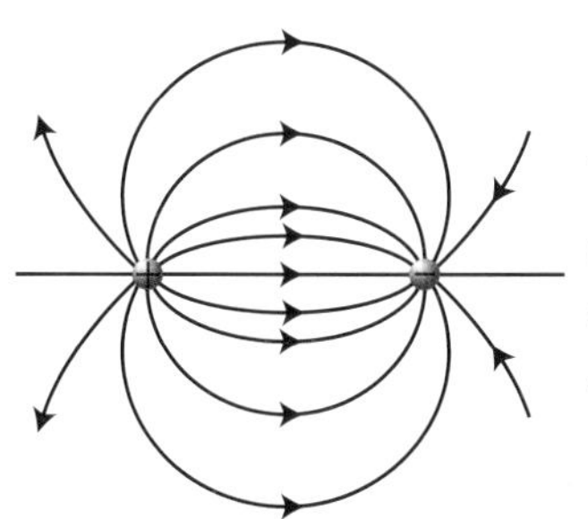

The field around adjacent positive and negative charges.

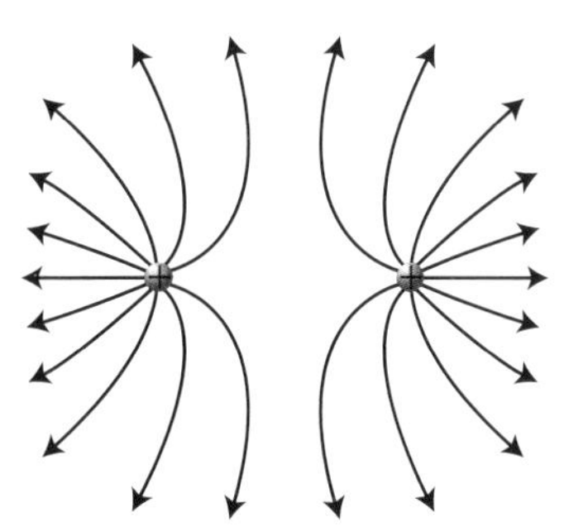

The field around a pair of adjacent positive charges. There is a neutral point between the charges.

The Uniform Electric Field

Consider a pair of parallel metal plates, a distance, **d**, apart, and suppose that between them we maintain a constant potential difference, V. Between such plates, and away from the edges, there is a uniform electrical field as illustrated below.

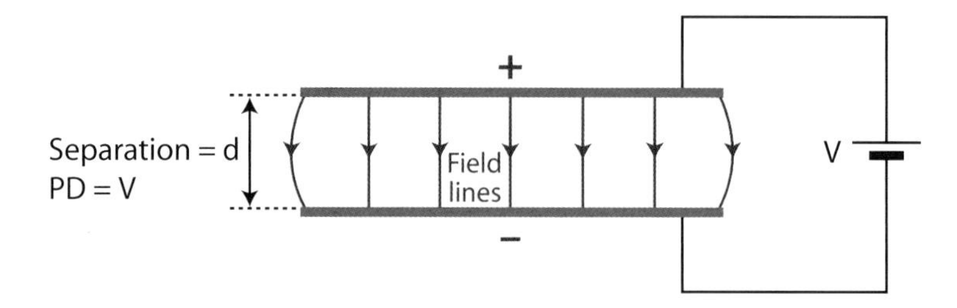

From the definitions of electrical potential and electric field strength, it is possible to establish the relationship:

$$\text{Field strength, } E = -\frac{V}{d} = -\text{Potential Gradient}$$

Note that **the minus sign** simply means that the electric field strength, E, is in the direction of decreasing potential and **is often omitted.**

The graph below illustrates how the potential varies as we move from the top (positive plate to the bottom (negative) plate.

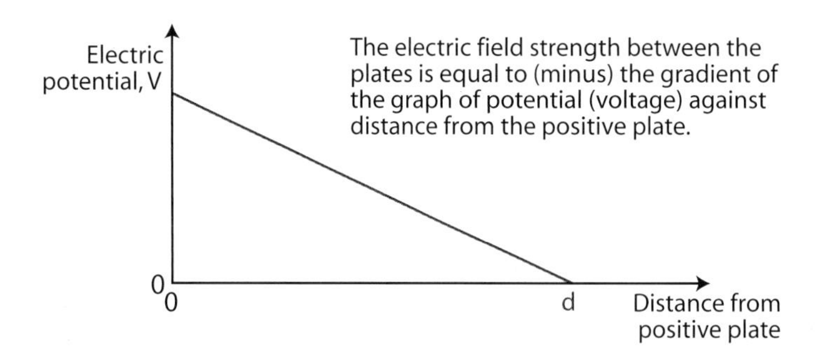

Differences and Similarities in Gravitational and Electric Fields

	Gravitational Fields	Electric Fields
Differences	• Acts on masses • Always produces an attractive force on a mass • Impossible to shield an object from a gravitational field	• Acts on charges • Can produce both attractive and repulsive forces because there are two types of charge (positive and negative) • Shielding is possible with a suitable material.
Similarities	• Field around a point mass decreases according to an inverse square law (falls off as $1/r^2$) • Field is of infinite range	• Field around a point charge decreases according to an inverse square law (falls off as $1/r^2$) • Field is of infinite range

Worked Examples

Example 1

Calculate the electrical force between a proton and an electron at a distance of 50 pm if the magnitude of the charge on each is 1.6×10^{-19} C

Solution

By Coulomb's Law,

$$F = \frac{Q_1Q_2}{4\pi\varepsilon_0 r^2} = \frac{\left(1.6 \times 10^{-19}\right)^2}{4\pi \times 8.85 \times 10^{-12} \times \left(50 \times 10^{-12}\right)^2} = 9.2 \times 10^{-8}\ \text{N}$$

Example 2

Calculate the size and direction of the electric field at a point 25 cm from an isolated point charge of –4 C.

Solution

Since E is the force on a charge of + 1 C,

$$E = \frac{Q \times 1}{4\pi\varepsilon_0 r^2} = \frac{(4 \times 1)}{4\pi \times 8.85 \times 10^{-12} \times (0.25)^2} = 5.75 \times 10^{11}\ \text{NC}^{-1}$$

Since the charge is negative, the direction is from the point towards the charge.

Example 3

At a distance of 60 pm from a certain charge the electric field strength is 400 NC^{-1}. Calculate the field strength at distances of (a) 120 pm (b) 150 pm and (c) 20 pm from the same charge.

Solution

(a) Around an isolated point charge the field strength obeys an inverse square law

120 pm is 2×60 pm, so the field strength falls to $\frac{1}{4}$ of its value at 60 pm

So, $E \text{ (at 120 pm)} = \frac{1}{4}(E \text{ at 60 pm}) = \frac{1}{4} \times 400 = 100\ \text{NC}^{-1}$

(b) $150 \text{ pm} = 2.5 \times 60 \text{ pm}$, so $E \text{ (at 150 pm)} = \frac{1}{2.5^2} \times E \text{ (at 60 pm)} = 0.16 \times 400 = 64\ \text{NC}^{-1}$

(c) $20 \text{ pm} = \frac{1}{3} \times 60 \text{ pm}$, so $E \text{ (at 20 pm)} = 3^2 \times E \text{ (at 60 pm)} = 9 \times 400 = 3600\ \text{NC}^{-1}$

Example 4

Charges of +4 C and +8 C are placed 1.00 m apart. At what distance from the +4 C charge is the electric field strength zero?

Solution

Suppose the distance from the +4 C to the neutral point is x, then the distance from the neutral point to the +8 C is $(1 - x)$, then:

Field strength due to +4 C = Field strength due +8 C

$$\frac{4}{(4\pi\varepsilon_0)x^2} = \frac{8}{(4\pi\varepsilon_0)\cdot(1-x)^2}$$

Cancelling $(4\pi\varepsilon_0)$, and taking the square root of both sides,

$$\frac{2}{x} = \frac{\sqrt{8}}{(1-x)}$$

$$2(1-x) = (\sqrt{8})x$$

which simplifies to $x = \frac{2}{(\sqrt{8}+2)}$ = 0.41 m from the charge +4 C

Example 5

Two parallel metal plates, 4 cm apart, are connected to a 12 V battery as shown in the diagram opposite.

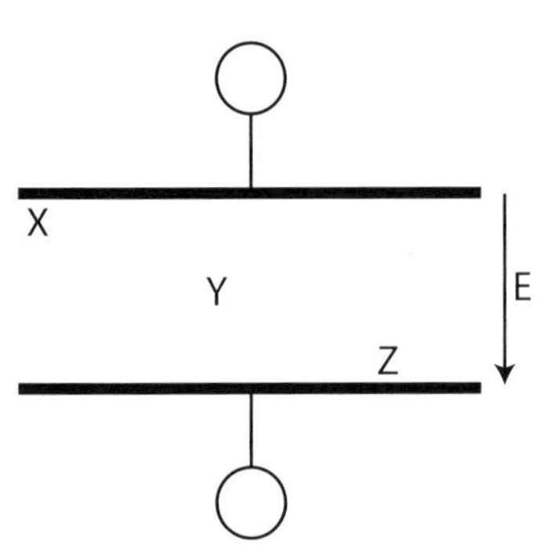

(i) Redraw the diagram showing the polarity of the plates, by writing "+" and "–" in the appropriate circles.

(ii) Calculate the size of the electric field between the plates.

Identical charges are placed at points X, Y and Z.

(iii) At which point (if any) will the charge experience the greatest force?

(iv) At which point (if any) is the electrical potential greatest?

Solution

(i) The upper plate is positive, the lower plate is negative

(ii) $E = \frac{V}{d} = \frac{12}{0.04} = 300 \text{ Vm}^{-1}$

(iii) The force is the same at all points because the field is uniform.

(iv) Potential is greatest at X

Exercise 26

Examination Questions

1 (a) State the purpose of the constant ε_0 (the permittivity of a vacuum) in the law for the force between two point charges.

(b) A point charge of +2.0 μC is subjected simultaneously to two electric fields E_1 and E_2 as indicated in the diagram. E_1 is a vertical field of magnitude of 6.0 Vm^{-1} and E_2 is a horizontal field of magnitude 4.5Vm^{-1}.

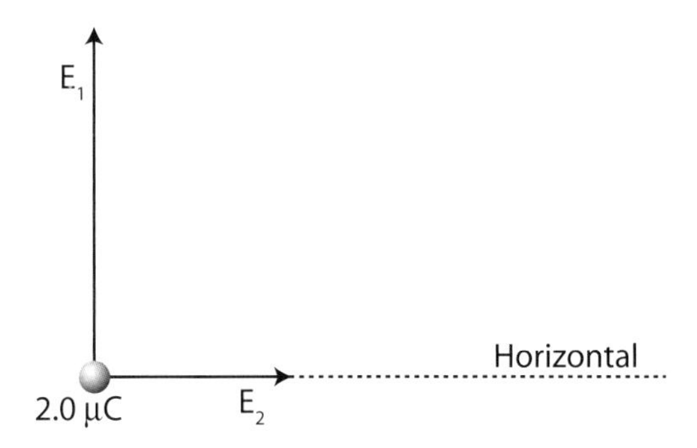

Calculate the resultant force on the point charge and find its direction relative to the horizontal.

(c) Three point charges are on a straight line. The magnitude of the charges and their separations are shown below. The two outer charges are both positive but the sign and the magnitude of the charge X are unknown. The resultant force on the +2.0 μC charge is 3.0×10^{-3} N to the right.

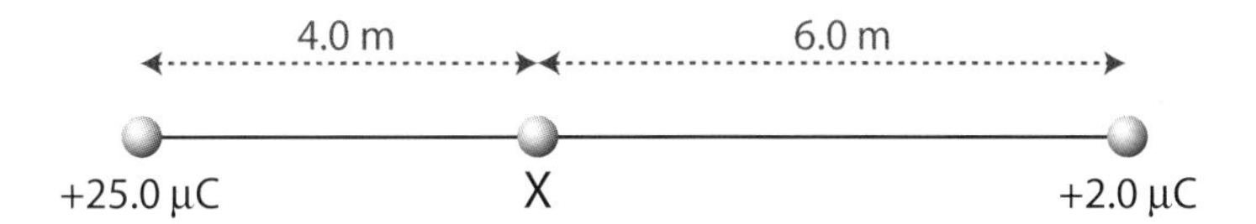

Find the sign and the magnitude of the charge X.
For simplification of calculations you may assume $\frac{1}{4\pi\varepsilon_0} = 9.0\times10^9\ \text{Fm}^{-1}$

[CCEA Summer 2007]

2 (a) Two point charges of magnitude 3 μC and 6 μC are placed 30 mm apart in a vacuum.

(i) Calculate the magnitude of the force between them. For simplification of calculations you may assume

$$\frac{1}{4\pi\varepsilon_0} = 9.0\times10^9\ \text{Fm}^{-1}$$

(ii) When they are free to move the charges accelerate towards each other. State what sign each charge could have.

(b) There are a number of differences between

1. the electric field produced between two parallel plates with a potential difference across them and

2. the field around a point charge.

Describe the difference between the electric field **strengths** in each case.

Use diagrams to help to explain your answer.

[CCEA Winter 2007]

3 A student, asked to explain what is meant by a **field** of force, gave the answer "A field of force is an area where a unit charge experiences a force".

(a) Identify **two** errors, omissions or irrelevant details in the student's explanation.

(b) It seems that the student may have been confusing the explanation of a field of force with the definition of electric field strength.

Define **electric field strength** and state how the direction of the electric field is obtained.

[CCEA June 2006]

4 (a) Two point charges in a vacuum are separated by a distance r. The magnitude of the force between them is F. On the diagram opposite sketch graphs to show how F depends on r when

(i) the charges have values +Q and +Q. (Label this graph 1)

(ii) the charges have values +2Q and –2Q. (Label this graph 2)

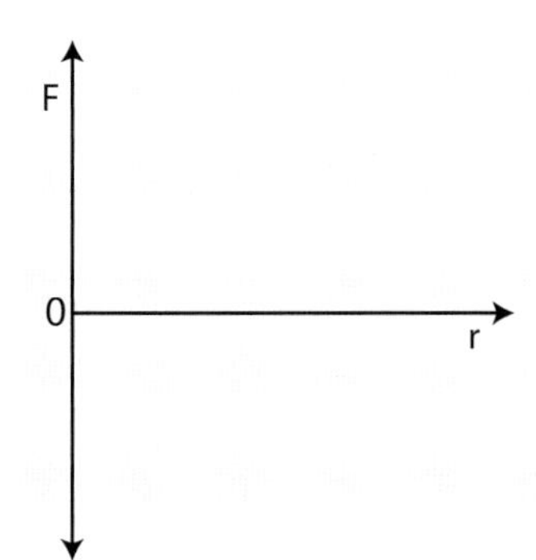

(b) Two charges –Q and +Q in a vacuum are a short distance apart, as shown opposite. X, Y and Z are points on the perpendicular bisector of the line joining the charges.

(i) On the diagram mark the directions of the electric field at the points X, Y and Z.

(ii) At which of the points X, Y or Z does the electric field have its largest value? Indicate your answer by placing a tick in the appropriate box. The electric field has its largest value at:

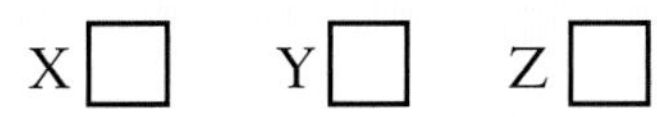

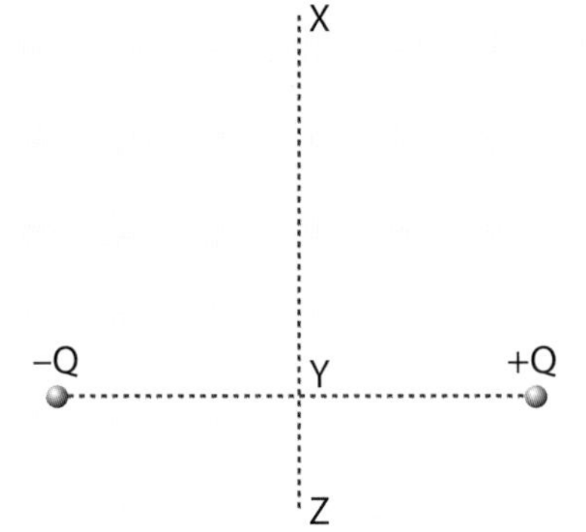

(c) A uniform electric field is set up by applying a potential difference between two parallel metal plates a distance, d, apart.

(i) On the diagram opposite sketch a graph to show how the electric field strength E between the plates depends on the inverse of the separation (i.e. 1/d) of the plates when the potential difference between the plates remains constant.

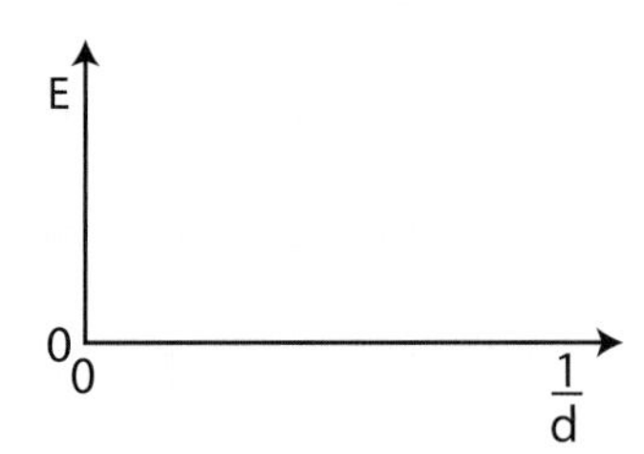

This diagram shows the arrangement of parallel plates described in (c)(i). The direction of the uniform field is marked on the sketch.

(ii) On the diagram, label the polarity of the plates required to obtain this field direction.

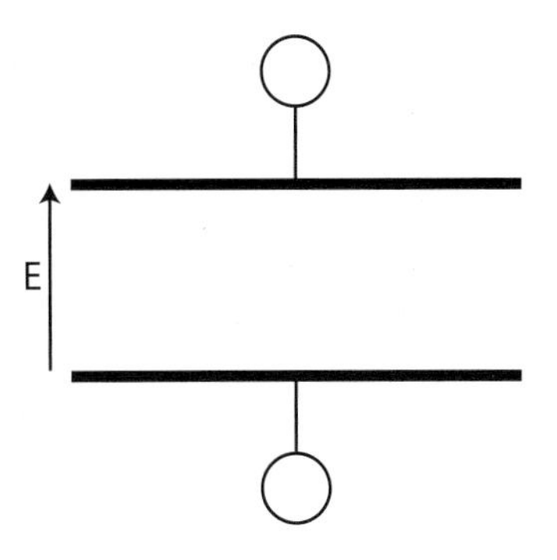

[CCEA January 2006]

5.4 Capacitors

Students should be able to:

5.4.1 Define capacitance;

5.4.2 Recall and use the equation $C = \frac{Q}{V}$;

5.4.3 Define the unit of capacitance, the farad;

5.4.4 Recall and use ½ QV for calculating the energy of a charged capacitor;

5.4.5 Use the equations for the equivalent capacitance for capacitors in series and in parallel;

5.4.6 Describe experiments to demonstrate the charge and discharge of a capacitor;

5.4.7 Explain exponential decay using discharge curves;

5.4.8 Define time constant and use the equation $\tau = CR$;

5.4.9 Describe an experiment to determine the time constant for R-C circuits;

5.4.10 Apply knowledge and understanding of time constants and stored energy to electronic flash guns;

A capacitor is an electrical component that can store energy in the electric field between a pair of metal plates separated by an insulator. Charging is the process of storing energy in the capacitor and involves electric charges of equal magnitude, but opposite polarity, building up on each plate. Capacitors come in many different types as shown in the picture.

The simplest arrangement of a capacitor is two parallel metal plates separated by an insulator.

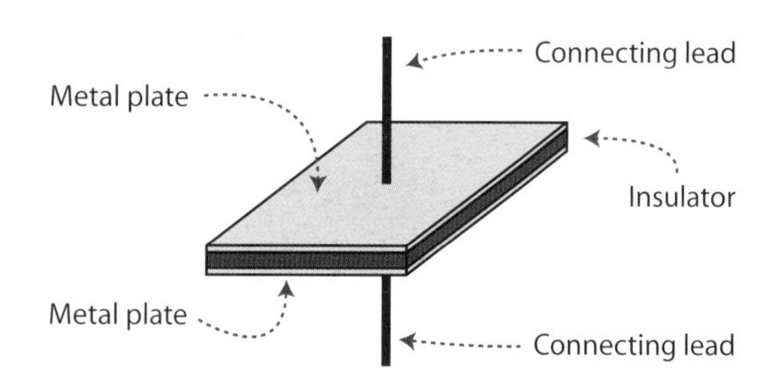

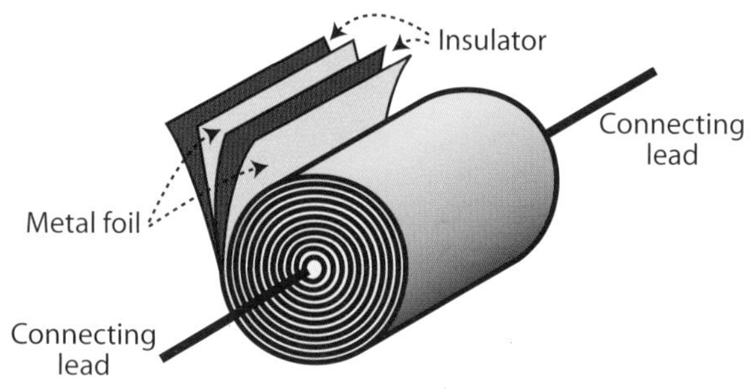

The diagram shows the parts that make up some of the capacitors shown in the photograph. The metals plates consist of thin metal foil and the insulator is either a plastic strip or waxed paper sandwiched between the metal foils. These layers are often rolled up into a cylinder to make them easy to handle.

A capacitor's ability to store charge is measured by its capacitance, in units of farads.

Capacitance is defined as the charge stored per volt.

$$C = \frac{Q}{V}$$

where C is the capacitance in farads (F)
Q is the charge stored in coulombs (C)
V is the potential difference between the plates in volts (V)

The farad is a very large unit and microfarads (μF) and picofarads (pF) are more common.

$$1\ \mu F = 1 \times 10^{-6}\ F \qquad 1\ pF = 1 \times 10^{-12}\ F$$

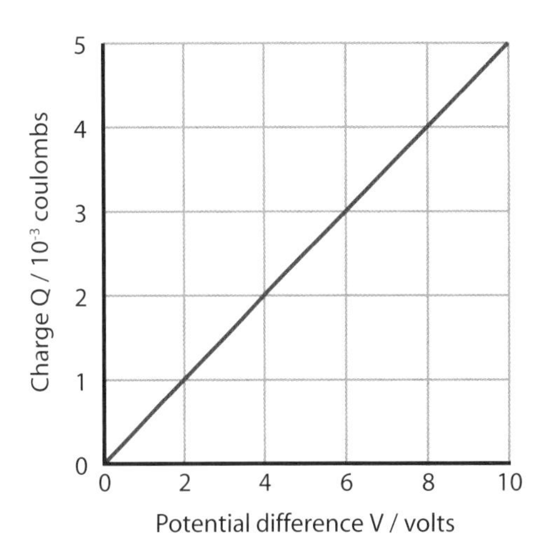

When electric charge is stored on the plates of a capacitor the potential difference between the plates rises.

The potential difference is proportional to the amount of charge stored. Therefore a graph of charge Q against the potential difference V is a straight line that passes through the origin.

The capacitance is equal to the gradient of the graph shown. The gradient of the graph is:

$$\frac{4 \times 10^{-3}}{8} = 5 \times 10^{-4} = 500\ \mu F$$

Capacitors in Parallel

The three capacitors below, C_1, C_2 and C_3 when connected to a potential difference V store electric charge Q_1, Q_2 and Q_3 respectively. Since they are connected in parallel the potential difference across each capacitor is the same and equal to V. They are equivalent to a single capacitor of value C which will store a charge Q, the sum of the individual charges stored on each capacitor ($Q = Q_1 + Q_2 + Q_3$), when connected to the same potential difference.

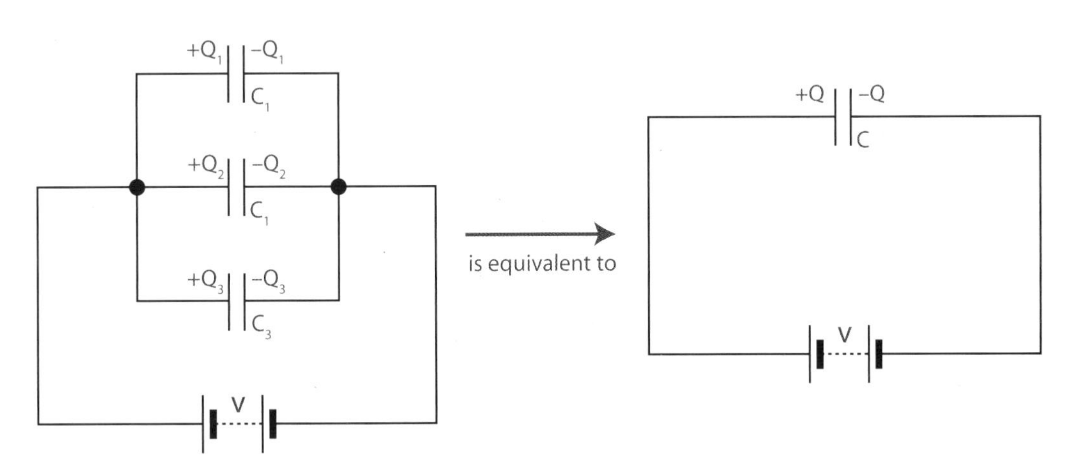

The charge on each capacitor is $Q_1 = C_1V$, $Q_2 = C_2V$ and $Q_3 = C_3V$

For the single capacitor that is equivalent to these three in parallel the charge $Q = CV$

$Q = Q_1 + Q_2 + Q_3$ $\quad CV = C_1V + C_2V + C_3V$ this simplifies to $C = C_1 + C_2 + C_3$

The total capacitance of any number, N, of capacitors in parallel is therefore just the sum of the capacitors:

$C = C_1 + C_2 + C_3 + \ldots\ldots\ C_N$

A larger capacitance can be made by connecting a number of smaller capacitors in parallel.

Capacitors in Series

When the capacitors are connected to a potential difference the electrons that move on to plate G of capacitor C_3 cause the same number of electrons to move from plate F of C_3 to plate E of capacitor C_2 and in turn this causes the same number of electrons to leave plate D and move to plate B of capacitor C_1.

This means that the charge that has moved around the circuit when the potential difference is applied to Q. It is not the 3Q that you might think when looking at the circuit. Another way to regard this is to think of plates A and G as the plates of the single capacitor that can replace the three in series.

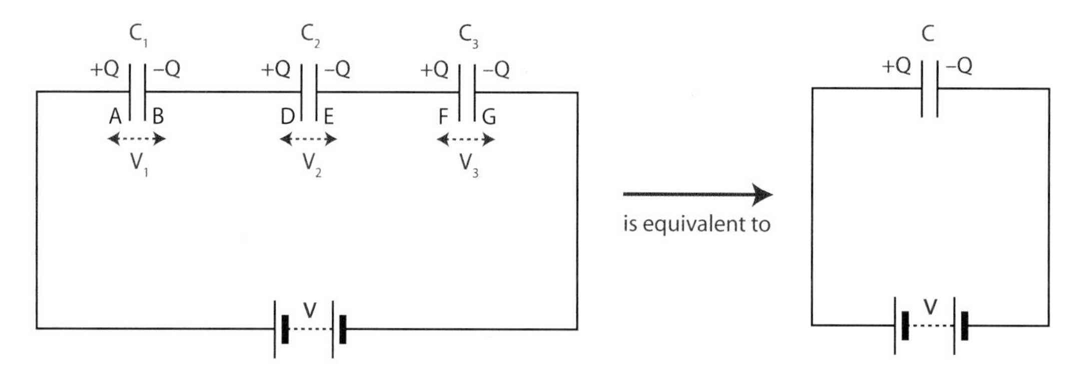

The potential differences across each capacitor are not equal. They would be equal if all the capacitors had the same capacitance. However $V_1 + V_2 + V_3 = V$.

From the defining equation for capacitance we have that

$$V_1 = \frac{Q}{C_1} \quad V_2 = \frac{Q}{C_2} \quad V_3 = \frac{Q}{C_3}$$

$$V = \frac{Q}{C} = \frac{Q}{C_1} + \frac{Q}{C_2} + \frac{Q}{C_3}$$

The total capacitance of any number, N, of capacitors in series is therefore given by:

$$\frac{1}{C} = \frac{1}{C_1} + \frac{1}{C_2} + \frac{1}{C_3} + \ldots \frac{1}{C_N}$$

A smaller capacitance can be made by connecting a number of larger capacitors in series.

Worked Example

Find the capacitance of circuits shown between the points A and B.

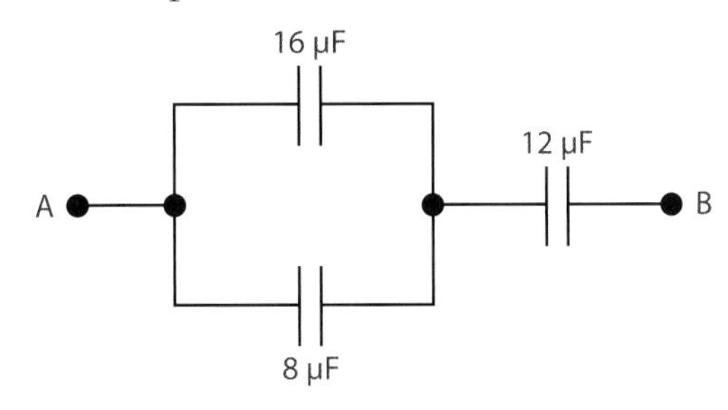

Solution

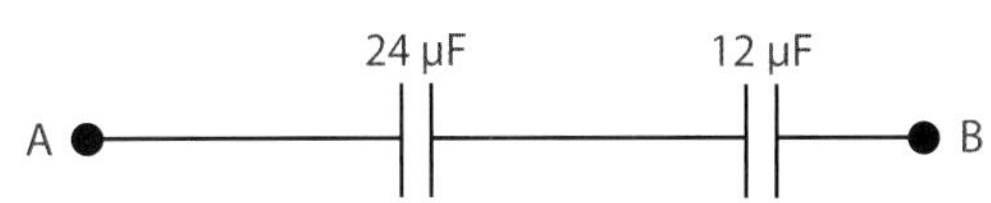

The 8 μF and the 16 μF capacitors are in parallel so their total capacitance is 24 μF. The circuit is then effectively a 24 μF capacitor in series with a 12 μF capacitor.

The total capacitance is then calculated using the rule for capacitors in series.

$$\frac{1}{C} = \frac{1}{24} + \frac{1}{12} = \frac{3}{24} = \frac{1}{8}$$

C = 8 μF

Exercise 27

1 Show how you would connect three 4 μF capacitors to form a network with a total capacitance of:

(a) 1.33 μF

(b) 6.0 μF

In each case draw a circuit diagram and provide suitable calculations for your answer.

2 An arrangement of capacitors is shown in the diagram.

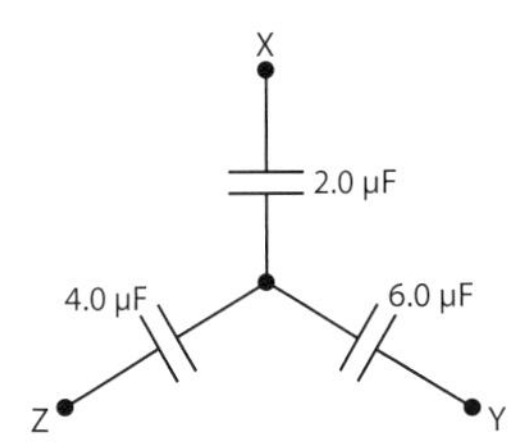

The only accessible terminals are X, Y and Z. The central common point is not accessible. State the pair of terminals between which the maximum capacitance is obtained. Make suitable calculations to support your answer.

3 A number of capacitors, each of value 10 μF, are connected in a network as shown below.

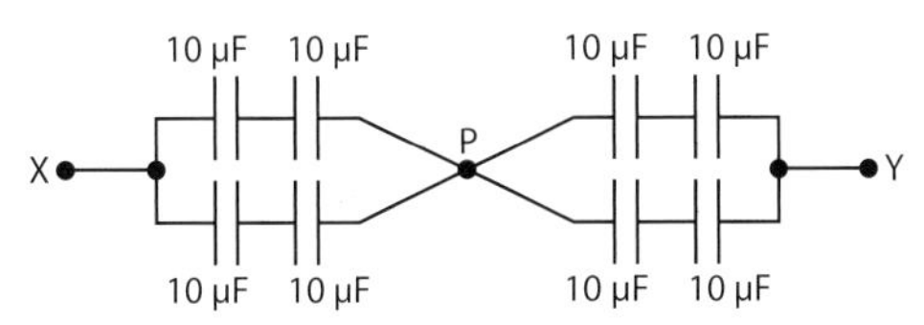

(a) Calculate the capacitance of the network between the points X and Y.

Later, the wires crossing at P become separated as shown below.

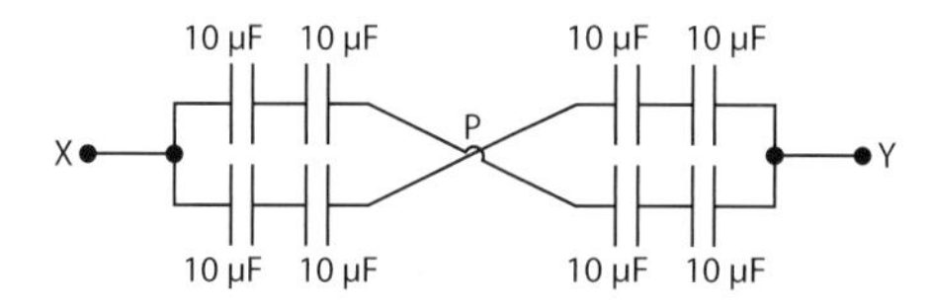

(b) Calculate the new capacitance of the network between X and Y.

4 The diagram shows six identical capacitors. The total capacitance of the network is 33 μF.

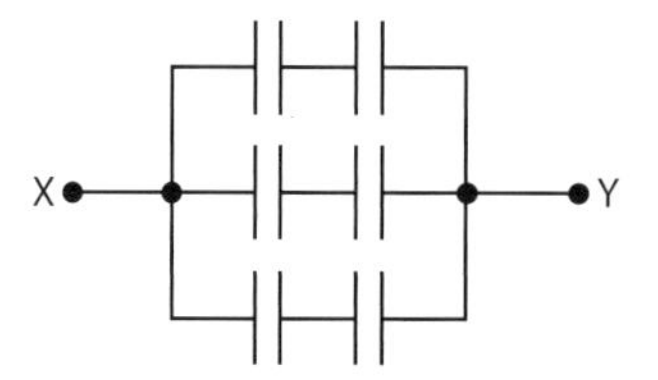

What is the value of each capacitance?

Energy of a Charged Capacitor

The charge stored by a capacitor is proportional to the potential difference across the plates. Therefore a graph of charge against the potential difference is a straight line through the origin.

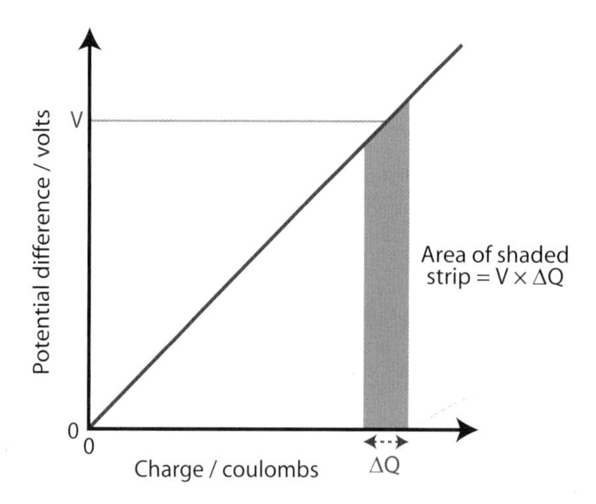

Suppose the potential difference across the capacitor is V when it stores a charge Q.

To move a small amount of charge Δq from the negative plate to the positive plate requires work to be done against the repulsive force of the charges already on that plate.

Potential difference = work done per unit charge

Work done or energy required = V × ΔQ = area of shaded strip

The total energy = area between the graph and the charge axis

The energy stored = $\frac{1}{2}QV$

Using C = Q/V also gives the energy stored as $\frac{1}{2}CV^2$

Eliminating the voltage gives the energy stored as $\frac{1}{2}\frac{Q^2}{C}$

The relationship energy stored = $\frac{1}{2}CV^2$ can be verified using the equipment arranged as shown below.

The capacitor is charged by closing the switch for a short time and then opening it. The joule meter which measures the energy supplied by the capacitor is reset to read zero. The capacitor is discharged, the bulb lights and then dims as the energy stored in the capacitor is dissipated as heat and light by the bulb.

The joule meter measures this energy. The discharging switch is kept closed until the reading no longer changes.

This is repeated several times and an average obtained.

The voltage is now changed and the procedure repeated. Typically, the voltage might be varied from 5 to 25 in steps of 5 volts. A set of readings is shown below.

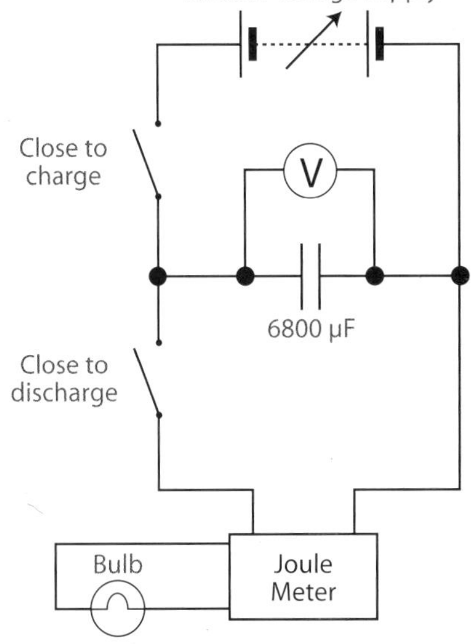

	Energy readings/J			
Potential difference/V	1st reading	2nd reading	3rd reading	Average energy/J
5	0.1	0.1	0.1	0.1
10	0.4	0.4	0.4	0.4
15	0.8	0.8	0.9	0.83
20	1.7	1.7	1.7	1.7
25	2.5	2.5	2.5	2.5

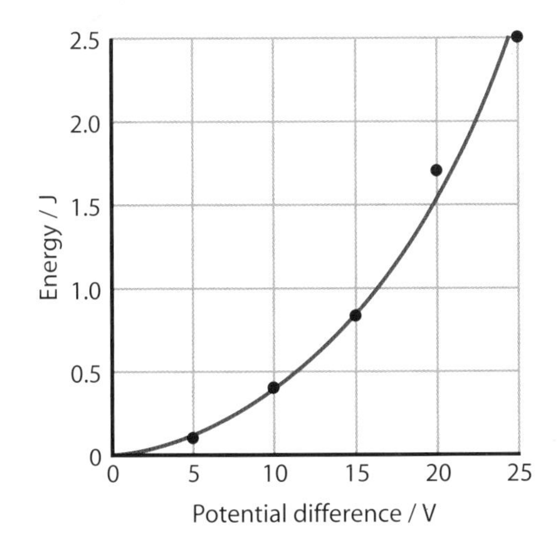

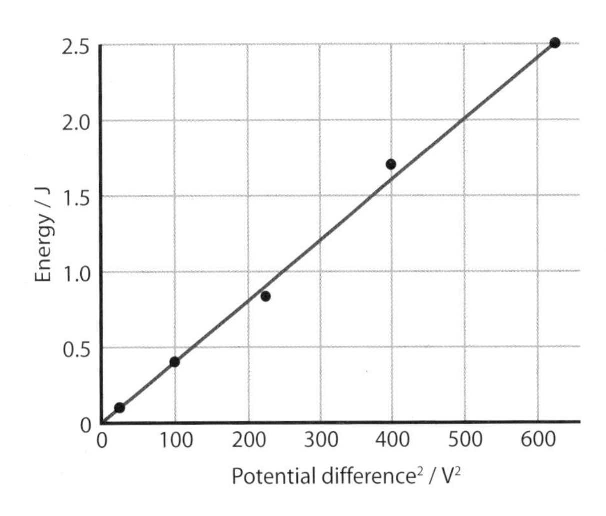

The graph of E against V^2 is a straight line that passes through the origin. This means that the energy E is directly proportional to the square of the potential difference V^2. The general relationship for a straight line graph that passes through the origin is y = mx. Applying it to the energy relationship the gradient of the energy stored against V^2 graph is equal to $\frac{1}{2}C$.

Worked Examples

(a) A capacitor of capacitance 100 μF is charged by a battery of e.m.f. 6.0 V. Calculate the charge on the capacitor.

Solution $Q = CV = 100 \times 10^{-6} \times 6.0 = 6.0 \times 10^{-4}$ C

(b) Calculate the energy stored in the capacitor when it is fully charged.

Solution Energy stored $= \frac{1}{2}QV = \frac{1}{2}(6.0 \times 10^{-4} \times 6) = 1.8 \times 10^{-3}$ J

Alternatively you could use energy stored $= \frac{1}{2}CV^2 = \frac{1}{2}(100 \times 10^{-6} \times 6^2) = 1.8 \times 10^{-3}$ J

In this case it is also possible to use $\frac{Q^2}{2C}$ to calculate the energy stored.

(c) The charged capacitor in (a) and (b) is now disconnected from the battery and connected in parallel with an uncharged capacitor of capacitance 50 μF.

Calculate the potential difference across the capacitors.

Solution

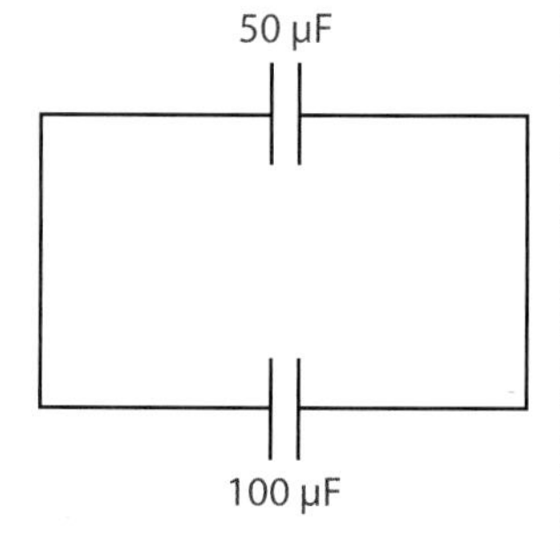

The total capacitance of the two in parallel is 150 μF.

The electric charge that was solely on the 100 μF capacitor is now shared between it and the 50 μF capacitor.

Electric charge is conserved.

$$Q = CV$$

$$6.0 \times 10^{-4} = 150 \times 10^{-6} \times V$$

$$V = 4.0 \text{ V}$$

(d) Calculate the energy now stored in the two capacitors in (c). Account for any difference between your answer and the initial energy stored in the 100 μF capacitor.

Solution

The energy stored in the combination $= \frac{1}{2}CV^2 = \frac{1}{2}(150 \times 10^{-6} \times 4^2) = 1.2 \times 10^{-3}$ J

The initial energy stored in the 100 μF capacitor was 1.8×10^{-3} J.

The loss of energy is due to transfer of electrons between the two capacitors when they are connected together. This movement of electrons is a current and the connecting leads have a small amount of resistance so some of the stored energy is dissipated as heat.

Exercise 28

1 Two capacitors of values 1 μF and 100 pF are each given a charge of 1.5 nC.

(a) Calculate the potential difference across each capacitor.

(b) Calculate the energy stored in each capacitor.

2 A capacitor C1 of value 12.0 μF is charged from a 400 V supply. It is then disconnected from the supply and connected across a second capacitor of value 4.0 μF , which is initially uncharged. C2 is then disconnected from C1 and discharged.

(a) What charge remains on C1?

(b) Find the minimum number of times that C2 has to be connected to C1, then disconnected from C1 and discharged, in order to reduce the charge on C1 to below 50% of its initial value.

3 Opposite is a circuit diagram showing three capacitors connected to a 12 V battery.

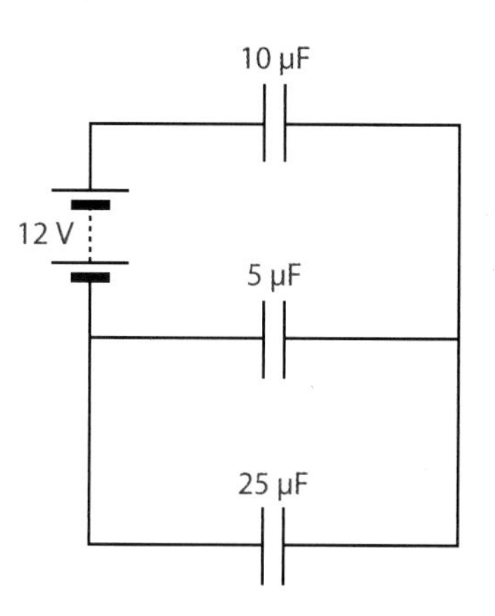

(a) Which capacitors are in parallel?

(b) Calculate the capacitance of the single capacitor, connected across the battery, which is equivalent to the network of the three capacitors.

(c) Calculate the energy stored in the network of capacitors.

4 Three capacitors are connected as shown below to a 12.0 V battery. The value of one of them is unknown and is labelled C.

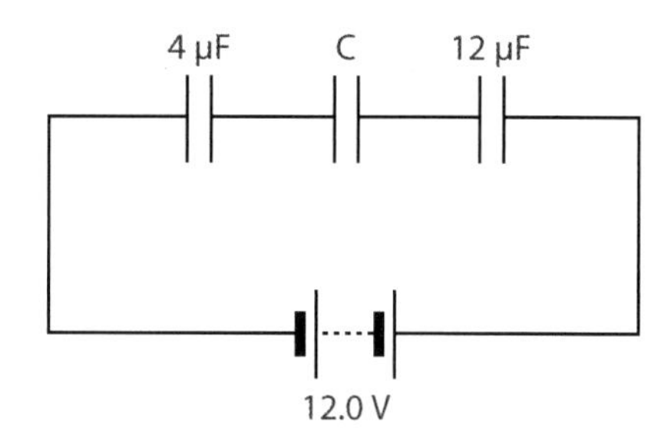

The total energy stored in the three capacitors is 1.44×10^{-4} J

(a) Calculate the total capacitance of the three capacitors when connected as shown.

(b) Calculate the value of the unknown capacitor.

[CCEA 2006]

Charging a Capacitor

The circuit shows a capacitor connected in series with a battery, a switch and a resistor. When the switch is closed the capacitor charges. Electrons flow onto one plate and which causes electrons to be repelled from the other plate. This movement of electrons constitutes a current.

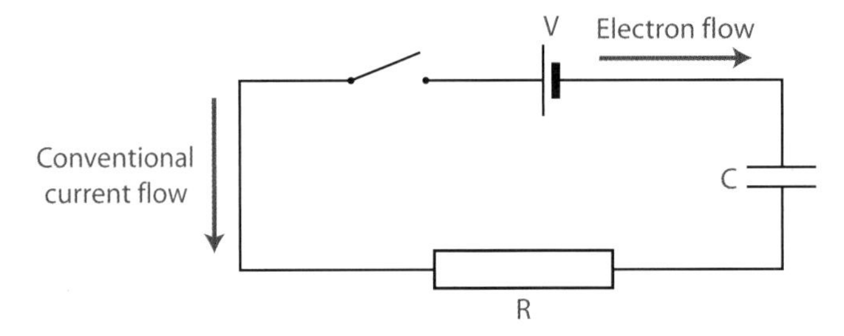

The rate at which the charging current decreases and the rate at which the potential differences across the capacitor rises depends on the capacitance of the capacitor and the resistance of the circuit.

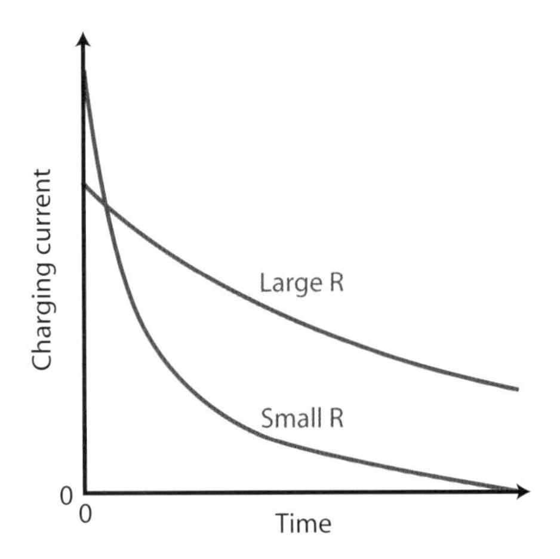

The charging current decreases with time. It decreases because it becomes more difficult to push electrons onto the plate because of the repulsion of the electrons already there.

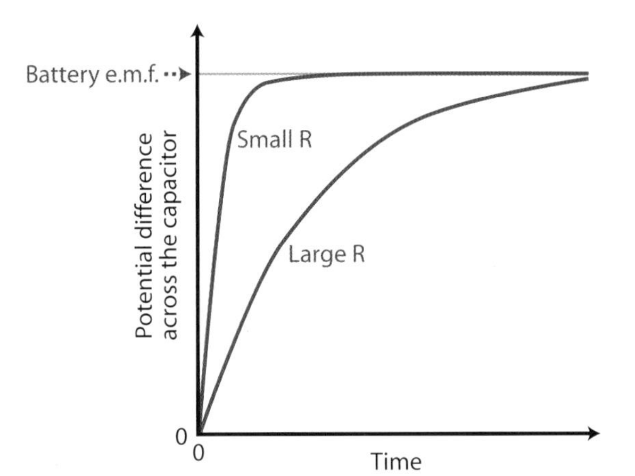

The voltage across the capacitor gradually increases until it is equal to the e.m.f. of the battery. However it is in the opposite sense, it opposes the further movement of electrons.

Discharging a Capacitor

The circuit below can be used to investigate the discharge of a capacitor.

Suitable values for this circuit are:

R = 100 kΩ (1×10^5 Ω)

C = 470 μF (470×10^{-6} F)

The microammeter should have range of 0 – 100 μA.

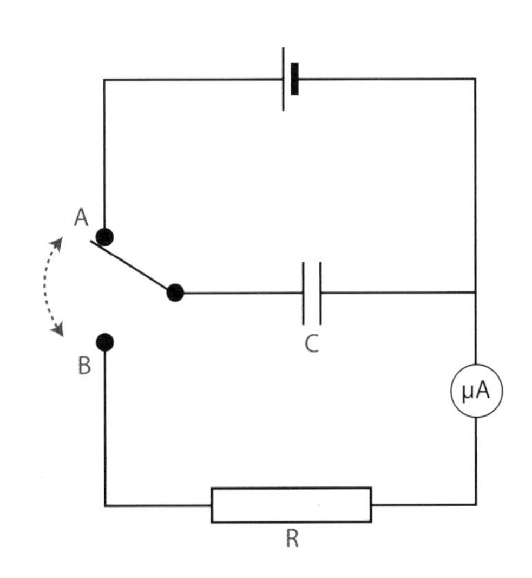

When the switch is moved to position A the capacitor is charged from the 9V battery. The capacitor becomes fully charged very quickly since the resistance of the charging circuit is very small. Moving the switch to position B will start the capacitor discharging through the resistor R.

The reading on the microammeter will start high and gradually fall. Values of the current can be recorded at 10 or 20 second intervals.

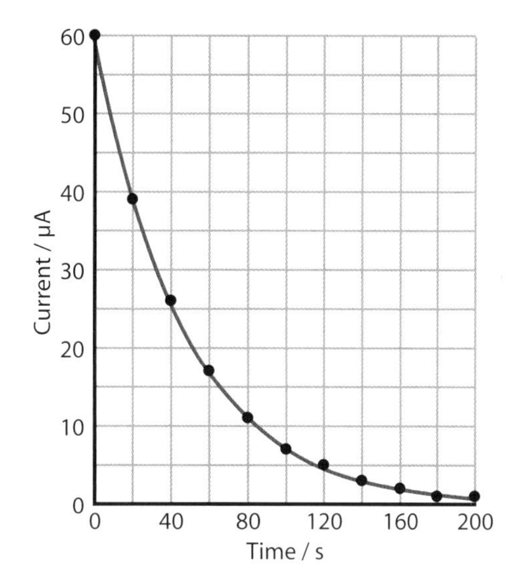

The product of the capacitance and the resistance is known as the time constant of the circuit.

For example a 470 μF capacitor in a circuit with a resistance of 100 kΩ has a time constant τ of 47 seconds.

$$\tau = CR$$

where C is the capacitance in farads (F)
R is the resistance in ohms (Ω)

The product of capacitance and resistance can be shown to have the units of time as follows;

$$C = \frac{Q}{V} \text{ and } R = \frac{V}{I} \text{ so } CR = \frac{Q}{V} \times \frac{V}{I} = \frac{Q}{I}$$

but $Q = \text{charge} = \text{current} \times \text{time} = I \times t$

$$\text{so } CR = \frac{Q}{I} = \frac{I \times t}{I} = t$$

The discharging current decreases exponentially. The equation that describes this decrease is explained below.

$$I = I_0 e^{\frac{-t}{\tau}}$$

where I_0 is the initial current
I is the current at a time t
t is the time in seconds
τ is the time constant

To explain the significance of the time constant τ consider what value the discharging current has after a period of time equal to the time constant has passed.

$$I = I_0 e^{\frac{-t}{\tau}} = I_0 e^{-1} = \frac{I_0}{e}$$

The value of e is 2.7183

$$I = \frac{I_0}{2.7183} \approx 0.37 \times I_0$$

After a period of time equal to the time constant the discharging current falls to 0.37 of its value at the start of that period. For the circuit shown on the previous page the time constant is 47 s. The initial discharging current was 60 μA, after 47 s it will have fallen to 0.37 × 60 = 22.2 μA and after another 47 s it will have fallen to 0.37 × 22 = 8.2 μA.

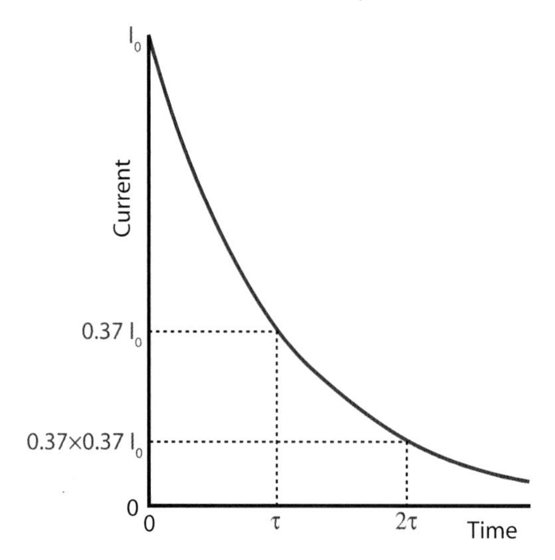

This approach can be used determine the time constant and therefore the capacitance if the resistance of the circuit is known.

The charge stored by the capacitor and the potential difference across the capacitor also decrease in a exponential way.

$$Q = Q_0 e^{\frac{-t}{\tau}}$$

where Q_0 is the initial charge
Q is the charge at a time t
t is the time in seconds
τ is the time constant

$$V = V_0 e^{\frac{-t}{\tau}}$$

where V_0 is the initial potential difference
V is the potential difference at a time t
t is the time in seconds
τ is the time constant

Some Points to Remember

How do you find V_0?

It equals the e.m.f. of the battery or power supply used to charge the capacitor.

How do you find Q_0?

Use the defining equation for capacitance $C = \frac{Q}{V}$ re-arranging gives $Q_0 = CV_0$

How do you find I_0?

Use Ohm's law $I = \frac{V}{R}$ so we have $I_0 = \frac{V}{R}$

R is the resistor through which the capacitor discharges.

Measuring the Time Constant τ for C R Circuits

One method of doing this has already been mentioned. The graph of current against time can be used to find the time constant directly. The time constant τ is the time taken for the current to fall to $\frac{1}{e}$ (0.37) of the initial value. Determining a number of values of τ from the exponential curve allows an average value to be found.

One problem with this approach is the difficulty of drawing such a curve by hand.

A straight line graph is much easier to draw and can be achieved as follows.

$$I = I_0 e^{\frac{-t}{\tau}}$$

Taking natural logarithms (ln) (base e) of both sides of the equation gives:

$$\ln I = \ln I_0 - \frac{t}{\tau}$$

Plotting a graph of ln I on the y–axis and t on the x–axis will produce a straight line. The gradient of the line will give 1/τ. Note that logarithms, natural or to the base 10, do not have a unit as you can see from the graph below.

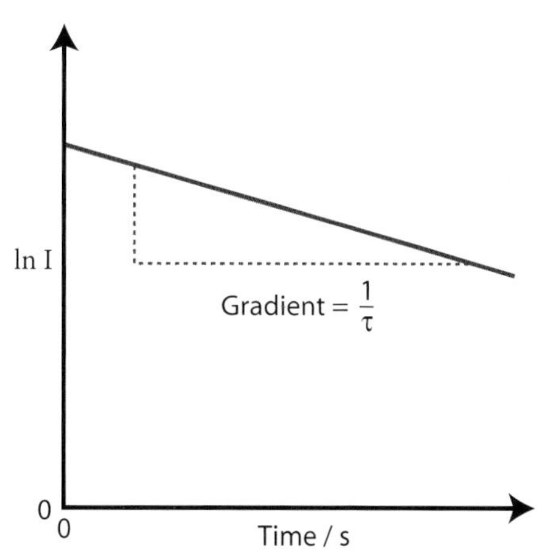

Worked Example

(a) A capacitor of value 1 μF may be connected either to a 100 V d.c. supply or to a resistor R of value 22 MΩ by means of a two-way switch as shown below.

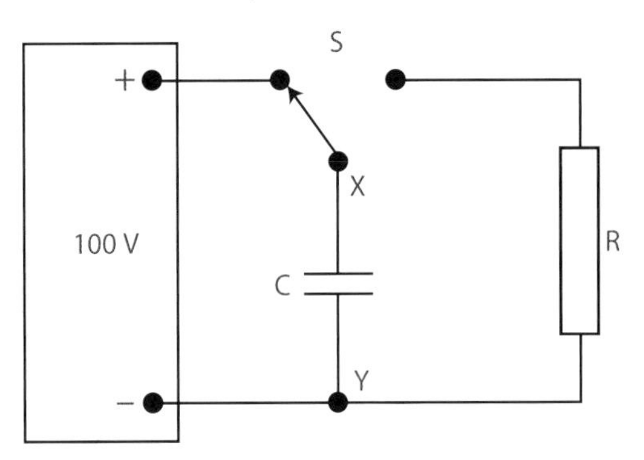

Explain the meaning of the time constant.

Solution

The time constant τ is the product of the capacitance and resistance.

When the switch is changed over so that the capacitor is connected to the resistor the charge stored on the capacitor will fall to a value of 1/e or 0.37, in one time constant period, of the charge stored before the switch was changed over.

Alternatively you explain the significance in terms of the discharge current, after a period equal to τ the discharge current will have decreased to 0.37 I_0, where I_0 is the value of the current at the moment the switch was changed over.

(b) Calculate the value of the time constant for this circuit.

Solution

$$\tau = C \times R = 1 \times 10^{-6} \times 22 \times 10^{6} = 22 \text{ s}$$

(c) The capacitor is fully charged by moving the switch to the position shown. The switch is then moved so that capacitor discharges through the resistor. Determine the time taken for the voltage across the capacitor to fall to 10V.

Solution

$$V = V_0 e^{\frac{-t}{\tau}}$$

t = time in seconds, $\tau = 22$ s and $V_0 = 100$ V so:

$$10 = 100 \times e^{\frac{-t}{22}} \quad \text{rearranging gives} \quad 0.1 = e^{\frac{-t}{22}}$$

Take natural logs

$$\ln(0.1) = \frac{-t}{22} \quad \text{which leads to}$$

$$-2.306 = \frac{-t}{22}$$

$$t = 2.306 \times 22 = 50.7 \text{ s}$$

Exercise 29

1 A charged capacitor is connected in series with a switch and a resistor. Before the switch is closed the potential difference across the capacitor is V_0.

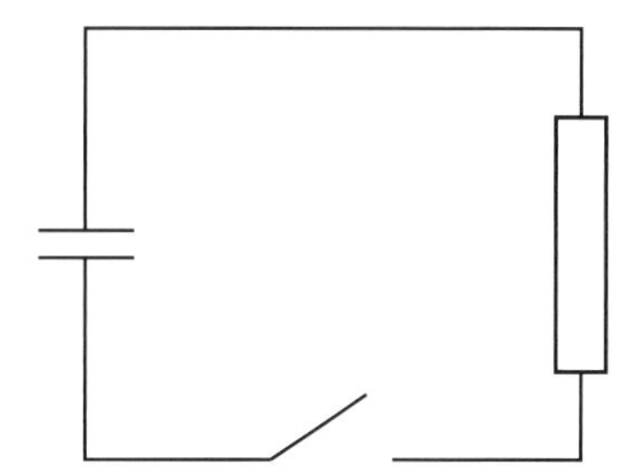

(i) Sketch a graph to show how the potential difference across the capacitor (y–axis) varies with time (x–axis) after the switch is closed.

(ii) Explain what is meant by the time constant of the circuit. Mark the time constant τ on the horizontal axis of your graph and label the corresponding voltage on the vertical axis.

(iii) The capacitor has a capacitance of 47 pF and the resistor has a resistance of 22 MΩ. How long after the switch is closed will it take for the potential difference across the capacitor fall to 14% of its initial value.

2 An uncharged capacitor is connected in series with a resistor, a battery and a switch as shown below. Initially, the switch is open.

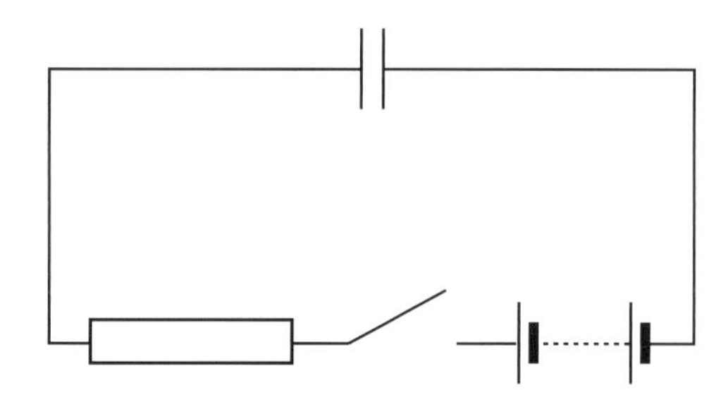

(a) The switch is now closed. Sketch a graph to show the variation of the charge (y–axis) on the capacitor with time (x–axis).

(b) Sketch a graph to show the variation of the current I (y–axis) drawn from the battery with time (x–axis) after the switch is closed.

3 A capacitor of value 470 μF is connected in series with a battery of e.m.f. 20 V, a resistor of value 10 kΩ, a milliammeter and a switch.

(a) Draw a circuit diagram of this arrangement.

(b) The switch is closed. Calculate the initial current shown by the milliammeter.

(c) Calculate the time constant of the circuit.

(d) Calculate how long it is before the discharge current has fallen to ¼ of its initial value.

4 A charged capacitor of value 8 μF is discharged through a milliammeter and resistor. The current is recorded at 5 second intervals. The measurements are shown below.

Time/s	0	5	10	15	20	25	30	35
Current/mA	72	54	41	31	23	17	13	10

(a) Use the results to plot a suitable linear graph from which the time constant of the circuit can be found.

(b) Calculate the resistance of the resistor used in the circuit.

Use of Capacitors

Defibrillators

For proper functioning the heart needs regularly timed electrical signals transmitted to the heart muscles, causing it to contract (and relax) in a delicate rhythm. The heart muscle beats at around 70 beats per minute throughout our life. It has its own pacemaker which produces electrical signals at around 70 pulses per minute. If this rhythm is disturbed, in a heart attack, the heart may undergo ventricular fibrillation which is a rapid, uncoordinated twitching of the heart muscles. If this happens the heart can not properly pump blood, which means that fresh oxygen will not be delivered to the cells of the body. A defibrillator is used to deliver a carefully controlled shock to the victim's heart. This is designed to stop the ventricular fibrillation and starts the normal heart rhythm again.

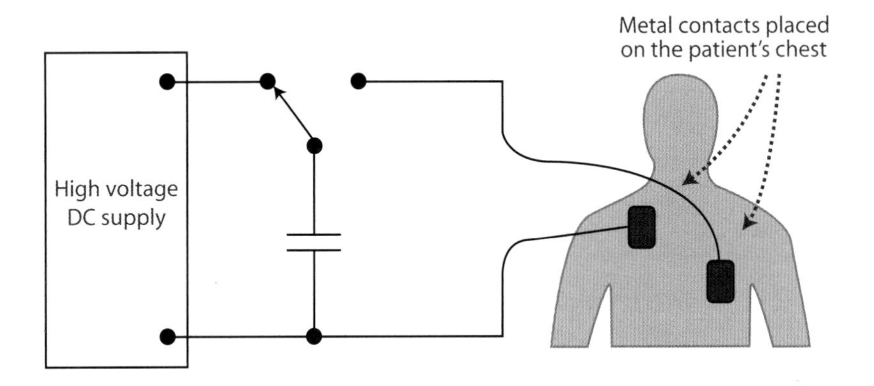

A defibrillator needs to transfer a precise amount of energy to a patient. The best way to do this is to use a capacitor. The capacitor stores electric charge on its plates. The energy that it stores is in form of the electric field that it created between its plates. Recent designs of defibrillators use energies of 120 – 200 J to stop fibrillation, these are much lower energies than earlier designs. This energy is delivered as a pulse lasting a few milliseconds.

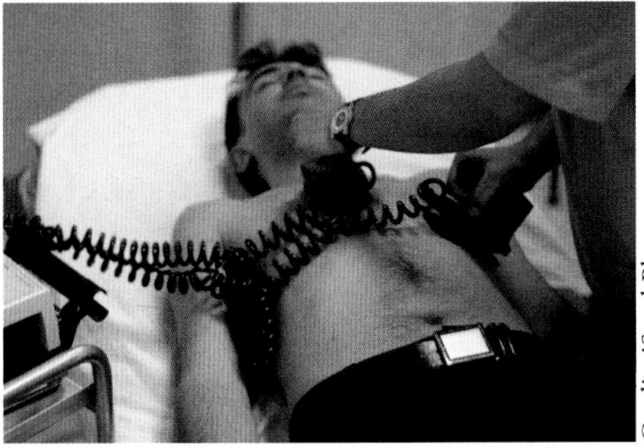
Credit: iStockPhoto

Exercise 30

Defibrillators use capacitors specially designed for use in portable devices.

1. What charging potential difference would be needed to store 100 J in a 50 μF capacitor?
2. Doctors need to use the defibrillator over and over again so they require a charging time of 3 seconds. What average charging current is needed?
3. Calculate the mean charging power.
4. A defibrillator uses a 50 μF capacitor charged to 1500 V. How would you use a number of 50 μF capacitors that can withstand a maximum of 500 V to construct a circuit with a capacitance of 50 μF and that can withstand the charging voltage of 1500 V?

   Draw a circuit diagram to illustrate your answer.

Flash Gun

Electronic flash guns use an electric discharge in a suitable gas such as xenon to produce an intense flash of light that lasts a short time. In the case of the flash gun used with a camera the flash gun is operated from a small battery, say, 6V. To achieve this, electronic circuitry has to be used to generate a high voltage, several hundred volts. When the shutter on the camera is pressed the charged capacitor is discharged through the gas filled flash tube so producing the intense flash of light.

The diagram below shows a simplified version of the circuit used.

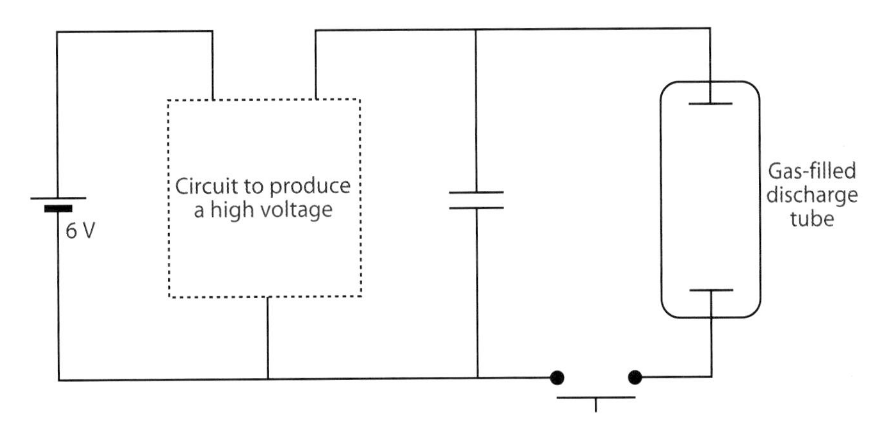

Exercise 31

1. A 47 μF capacitor in a flash gun supplies an average power of 1.5×10^4 W for 20 μs. What is the potential difference across the capacitor before it is discharged?

Exercise 32

1 An uncharged capacitor C_1 of capacitance 200 μF is connected in series with a resistor R of value 470 kΩ, a 1.50 V cell and switch S as shown below.

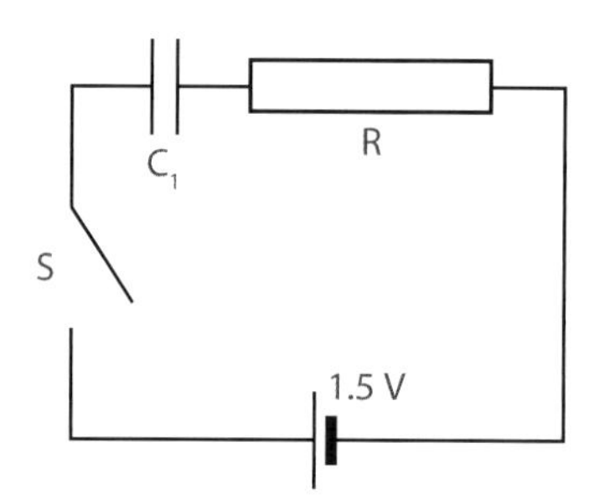

The switch is closed and the capacitor is charged.

(a) Sketch a graph to show the variation of the energy E stored in the capacitor (y–axis) with the voltage V (x–axis) as the capacitor is charged.

(b) Calculate the energy E_{max} stored in the fully charged capacitor.

The switch S is opened and then a second, uncharged, capacitor C_2 of value 400 μF is connected as shown below.

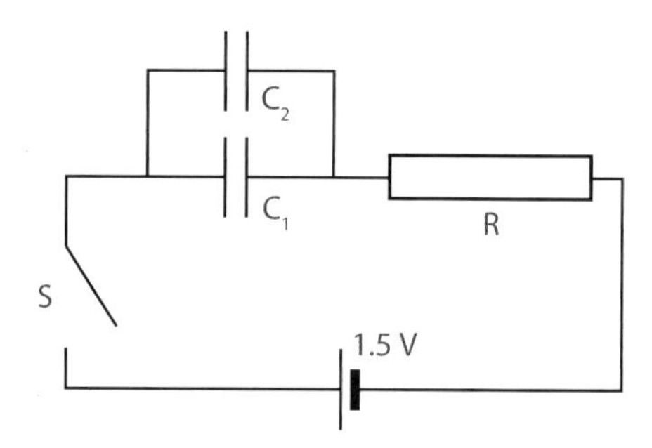

(c) Calculate the voltage across the combination of capacitors after they have been joined and the switch is still open.

(d) Calculate the total energy stored in the combination of capacitors.

[CCEA May 2008]

2 (a) Define the capacitance of a capacitor.

(b) A capacitor of value 47 μF is charged by a battery of e.m.f. 6.0 V.

(i) Show that the charge on the capacitor is 2.8×10^{-4} C.

(ii) Calculate the energy stored in the charged capacitor

(c) The charged 47 μF capacitor is then disconnected from the battery and connected to an initially uncharged capacitor C as shown. It is found that the potential difference across the 47 μF falls to 4.1 V as a result of this reconnection.

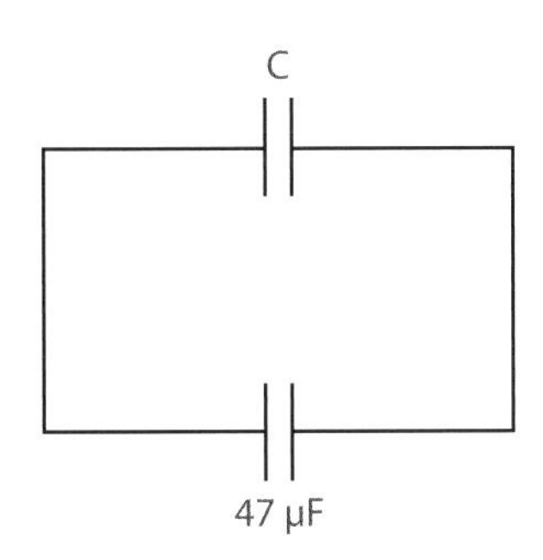

(i) Calculate the value of the capacitance C.

(ii) Calculate the energy now stored in this combination of capacitors.

(iii) Account for any difference between your answer and the answer to (b) (ii).

[CCEA June 2005]

3 (a) When a battery of e.m.f. V is connected across an initially uncharged capacitor, charge moves onto one plate of the capacitor and away from the other until the potential difference between the plates equals V. The process can be thought of as the transfer of a large number of equal amounts Δq of charge, amounting to a total charge Q.

(i) Explain why work is needed to transfer charge.

(ii) Why is zero work needed to transfer the first Δq of charge?

(iii) Write down an expression for the work ΔW needed to transfer the last Δq of charge.

(iv) Hence explain why the work required to transfer the total charge Q is given by

$$W = \frac{1}{2}QV$$

(b) A network of initially uncharged capacitors is shown below.

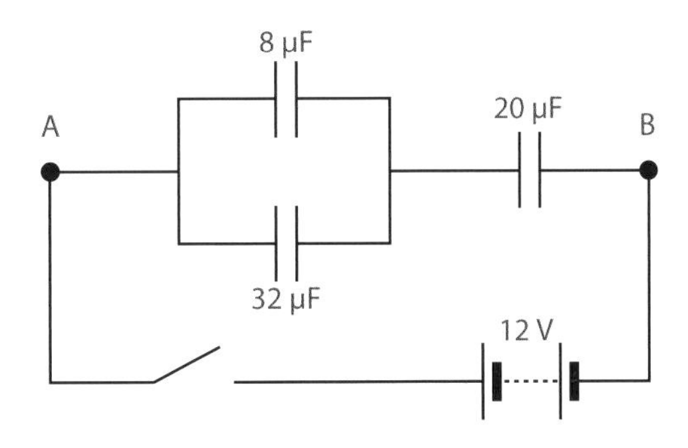

(i) Calculate the total capacitance of the network between the points A and B.

(ii) The switch is closed. Calculate the potential difference across the 20 μF capacitor when the capacitors are fully charged.

[CCEA May 2002]

4 An uncharged capacitor C is connected in series with a resistor R, a battery and a switch as shown below.

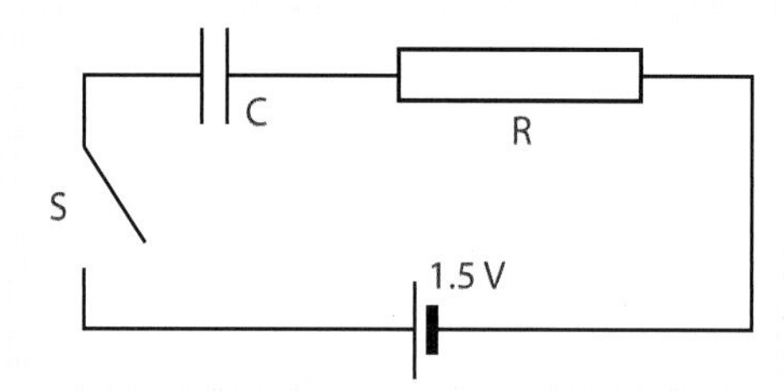

(a) Sketch a graph to show how the potential difference V_R across the resistor varies with time after the switch is closed.

(b) Sketch a graph to show how the charge Q on the capacitor varies with time after the switch is closed.

[CCEA 2006]

5.5 Magnetic Fields

Students should be able to:

5.5.1 Explain the concept of a magnetic field;

5.5.2 Understand that there is a force on a current-carrying conductor in a perpendicular magnetic field and be able to predict the direction of the force;

5.5.3 Define magnetic flux density using the equation $F = BI\ell$;

5.5.4 Define the unit of magnetic flux density, the tesla;

5.5.5 Understand the concepts of magnetic flux and magnetic flux linkage;

5.5.6 Recall and use the equations for magnetic flux, $\phi = BA$, and flux linkage, $N\phi = BAN$;

5.5.7 Define the unit of magnetic flux, the weber;

5.5.8 State, use and demonstrate experimentally Faraday's and Lenz's laws of electromagnetic induction;

5.5.9 Recall and calculate induced e.m.f. as rate of change of flux linkage with time;

5.5.10 Describe how a transformer works;

5.5.11 Recall and use the equation $\frac{V_s}{V_p} = \frac{N_s}{N_p} = \frac{I_p}{I_s}$ for transformers;

5.5.12 Explain power losses in transformers and the advantages of high voltage transmission of electricity;

The space surrounding a magnet where a magnetic force is experienced is called a magnetic field. This magnetic force is experienced by other magnets, objects made of metals such as iron and steel and also by current carrying conductors. The direction of a magnetic field at a point is taken as the direction of the force that acts on a north pole placed at that point.

The shape of a magnetic field can be represented by a magnetic field lines (lines of magnetic force). Arrows on the lines show the direction of the magnetic force. Since a north pole is repelled from another north pole and attracted by a south pole the direction of a magnetic field is from North to South.

Magnadur magnets are commonly used in school laboratories to produce a strong magnetic field. Two are arranged on a U-shaped steel yoke as shown in the diagram. The diagram also shows the shape and direction of the magnetic field that is created by this arrangement.

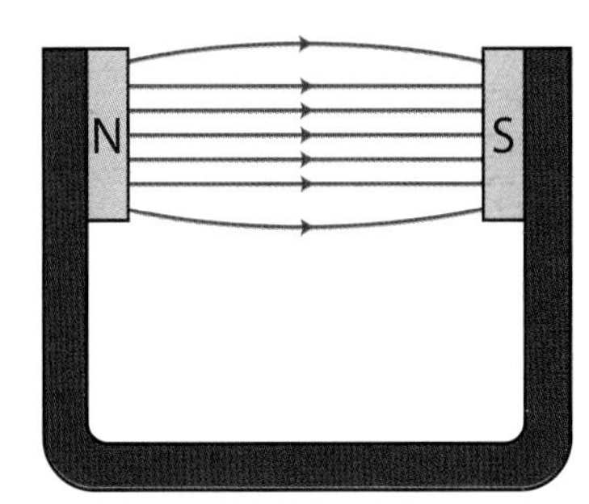

The field lines in the central portion are parallel and uniformly spaced this indicates a magnetic field in this region that is uniform, i.e. it has the same strength at all points in the region. The field weakens at the edges as indicated by the increased spacing between the lines of force.

Electromagnetism

A conductor carrying an electric current is surrounded by a magnetic field. Different shaped conductors produce magnetic fields of various shapes. The magnetic field lines due to the current in a straight wire are circles concentric with the wire. The shape can be revealed by sprinkling iron filings around the wire. The direction can be obtained by using the right hand screw rule. If a right handed screw moves forward in the same direction as the current, then the direction of rotation of the screw gives the direction of the magnetic field lines.

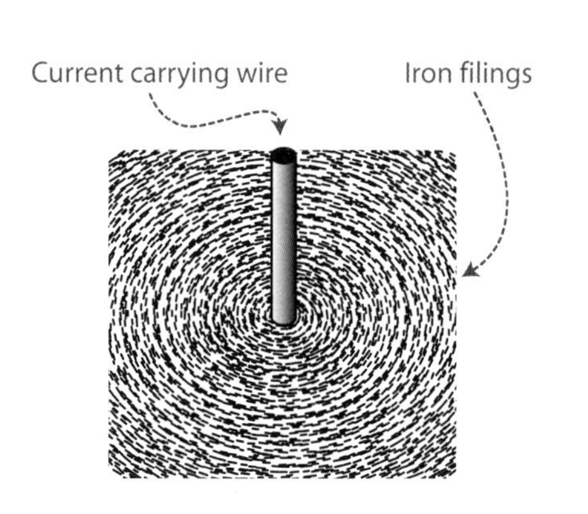

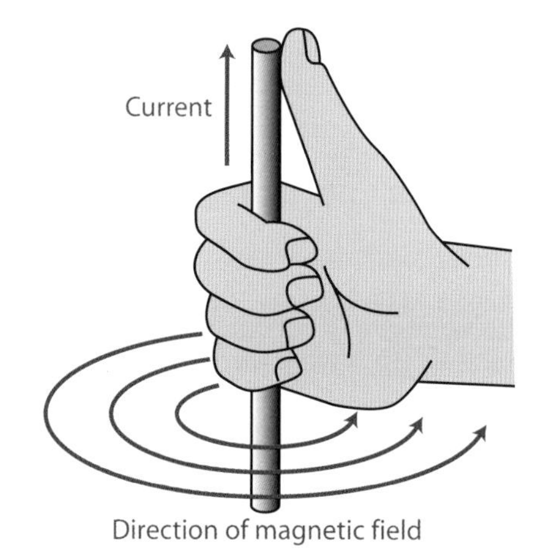

The direction can also be obtained using the right hand grip rule, this is illustrated in the diagram above.

Pretend to grab the wire with your right hand, your thumb pointing in the direction of the current. Your fingers curl in the direction of the magnetic field.

The diagram below shows a view looking along the current carrying wire.

The symbol ⊗ represents a current flowing into the page (away from you) and the symbol ⊙ represents a current flowing out of the page (towards you). You should apply the right hand grip rule to the diagrams below to confirm the magnetic field direction in each case.

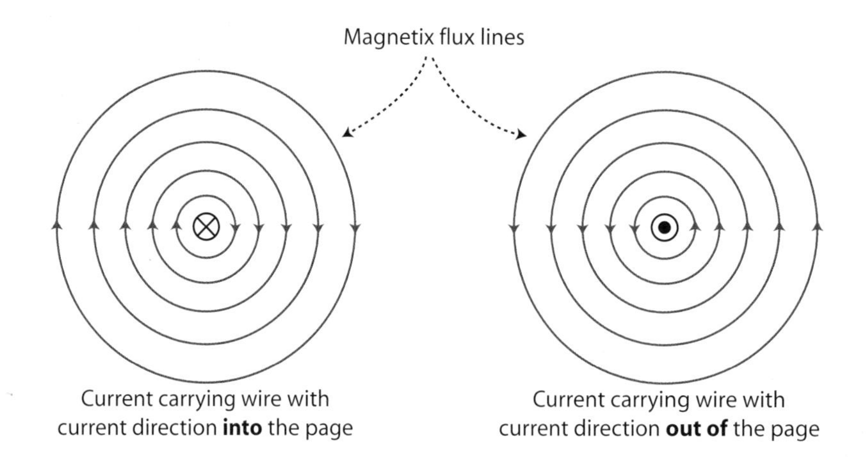

Force on a Current in a Magnetic Field

When a current flows through a conductor in a magnetic field the conductor experiences a force. This can be demonstrated using the apparatus shown. When a current is passed along the flexible wire, the wire moves up. The force acts at right angles to both the current and the magnetic field direction.

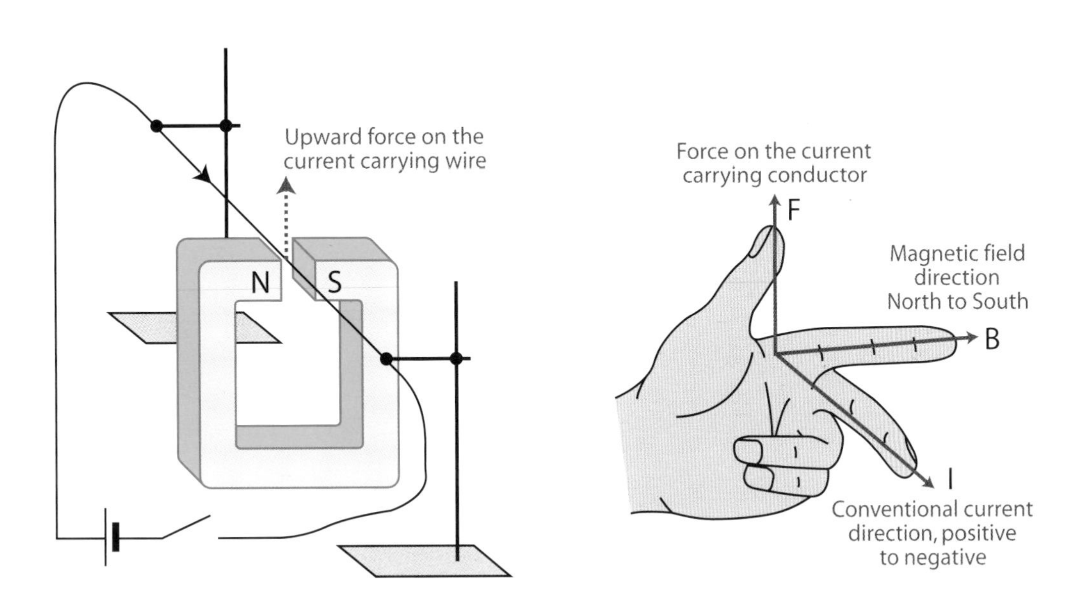

The direction of the force (movement) of the wire is obtained from Fleming's Left Hand rule. Extend the thumb, first and second fingers of the left hand so that they are at right angles to each as shown in the diagram above. The first finger represents the magnetic field direction. The magnetic field direction is from the north pole (N) to the south pole (S). The second finger represents the conventional current direction, from positive to negative. The thumb represents the force acting on the current carrying wire.

This force arises because of the interaction of the two magnetic fields: the uniform field of the permanent magnet and the field due to the current in the wire. The magnetic field lines of force are vectors and the field lines due to two fields have to be combined vectorially. This interaction of magnetic fields is illustrated below and the resultant field is some times called a catapult field. The field lines resemble the stretched rubber band in a catapult, the conductor experiencing a force in the direction shown.

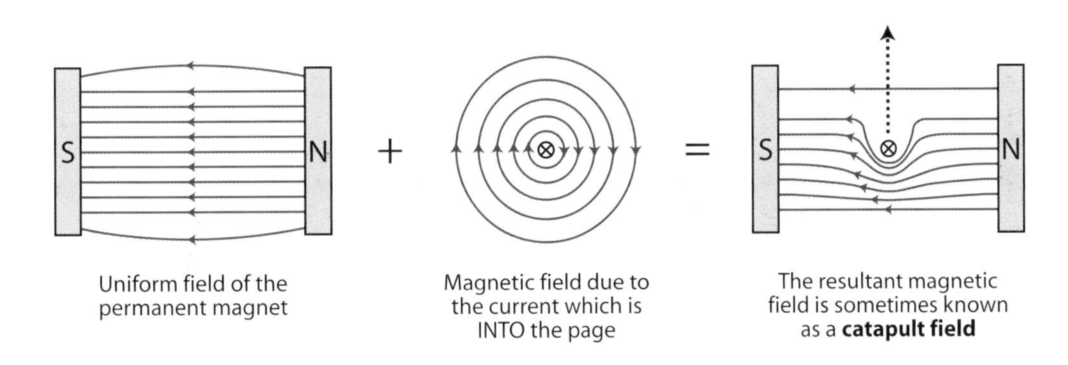

Uniform field of the permanent magnet

Magnetic field due to the current which is INTO the page

The resultant magnetic field is sometimes known as a **catapult field**

When the current and the magnetic field are in the same direction or act at 180° to each other the force is zero. It is a maximum when the current and the magnetic field are perpendicular to each other.

In the case when the magnetic field and the current are at right angles the force is given by:

$$F = BI\ell$$

where B is the magnetic flux density in tesla (T)
I is the current in the conductor in amperes (A)
ℓ is the length of the conductor in the magnetic field in metres (m)

The term magnetic flux density is used to indicate the strength of a magnetic field. Magnetic flux density is measured in units known as the tesla, symbol T. The tesla is defined from the equation shown above. Re-arranging the equation gives:

$$B = \frac{F}{I\ell}$$

The magnetic flux density is defined as the force per unit current carrying length. A current carrying length is the product of the current and the length of the conductor in the magnetic field. A current of 2.0 A and a length of 0.5 m is a current carrying length of 1.0 A m.

A magnetic field of flux density 1.0 T will exert a force of 1.0 N on a current carrying length of 1.0 Am when the current and magnetic field directions are perpendicular to each other.

Worked Examples

Example 1

A straight wire, of length 60 cm, carries a current of 1.25 A. Calculate the value of the force that acts on this wire when a length of 25 cm of the wire is placed at right angles to a magnetic field of flux density 4.5×10^{-2} T.

Solution

$$F = BI\ell$$
$$= 4.5\times10^{-2} \times 1.25 \times 0.25$$
$$= 0.014 \text{ N}$$

Only 0.25 m of the wire is in the magnetic field.

Application of Fleming's Left Hand Rule tells us that the force acts on the wire to the left.

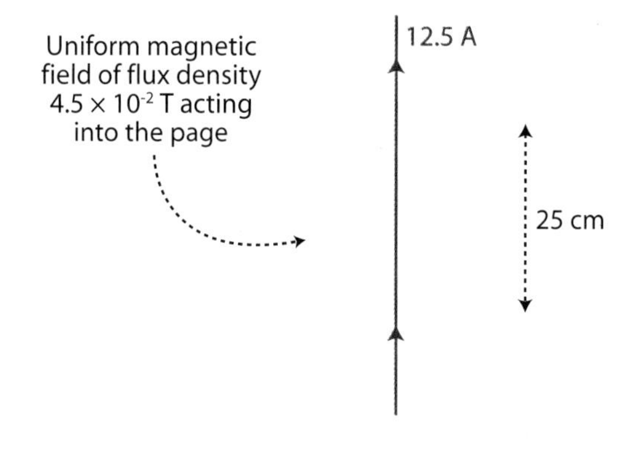

Example 2

A straight wire carrying a current of 500 mA is placed in a magnetic field of flux density 0.2 T. The length of the wire in the magnetic field is 30 cm and the wire makes an angle of 50° with the magnetic field direction. Calculate the force on the current carrying wire due to the magnetic field.

Solution

The magnetic flux density is resolved into a component perpendicular to the wire $B_\perp$ and a component parallel to the wire, $B_=$. Only the perpendicular component produces a force on the wire..

$$F = B_\perp I\ell$$
$$B_\perp = 0.2 \times \sin 50^\circ$$
$$= 0.153 \text{ T}$$
$$F = 0.153 \times 0.5 \times 0.3$$
$$= 0.023 \text{ N}$$

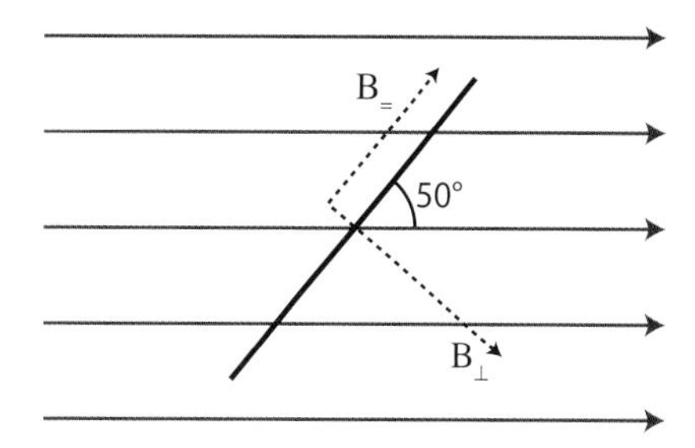

Verification of $F = BI\ell$

The apparatus shown below can be used to investigate how the force acting on a current carrying conductor depends on the current flowing in the conductor and the length of the conductor in the magnetic field.

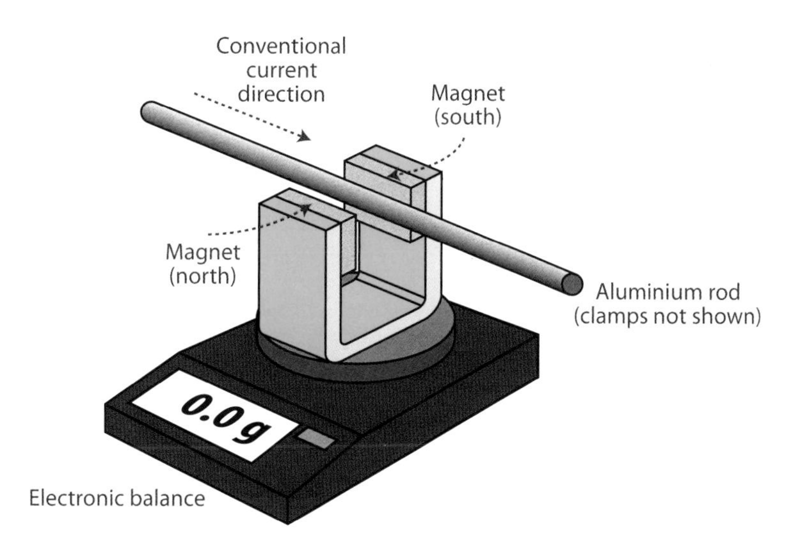

An aluminium rod (e.g. the upright from a retort stand) is clamped horizontally above a sensitive electronic balance. The rod is connected to a variable low voltage supply. An ammeter connected in series with both will allow the current to be measured.

A permanent magnet is placed on the balance and the aluminium rod positioned so that it is located in the centre of the magnetic field. The balance is set to read zero after the magnets have been placed on it. When a current is then passed along the clamped aluminium rod the rod experiences a force due to the interaction of the permanent magnetic field and the magnetic field due to the current in the aluminium rod. In the case shown in the diagram, the force is upwards, use Fleming's Left Hand Rule to verify this.

Since the magnet exerts an upward force on the rod, then by Newton's third law, the rod must exert an equal but opposite force on the magnets. This downward force will cause the reading on the electronic balance to increase.

The current is varied with a single magnet in place. This ensures that the length of the conductor in the magnetic field remains constant.

A graph of the force (y–axis) against current (x–axis) produces a straight line through the origin. This is verification that the force on the current carrying conductor in the magnetic field is directly proportional to the current.

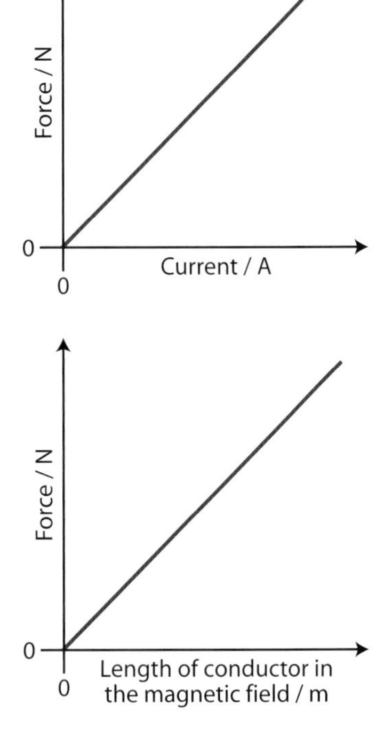

The current is then fixed and a number of identical magnets are placed side by side, this changes the length of the conductor in the magnetic field. A graph of force (y–axis) against the length of conductor in the magnetic field (x–axis) produces a straight line through the origin. This is verification that the force on the current carrying conductor in the magnetic field is directly proportional to the length of the conductor in the magnetic field.

Exercise 33

1 The horizontal component of the Earth's magnetic field has a flux density of 2.0×10^{-5} T. A straight piece of wire XY 1.2 m long of mass 0.8 g is resting on a wooden bench so that it is at right angles to the magnetic field direction. A current is passed through the wire which just causes the wire to lift off the bench.

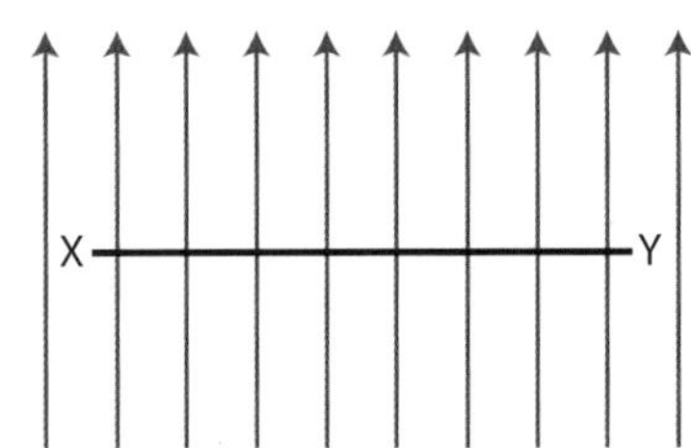

(a) State the direction of the current in the wire.

(b) Calculate the current.

2 A single-turn rectangular wire loop ABCD hangs from a sensitive balance so that its lower portion is in a region of uniform magnetic field. The direction of the magnetic field is at right angles to the plane of the loop. This arrangement is shown in the diagram. The upper portion of the loop is not in the magnetic field.

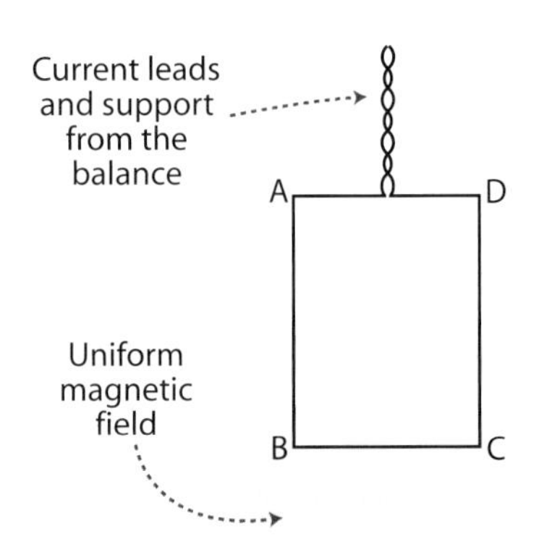

When a current passes round the loop in the direction ABCD, the balance reading gets less.

(a) Explain why the balance reading is less when there is a current in the loop in this direction than when it is not.

(b) The dimensions of the loop are AB = 150 mm, BC = 50 mm. The initial reading of the balance is 50.57 grams. When the current in the loop is 2.8 A in the direction ABCD, the balance reading is 50.035 grams. Calculate the magnitude of the magnetic flux density of the field.

(c) Is the direction of the magnetic field out of or into the plane of the page? Give a reason for your answer.

(d) Without making any other changes, the direction of the current is reversed so that it passes round the loop in the direction DCBA. What is the balance reading now?

[CCEA June 2005]

Magnetic Flux ϕ

Gravitational fields, electric fields and magnetic fields can be visualized using lines of force. Magnetic flux lines (magnetic field lines) show the direction of a magnetic field. Their spacing indicates the strength of the field, the closer the field lines, the stronger the magnetic field.

Magnetic flux ϕ, represents the total number of magnetic flux lines that pass at 90° through a given area. Magnetic flux is measured in webers (Wb).

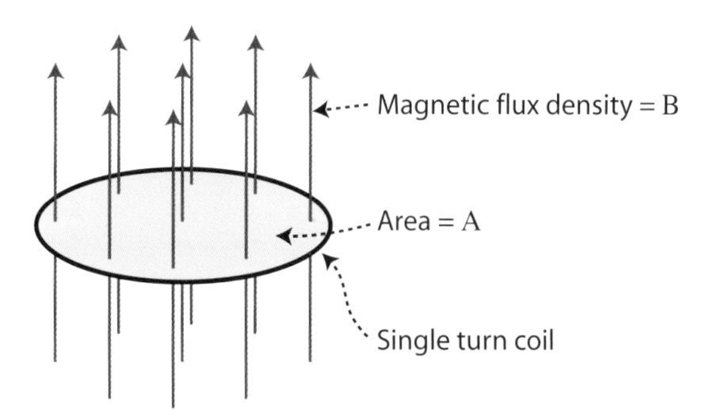

$$\phi = B \times A$$

where ϕ is the magnetic flux in Wb

B is the magnetic flux density, perpendicular to the plane of the coil, in T

A is the area in m^2

Magnetic Flux Linkage $N\Phi$

The term magnetic flux linkage is used when calculating the total magnetic flux passing through or linking a coil of N turn and area of cross section A. Magnetic flux linkage is also measured in webers (Wb).

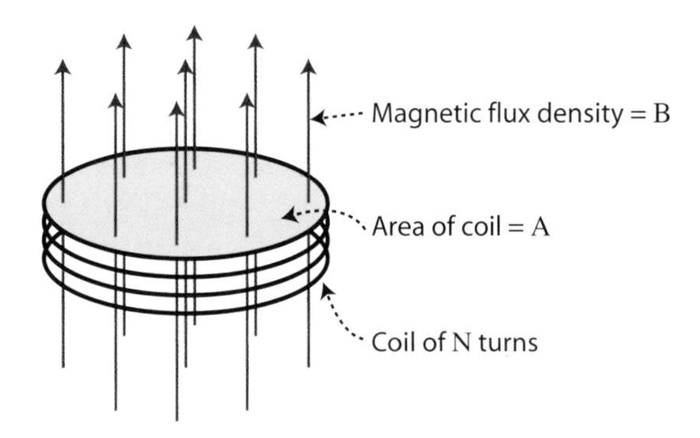

$$N\Phi = B \times A \times N$$

where $N\Phi$ is the magnetic flux linkage in Wb
B is the magnetic flux density, perpendicular to the plane of the coil, in T
A is the area in m^2
N is the number of turns in the coil

Worked Example

A coil of 250 turns each of area 80 cm^2 is placed in a uniform magnetic field of flux density 0.25 T. The magnetic field direction makes an angle of 30° with the normal to the plane of the coil. Calculate the magnetic fluxing the coil.

Solution

The magnetic flux density is resolved into a component perpendicular to the plane of the coil, $B_\perp$ and a component parallel to the plane of the coil, $B_=$. Only the perpendicular component passes through the coil.

$$B_\perp = 0.25 \cos 30^\circ = 0.217 \text{ T}$$

$$N\Phi = B_\perp AN = 0.217 \times 80 \times 10^{-4} \times 250$$

$$= 0.434 \text{ Wb}$$

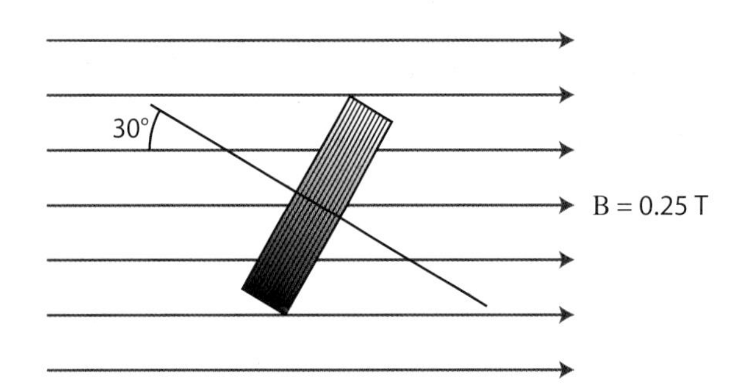

Electromagnetic Induction

An **e.m.f.** (electromotive force) can be induced in a coil of wire by moving a magnet towards or away from the coil or by moving a wire so that it cuts across the magnetic lines of flux. For an e.m.f. there must be relative motion between the magnet and the conductor, the wire or coil of wire.

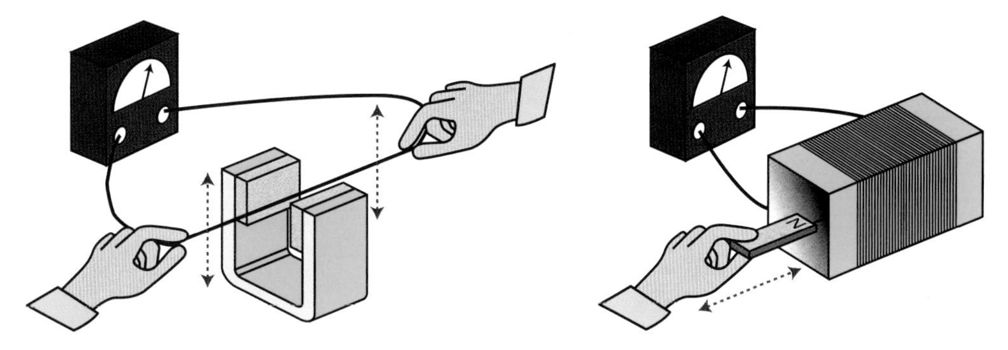

The magnitude of the e.m.f. is proportional to:

- the strength of the magnet
- the number of turns on the coil
- the speed of the moving magnet

The factors affecting the size of the induced e.m.f. can be stated formally as Faraday's Law of Electromagnetic Induction:

The magnitude of the induced e.m.f. is equal to the rate of change of magnetic flux linkage.

The direction of the induced e.m.f. depends on the direction in which the magnet is moving and on the type of magnetic pole nearest the coil. We can demonstrate this using the simple apparatus shown below.

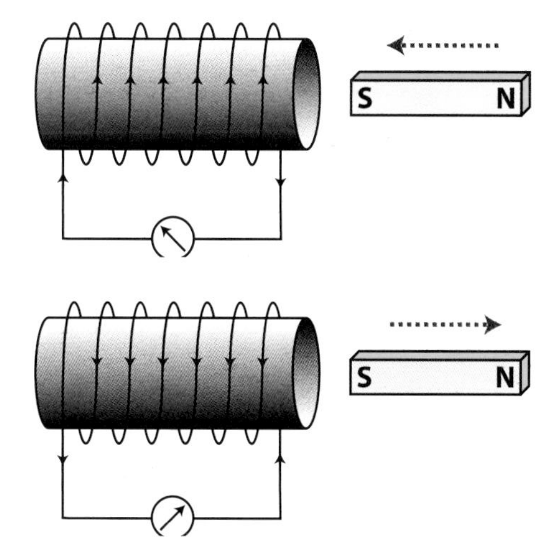

Moving the south pole of the magnet towards the coil causes the induced current to flow so that it creates a south magnetic pole in the coil opposing the incoming south pole of the magnet. Work has to be done against this opposing force.

Moving the south pole of the magnet away from the coil causes the induced current to reverse direction. It now flows so that it creates a north magnetic pole in the coil attracting the retreating pole of the magnet. Work again has to be done against this opposing force.

These observations can be stated as Lenz's Law:

The direction of the induced current is such that it opposes the change in the magnetic flux that is producing it.

Lenz's Law is the Principle of Conservation of Energy in action. The kinetic energy of the moving magnet is converted to electrical energy when work is done against the opposing force.

Calculation of Induced e.m.f.

Faraday's model of electromagnetic induction is based on magnetic field lines. He suggested that an e.m.f. was induced in the conductor when the number of magnetic field lines passing through the conductor changed.

Faraday's law states that the size of the induced e.m.f. is equal to the rate of change of the number of magnetic field lines passing through the coil. The number of magnetic flux lines that passes through a coil is calculated as the magnetic flux $\Phi = BA$ and for a coil of N turns it is calculated as the magnetic flux linkage $N\Phi = BAN$.

The diagram below illustrates the role magnetic flux linkage plays in Faraday's model of electromagnetic induction.

A bar magnet is moved closer to a coil of wire. At this instant the number of magnetic flux lines linking the coil is 3.

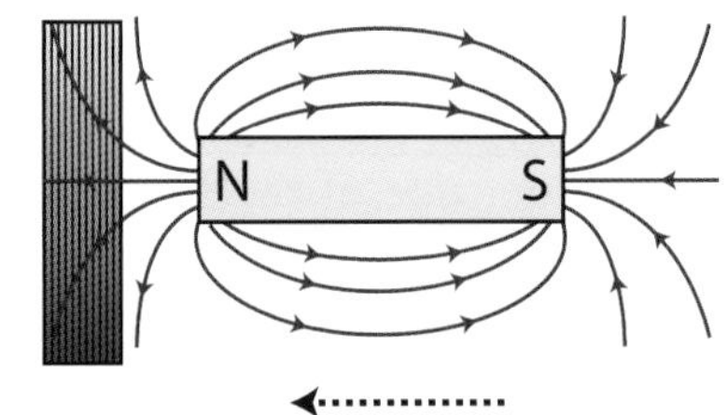

A short time later the magnet is closer to the coil and the number of magnetic flux lines linking the coil has increased to 5.

There has been a change of 2 in magnetic flux linking the coil, this results in an e.m.f. being induced in the coil. The more rapidly this change takes place the larger the induced e.m.f.

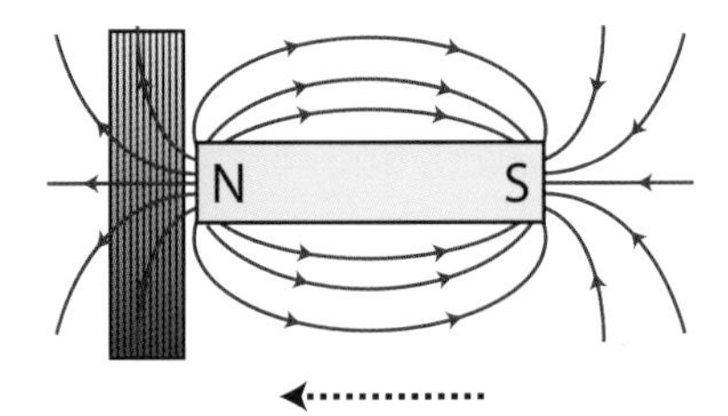

The induced e.m.f. can be calculated as follows:

Induced e.m.f. = Rate of change of magnetic flux linkage

$$E = -\frac{\Delta N\Phi}{\Delta t}$$

where E is the induced e.m.f. in V

$\Delta N\Phi$ is the change in the magnetic flux linkage in Wb

Δt is the time in which the change occurs in s

The minus is a consequence of Lenz's Law

A.C. Generator

This consists of a coil of wire that it rotated at a constant angular velocity in a magnetic field. As the coil turns the magnetic flux linking it changes. This change in magnetic flux linkage results in an alternating e.m.f. being induced in the coil. The output from the rotating coil is led to the outside by means of carbon brushes that rub against metal slip rings.

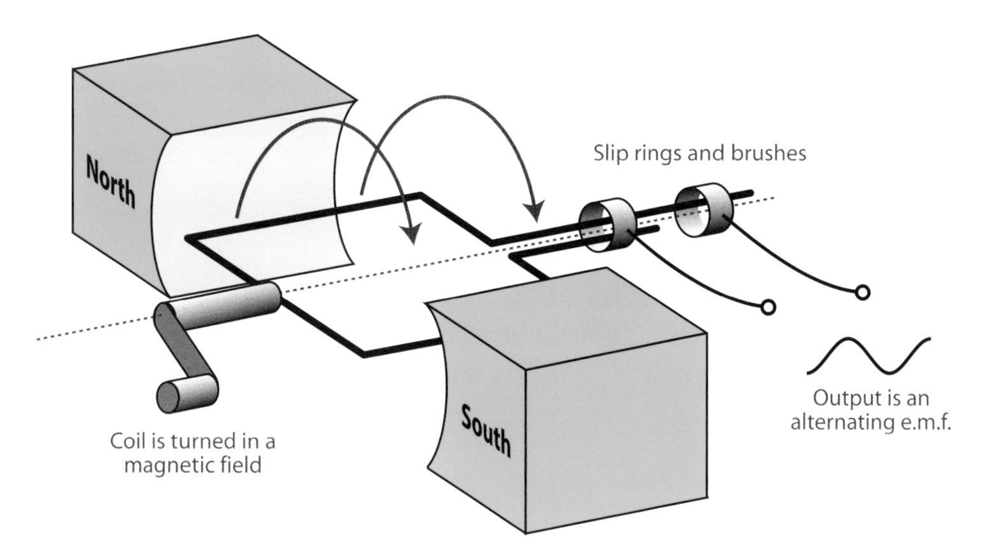

In a practical generator there would be a lot more turns on the coil. The coil would be formed around a soft iron core. These two changes produce an alternating output voltage with a much greater peak value.

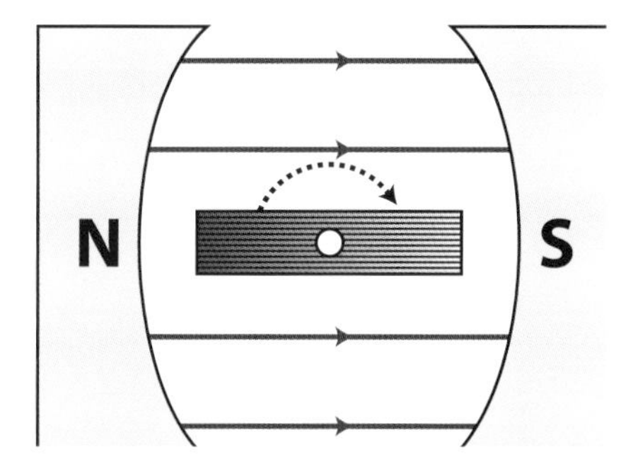

In this position the plane of the coil is parallel to the magnetic flux lines. The magnetic flux linking the coil is zero

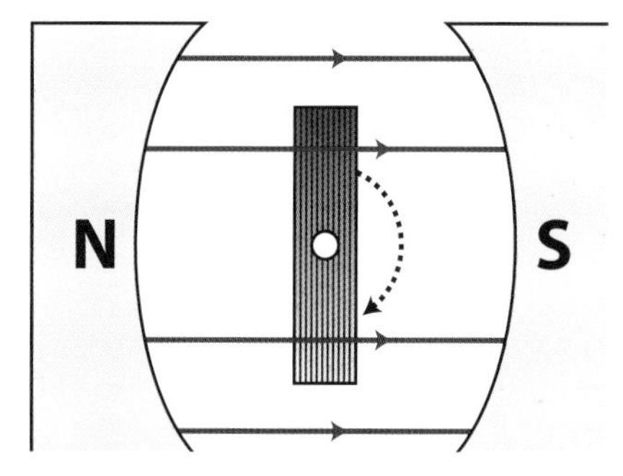

In this position the coil has turned through 90°. The plane of the coil is now perpendicular to the magnetic flux lines. The magnetic flux linkage is now a maximum.

$$N\Phi = BAN$$

B = magnetic flux density

A = area of coil

N = number of turns on the coil

The diagram on the next page shows the position of the coil after it has rotated through an angle θ. The dotted line is the normal to the plane of the coil. To calculate the magnetic flux linking the coil at this position it is necessary to resolve the magnetic flux density into two components. The component perpendicular to the plane of the coil is $B_{\perp}$, which is the one used to calculate the magnetic flux linking the coil.

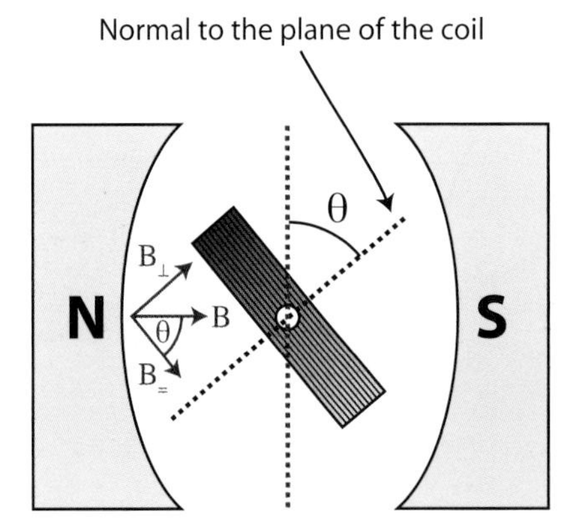

The coil is rotating with a constant angular velocity ω, the angle θ = ωt.

The flux linking the coil varies with time and at any instant is given by BAN cos ωt.

$$B_{\perp} = B \cos \theta$$

$$\text{so } N\phi = B_{\perp}AN = B \cos \theta \cdot AN$$

which is more conveniently written as

$$N\phi = BAN \cos \theta$$

The magnetic flux linking the coil varies as cosine function.

This diagram shows the position of the coil in relation to the magnetic field. Notice that when the flux linking the coil is momentarily zero the induced e.m.f. is a maximum.

Although at this instant the magnetic flux linkage is zero its rate of change is a maximum so the induced e.m.f. is a maximum.

The e.m.f. is obtained by taking the negative of the gradient of the flux linkage graph. This results in a sine function.

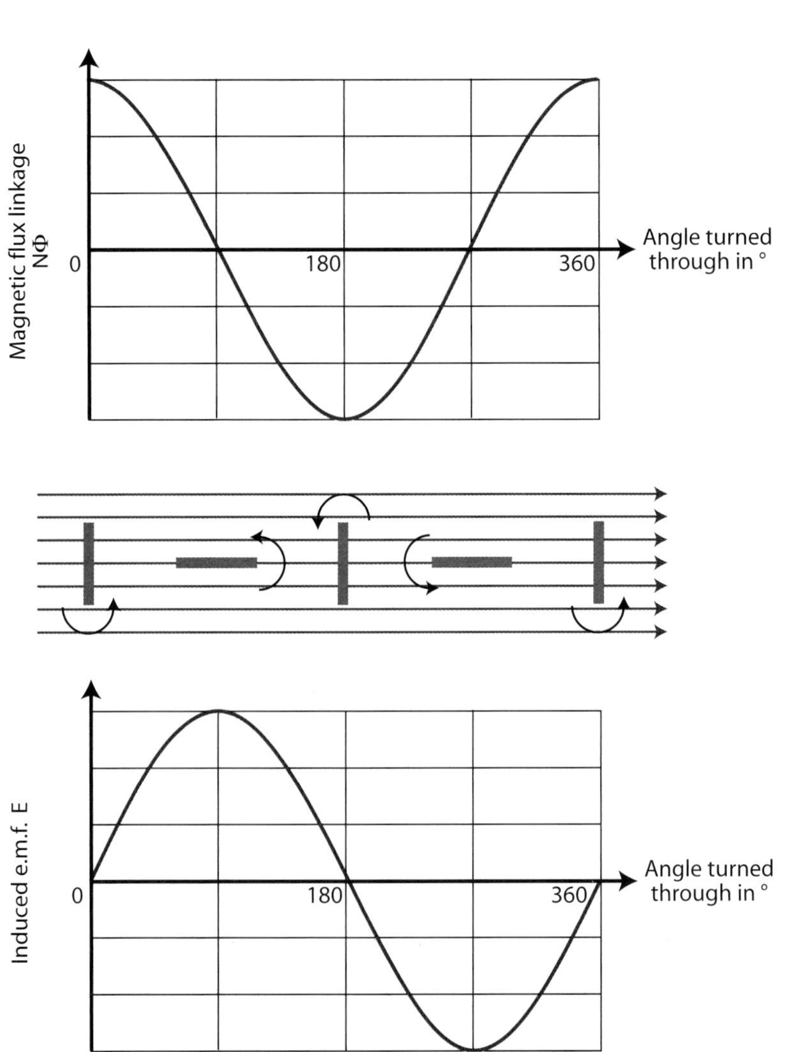

Worked Examples

Example 1

A circular coil of wire has a radius of 2 cm and 500 turns. It is situated in a magnetic field so that its plane is perpendicular to a magnetic field of flux density 20 mT. The magnetic field is then reduced in strength to zero and then increased to 20mT in the opposite direction, this change, which takes 60 ms, takes place at a constant rate. Calculate the magnitude of the e.m.f. induced in the coil.

Solution

$$\text{Area of coil} = \pi(2\times10^{-2})^2 = 1.26\times10^{-3}\ \text{m}^2$$

$$\text{Change of magnetic flux density} = 20-(-20) = 40\ \text{mT}$$

$$\text{Induced e.m.f. } E = \frac{\Delta N\Phi}{\Delta t} = \frac{500 \times 1.26\times10^{-3} \times 40\times10^{-3}}{60\times10^{-3}} = 0.42\ \text{V}$$

Example 2

A long solenoid is connected to a signal generator. A short coil is wound over the longer coil as shown below.

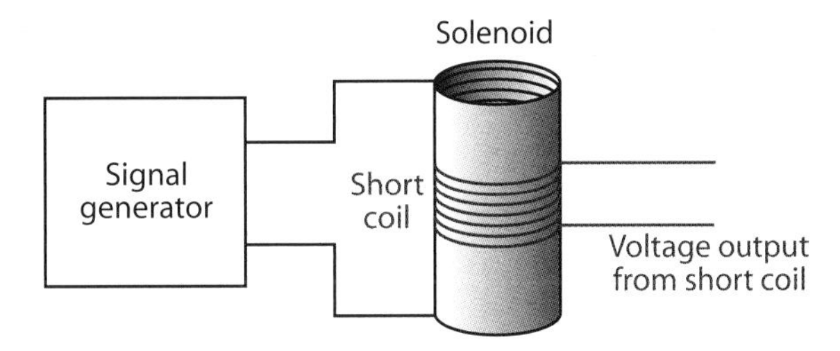

The voltage delivered by the signal generator varies with time as shown below.

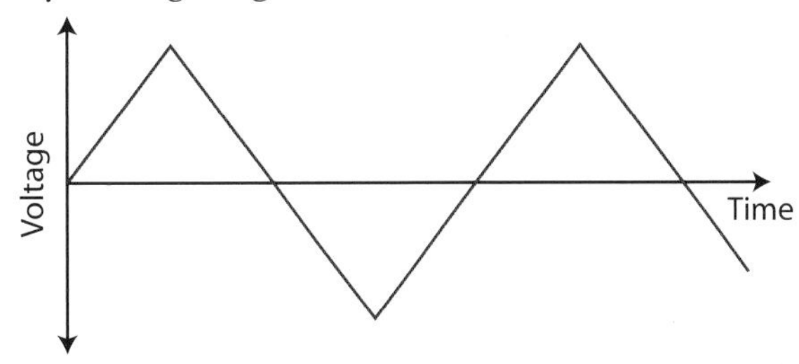

Sketch a graph to show the voltage induced in the short coil.

Solution

The voltage causes a current to flow in the longer coil. This current varies with time in the same way as the voltage does. This current creates a magnetic field which also varies in the same manner as shown in the graph. This means that the magnetic flux linking the shorter coil changes so an e.m.f. is induced in this coil. The size of the induced e.m.f. is proportional to the rate of change of magnetic flux linkage. The size of the induced e.m.f. is obtained by taking the gradient of the voltage-time graph and its direction is the negative of this gradient.

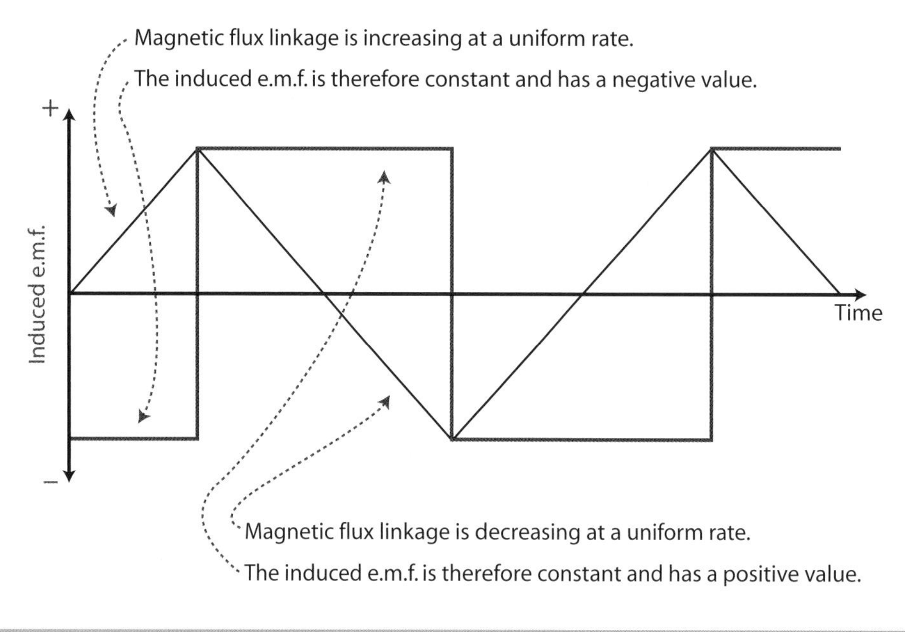

Exercise 34

1 The graph below shows how the magnetic flux linkage through a coil varies over a period of 13 seconds. Draw second set of axes, with induced e.m.f. on the y–axis and time of the x –axis. On this set of axes show how the e.m.f. induced in this coil varies during the same time. Place values on both of the axes.

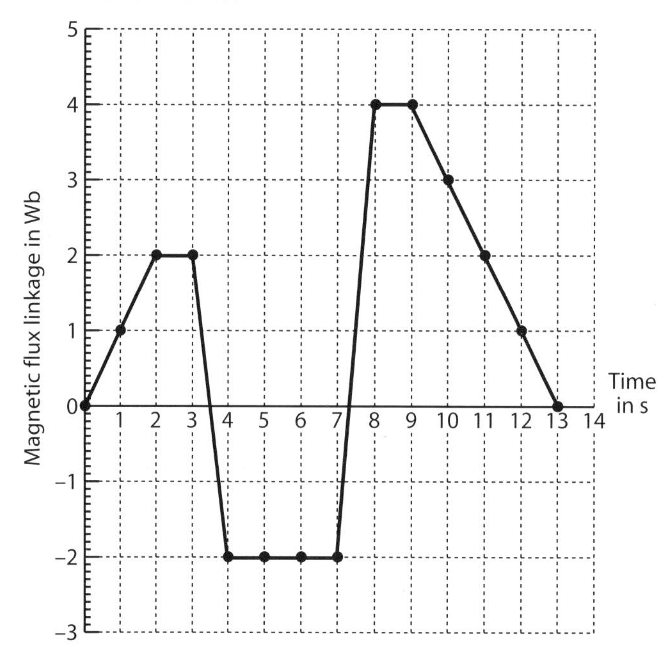

2 (a) State what is meant by electromagnetic induction.

(b) A coil of wire with an iron core is connected to a battery through a switch. A filament lamp is connected in parallel with the coil as shown below.

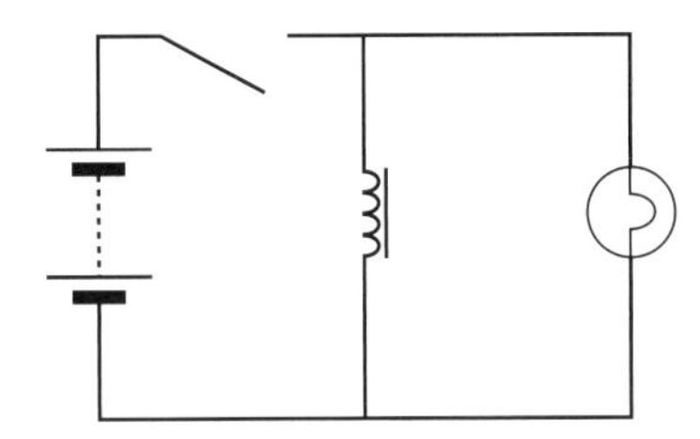

When the switch is closed, the intensity of the lamp increases to a steady glow.

When the switch is opened, the lamp produces a very bright flash before going out.

(i) Why does the lamp produce this very bright flash when the switch is opened?

(ii) Why is this bright flash much brighter than the lamp's initial steady glow?

[CCEA 2005]

3 (a) State Faraday's law of electromagnetic induction.

(b) Lenz's law is sometimes thought of as being simply a statement of the law of conservation of energy. The diagram shows a magnet being pushed into a coil of wire. The coil is connected to a resistor.

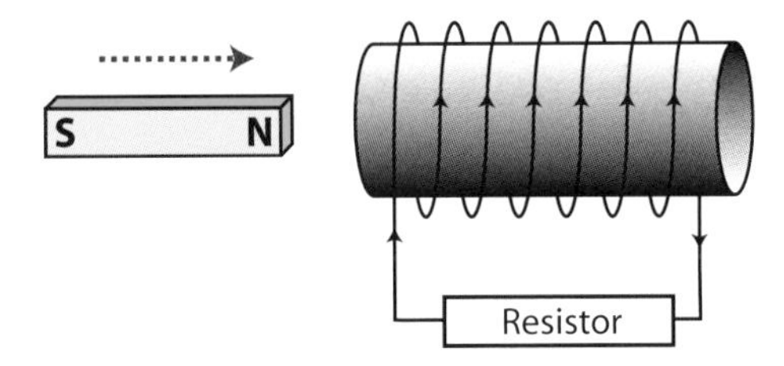

Explain how Lenz's law and the conservation of energy are related in this example. Make reference to Faraday's law and Lenz's law, the work done in pushing the magnet towards the coil and the thermal energy in the resistor.

[CCEA 2007]

4 Before the introduction of CDs and MP3 players music was stored in a continuous groove on the surface of vinyl discs. The music was recovered by rotating the vinyl disc on a turntable rotated at a constant angular velocity. A stylus placed in the groove will move from side to side as the record spins. This movement produces an em.f. The graph shows the variation of the angular position of the coil with time t as the stylus moves through a section of the groove.

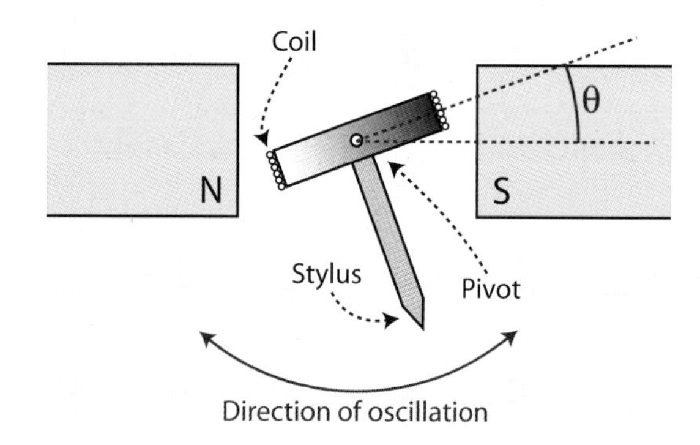

(a) Explain why the movement of the stylus creates an e.m.f.

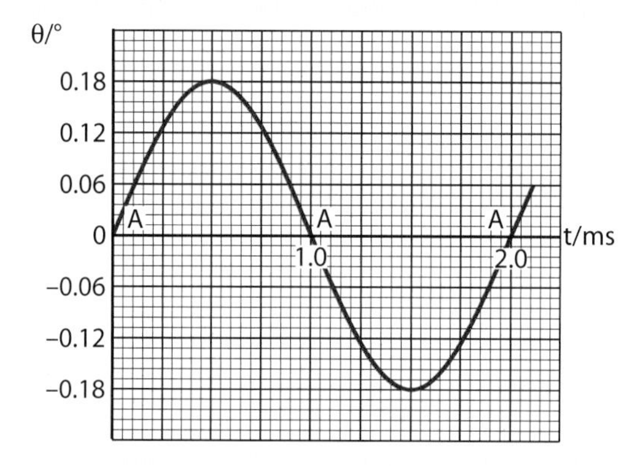

(b) The maximum e.m.f. in the coil is induced at the times marked A i.e. at 0, 1.0 ms and 2.0 ms. Making reference to the graph explain briefly why the induced e.m.f. is greatest at these times.

[CCEA 2008]

The transformer

The principle of the transformer can be demonstrated using two coils arranged as shown in the diagram below. Coil 1 is connected to a battery and a switch. Coil 2 is connected to a sensitive ammeter. A soft iron core passes through both coils.

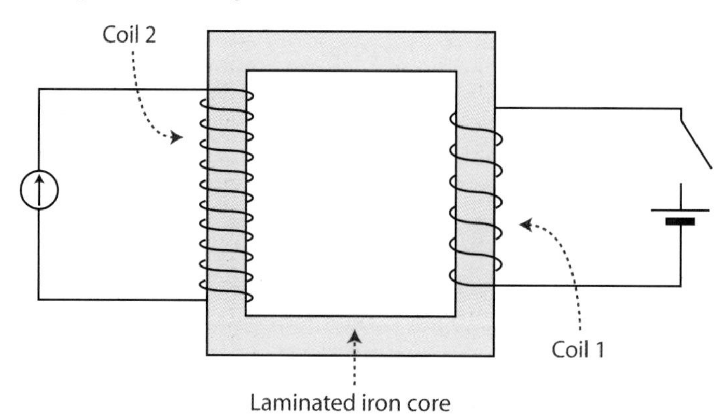

When the switch of coil 1 is closed the meter gives a momentary deflection to the right, and then goes back to zero. Closing the switch completes the circuit of coil 1, a current flows and a magnetic field quickly develops and the iron core ensures that magnetic flux lines link the turns of coil 2. This change in the magnetic flux linkage causes an e.m.f. to be induced in coil 2.

When the current in coil 1 reaches its final steady value, the magnetic field around it is steady, so the flux linkage is no longer changing and so there is no induced e.m.f.

When the switch is opened the current in coil 1 falls, its magnetic field decreases. The magnetic flux linkage through coil 2 decreases and e.m.f. is induced again, but in the opposite direction to the first one.

An em.f. is only induced in coil 2 if there is a changing current in coil 1, which produces a changing magnetic field and of course a changing magnetic flux linking coil 2.

If a low frequency alternating voltage (1 Hz) is applied to coil 1, you will observe the needle of the meter attached to coil 2 alternate from side to side.

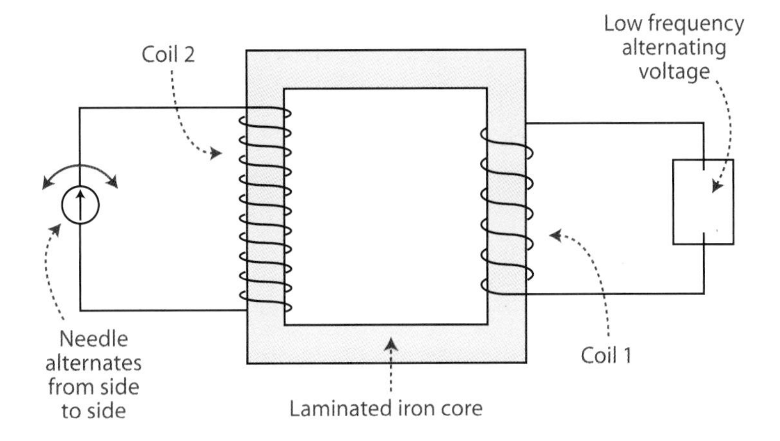

The apparatus shown in the diagram above is a simple transformer. The input voltage is applied to coil 1 which is called the primary coil. The output voltage is taken across coil 2 which is known as the secondary coil.

Transformers convert alternating current at one voltage to alternating current at a different voltage.

The diagram shows the structure of a typical transformer. In practice, for increased efficiency, the coils are wound on top of each other. They are shown separated here for clarity.

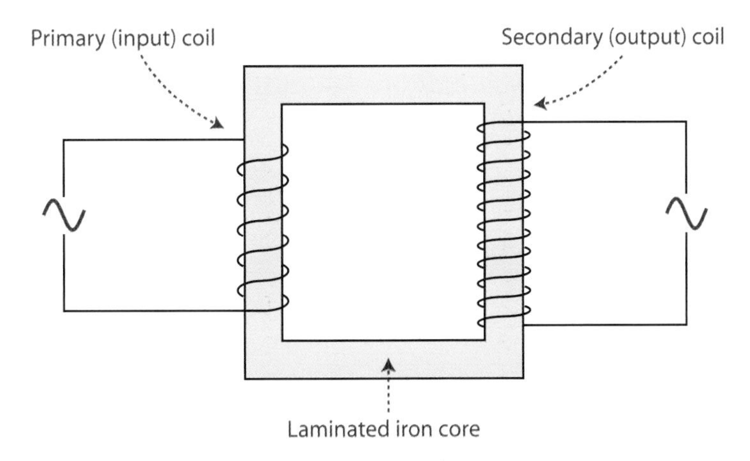

An alternating current in the primary coil produces an alternating magnetic field, a field whose strength varies continuously and whose direction reverses periodically. This changing magnetic field causes the magnetic flux linking the secondary coil to change so an e.m.f. is induced in the secondary coil. The iron core maximizes the magnetic flux linking both coils.

The ratio of the turns on each coil determines the ratio of the two voltages.

$$\frac{N_S}{N_P} = \frac{V_S}{V_P}$$

where N_P is the number of turns on the primary coil

N_S is the number of turns on the secondary coil

V_P is the voltage applied to the primary coil (input)

V_S is the voltage developed across the secondary coil (output)

If we assume that the efficiency of the transformer is 1 then we can write:

Input power = Output power

$$I_P \times V_P = I_S \times V_S$$

$$\frac{I_P}{I_S} = \frac{V_S}{V_P}$$

where I_P is the current in the primary coil

I_S is the current drawn from the secondary coil

V_P is the voltage applied to the primary coil (input)

V_S is the voltage developed across the secondary coil (output)

In a Step-up transformer, V_s is greater than V_p and in a Step-down transformer V_s is less than V_p.

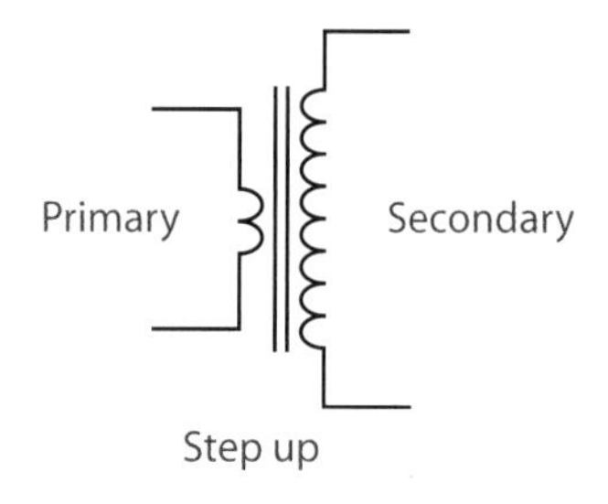

Step up

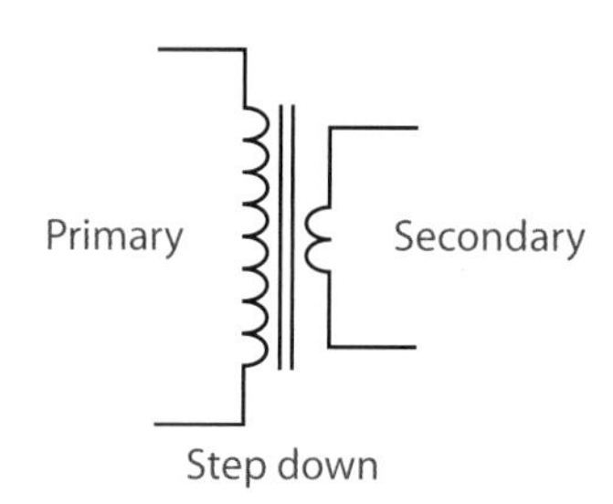

Step down

Worked Example

1 A transformer, assumed to be 100% efficient, is used with a supply voltage of 120 V. The primary winding has 50 turns. The required output voltage is 3000 V. The output power is 200 W.

(a) Name this type of transformer.

(b) Calculate the number of turns in the secondary coil.

(c) Calculate the current supplied to the primary coil.

[CCEA 2001]

Solution

(a) This is step up transformer since the output voltage is required to be greater than the input voltage.

(b) $\frac{N_S}{N_P} = \frac{V_S}{V_P}$ substituting values gives:

$\frac{N_S}{50} = \frac{3000}{120}$ so $N_S = 1250$

(c) Input power = Output power

$I_P \times V_P = P$

$I_P \times 120 = 200$ so $I_P = 1.7$ A

Power Losses in a Transformer

The efficiency of real transformers is less than 1(100%), not all of the input electrical energy appears as useful output electrical energy. However transformers are some of the most efficient electrical devices with some designs having an efficiency of around 99%. Some of the various ways in which energy is wasted are described below.

1 Real transformers have resistive heat losses due to the wires in the primary and secondary coils. As the coils heat up the resistance increases so the amount of energy lost also increases. Many large transformers use oil as a coolant to reduce this type of energy loss.

2 Not all of the magnetic flux of the primary passes through or links the secondary coil.

3 Repeatedly magnetising the iron core in one direction and then reversing the direction of magnetisation results in heating of the iron core.

4 The iron is a conductor in a changing magnetic field so currents are induced to flow in it. These are called **eddy currents** and are very large. They result in heating of the core. These eddy current are reduced by laminating the core.

Transmission of Electricity

Transformers play an important role in the transmission of electricity from the generating stations to the consumers. At the generating end they step the voltage up before it is connected to the transmission cables and at the consumer end they step the voltage down for use in appliances.

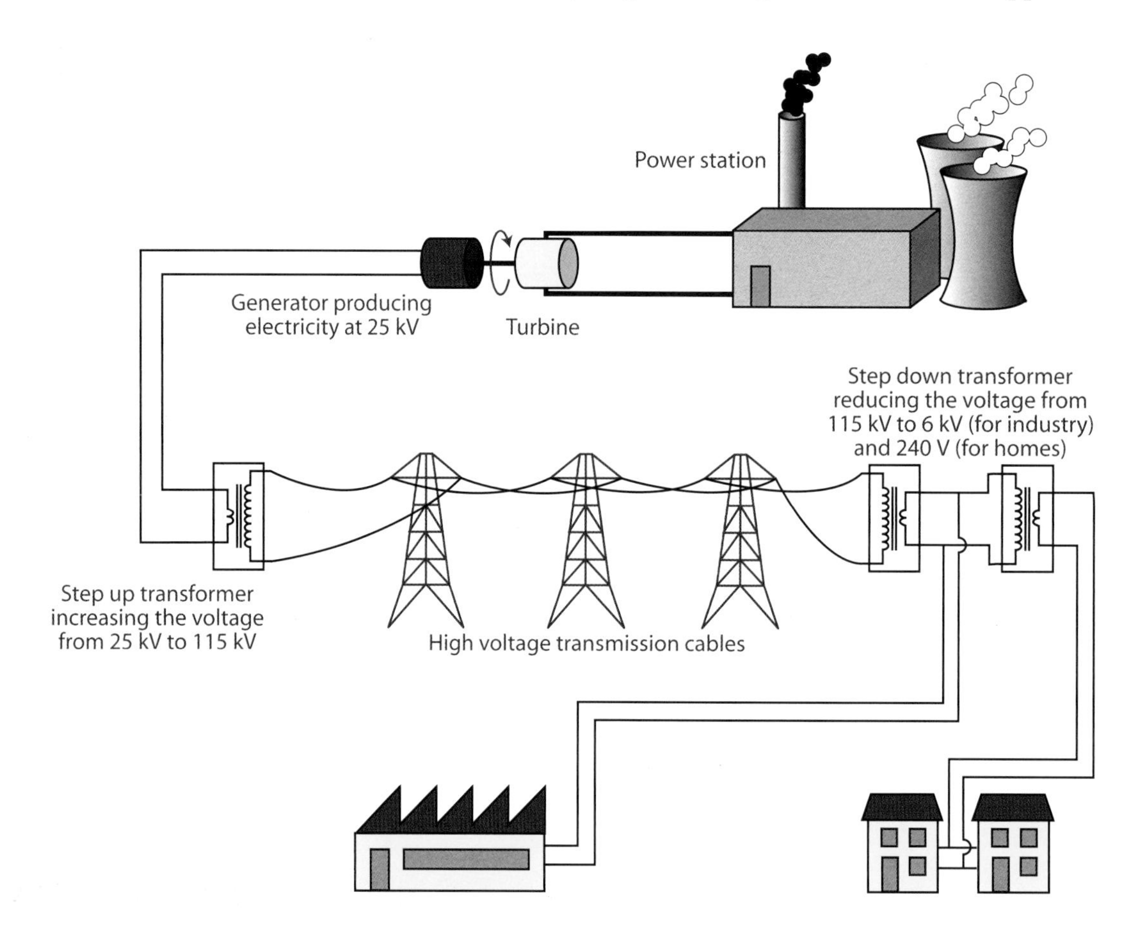

Advantages of High Voltage Electricity Transmission

The cables used to transmit the electrical power from the generator to the consumer have resistance. This means energy would be lost as heat due to resistive heating.

The diagram below is a simplified picture of the electricity generation and transmission system.

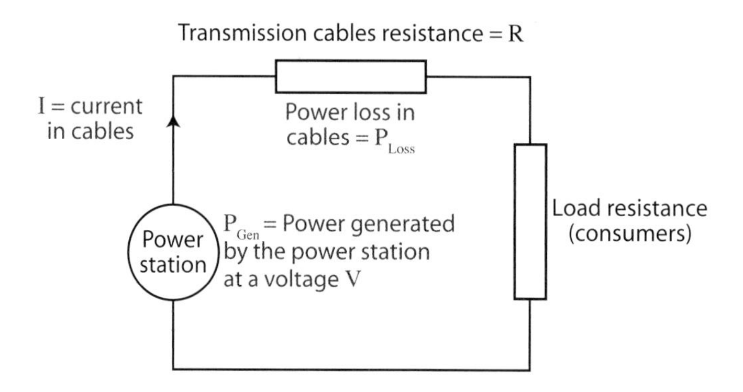

The power loss in the cables $P_{Loss} = I^2R$.

One way to reduce this power loss is to make the resistance of the cables smaller by using cables of a very large cross section area. This would considerably increase their weight and the cost.

Alternatively the current could be reduced. This is the function of the step up transformer at the generating station. The voltage is stepped up and the current is reduced. The electrical power is transmitted at a high voltage and a low current.

$$\text{Power generated } P_{Gen} = IV \quad \text{so } I = \frac{P_{Gen}}{V}$$

$$P_{Loss} = \frac{P_{Gen}^2 R}{V^2}$$

Since P_{Gen} and R are both constants this shows that the power loss in the cables P_{Loss} is inversely proportional to the square of the voltage at which the electricity is transmitted to the consumer. If the voltage is doubled the power loss is reduced by a factor of four.

The advantage of the high voltage transmission can be seen if we calculate the power loss for 1 km of transmission line at two different voltages.

A power station generates 300 MW of electrical power at a voltage of 25 000 V. The transmission lines have a resistance of 0.2 Ω per kilometre.

At 25 kV the current is $I = \frac{P}{V} = \frac{300 \times 10^6}{25 \times 10^3} = 1.2 \times 10^4$ A

The power loss in the cables = $P_{Loss} = I^2R = (1.2 \times 10^4)^2 \times 0.2 = 2.88 \times 10^7$ W or 28.8 MW

This represents 9.6% power loss for a 1km length of transmission line.

At 115 kV the current is $I = \frac{P}{V} = \frac{300 \times 10^6}{115 \times 10^3} = 2.6 \times 10^3$ A

The power loss in the cables = $P_{Loss} = I^2R = (2.6 \times 10^3)^2 \times 0.2 = 1.35 \times 10^6$ W or 1.35 MW

This represents 0.45% power loss for a 1 km length of transmission line.

Worked Example

A generator and transmission system is designed to transmit power of 200 kW. The total resistance of the transmission lines is 0.60 Ω. It is required that the power loss in the transmission lines should not exceed 0.015% of the total power. Calculate the minimum output voltage needed to achieve this.

[CCEA 2000]

Solution

$$P_{Loss} = 0.015\% \text{ of } 200 \text{ kW} = 0.03 \text{ kW or } 30 \text{ W}$$

$$P_{Loss} = \frac{P_{Gen}^2 R}{V^2}$$

$$30 = \frac{(200 \times 10^3)^2 \times 0.6}{V^2}$$

$$V^2 = \frac{(200 \times 10^3)^2 \times 0.6}{30} = 8 \times 10^8$$

$$V = 28\,284 = 28.3 \text{ kV}$$

Exercise 35

1 (a) Describe how a suitable input voltage, applied to the primary coil of a transformer, results in an output being obtained from the secondary coil.

(b) A supply of fixed peak voltage is available. A voltage of half this peak value is to be obtained, using a transformer. What type of transformer should be used? State the relation which must apply between the number of turns on the primary coil N_P and the number of turns on the secondary coil N_S.

2 A generator in a power station produces 176 MW of power at 11.0 kV. This is transformed to 275 kV for transmission to a nearby town.

(a) Calculate the ratio of turns in the primary coil to the number of turns in the secondary coil of the transformer.

(b) Calculate the currents in the primary and secondary coil, assuming the transformer to have an efficiency of 100%.

(c) Calculate the maximum resistance of the transmission lines if the power loss in the lines is not to exceed 2% of the power generated.

(d) In practice, transformers are less than 100% efficient. Name three sources of power loss in a transformer.

[CCEA 1997]

5.6 Deflection of Charged Particles in Electric and Magnetic Fields

5.6.1 Understand that a moving charge in a uniform electric field experiences a force;

5.6.2 Recall and use the equation F = Eq to calculate the magnitude of the force on a charged particle in an electric field, and determine the direction of the force;

5.6.3 Understand that a moving charge in a uniform, perpendicular magnetic field experiences a force;

5.6.4 Recall and use the equation F = Bqv to calculate the magnitude of the force, and determine the direction of the force;

5.6.5 Outline the structure of the cathode ray oscilloscope;

5.6.6 Explain how the cathode ray oscilloscope can be used as a measuring instrument for voltage

Motion of Electrons in an Electric Field

Television tubes, cathode ray oscilloscopes, X–rays tubes and electron microscopes are all devices that use beams of electrons. The electrons come from a heated wire filament of an electron gun. The electrons with sufficient energy escape from the surface of the filament by a process called thermionic emission.

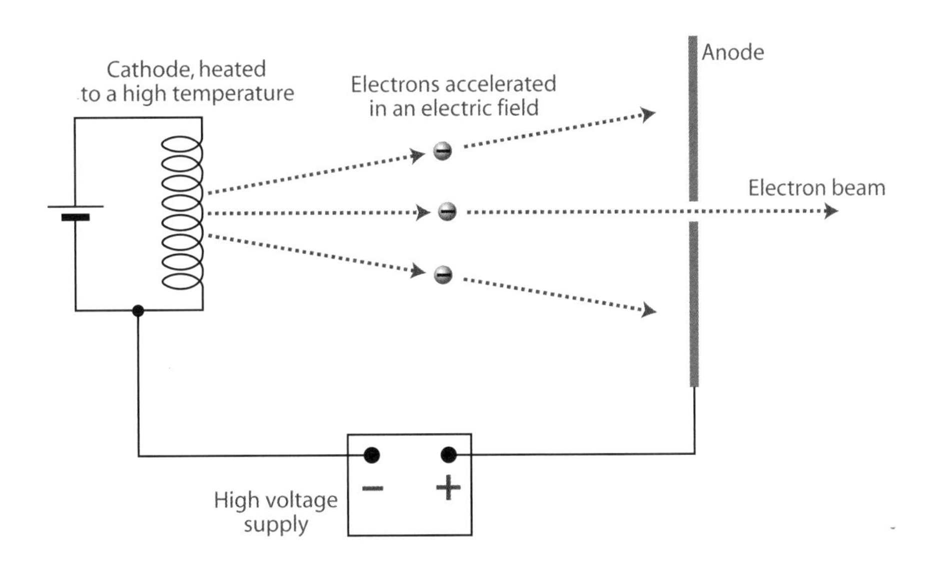

The electrons emerge into an electric field created by a large potential difference between the cathode and the anode. The electrons are accelerated by this electric field. They lose electric potential energy and gain kinetic energy. Applying the Principle of Conservation of energy we say that the kinetic energy gained equals the loss of electric potential energy. An electron beam emerges through a opening in the anode.

The diagrams below show a typical electron beam tube found in most school laboratories. The electron gun directs a beam of electrons towards the fluorescent screen, which emits green light when struck by electrons. When the Maltese cross electrode is made positive it deflects the electrons away from their original path preventing them reaching the fluorescent screen so a black shadow is seen on the screen.

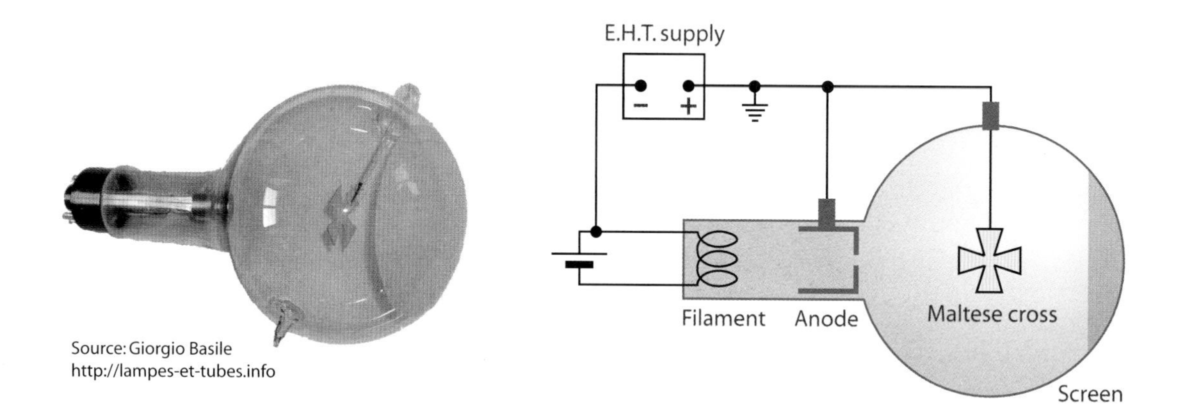

Energy Changes

Electric potential difference is measured in volts. The volt is defined as a joule per coulomb ($J\ C^{-1}$), which means that a charge of 1 coulomb accelerated through a potential difference of 1 volt will gain 1 joule of energy.

Using this definition of the volt and applying the principle of conservation of energy we have:

Loss of electrical potential energy = Gain of kinetic energy

$$eV = \frac{1}{2}m_e v^2$$

V = potential difference between anode and cathode
e = charge on the electron
m_e = mass of the electron
v = velocity of the electron

Re-arranging the above expression we obtain the equation below for the velocity of the electron after it has been accelerated from rest through a potential difference V.

$$v = \sqrt{\frac{2ev}{m_e}}$$

If the charged particles have a mass m and a charge q the expression for the velocity becomes:

$$v = \sqrt{\frac{2qV}{m}}$$

Worked Examples

Example 1

A potential difference of 50 V is applied between two electrodes. An electron is emitted from the negative electrode with negligible speed.

(a) Calculate the increase in kinetic energy of the electron when it reaches the positive electrode.

(b) Calculate the speed of the electron when it reaches the positive electrode.

Solution

(a) Gain in kinetic energy = change in electric potential energy

$$= eV$$

$$= 1.6\times10^{-19} \times 50 = 8.0\times10^{-18} \text{ J}$$

(b) $$v = \sqrt{\frac{2\times8\times10^{-18}}{9.1\times10^{-31}}} = 4.2\times10^{6} \text{ ms}^{-1}$$

Example 2

Positive ions, each of mass 8.35×10^{-27} kg and charge 3.20×10^{-19} C are accelerated in a vacuum from rest to a speed of 5.75×10^{4} ms^{-1}.

(a) Calculate the potential difference through which the ions are accelerated to give them this speed.

(b) If ions which have twice this charge but of the same mass were accelerated through the same potential difference how would this affect their final kinetic energy and their final speed?

Solution

(a) Gain of kinetic energy = loss of electric potential energy

$$\tfrac{1}{2}mv^2 = qV$$

Rearranging the above gives $v = \sqrt{\frac{2qV}{m}}$

$$5.75\times10^{4} = \sqrt{\frac{2\times3.2\times10^{-19}\ V}{8.35\times10^{-27}}}$$

$$V = 43.1 \text{ V}$$

(b) Gain of kinetic energy = qV

Twice the charge and same potential difference means twice the kinetic energy.

With twice the charge, $v = \sqrt{\frac{4qV}{m}} = \sqrt{2}\sqrt{\frac{2qV}{m}} = 1.41v = 8.13\times10^{4} \text{ ms}^{-1}$

Example 3

Charged particles of different charges and masses are accelerated through the same potential difference. Sketch graphs to show how:

(a) The kinetic energy of the particles depends on their mass.

(b) The velocity of the particles depends on the charge of the particle.

Solution

(a) kinetic energy = loss of electric potential energy

$$E_k = qV$$

E_k does not depend on the mass, only on q and V

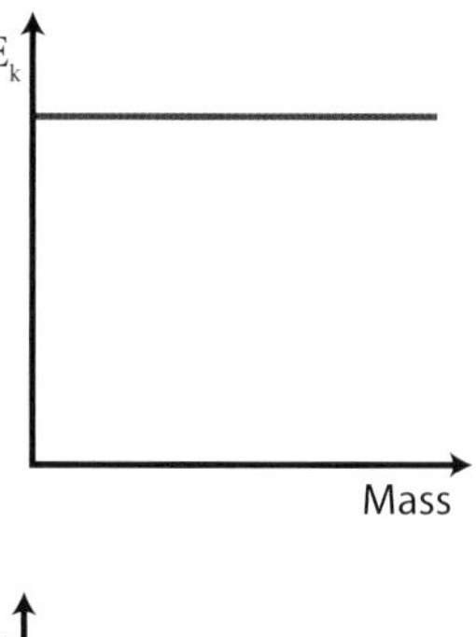

(b) $v = \sqrt{\dfrac{2qV}{m}}$

$v \propto \sqrt{q}$

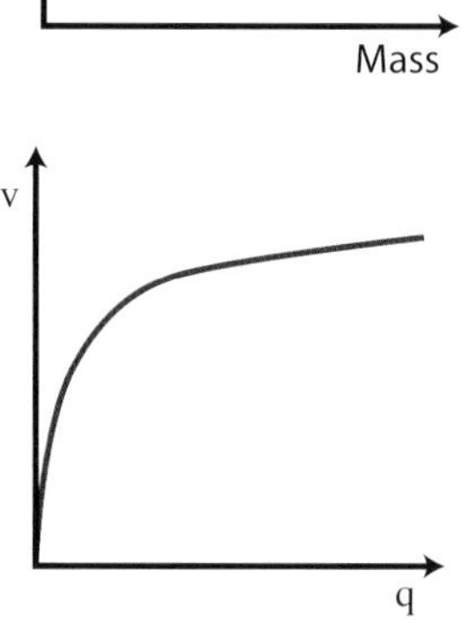

Force on a Charged Particle

The force on a charged particle in an electric field is given by:

$$F = qE$$

where F is the force in N

q is the charge in C

E is electric field strength in Vm^{-1} or NC^{-1}

If the field is non-uniform this force varies from point to point within the field. However in the case of a uniform electric field, which has the same field strength throughout, the force is constant.

Here a beam of positively charged particles is directed into an electric field so that its direction is parallel to the electric field lines.

The particles in the beam experience a force in the same direction as the electric field.

The particles will accelerate to the right.

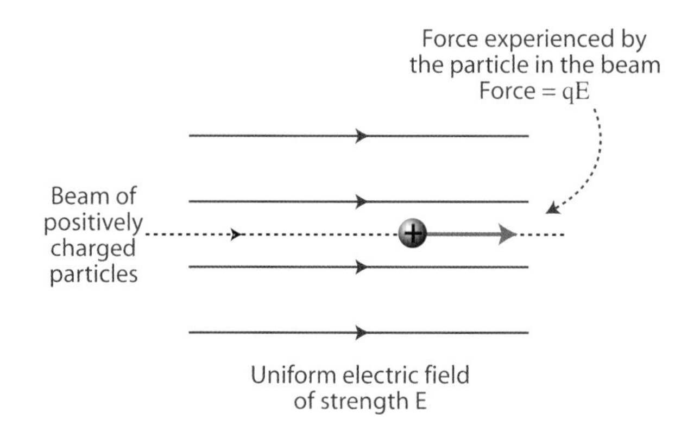

Here a beam of positively charged particles is directed into an electric field so that its direction is parallel to but in the opposite direction to the electric field lines.

The particles in the beam experience a force in the same direction as that of the electric field, but opposite to the direction of motion of the particles

The particles will experience a deceleration to the left.

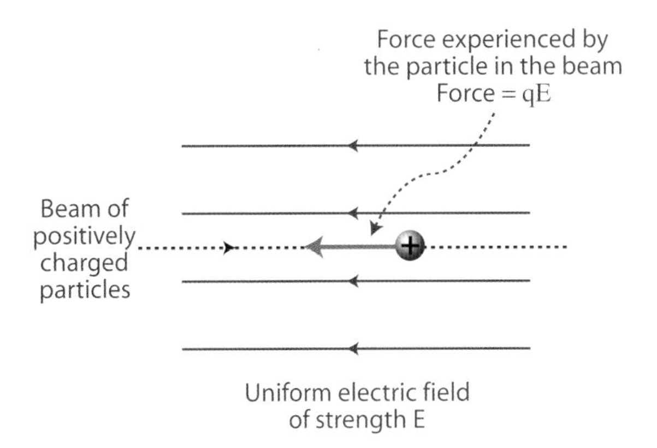

Deflection of Electrons by an Electric field

The diagram shows a beam of charged particles moving with velocity v entering a uniform electric field of strength E. The charged particles enter the electric field at right angles to the electric field lines. A uniform electric field has the same field strength throughout its region. The parallel, uniformly spaced field lines indicate this uniformity. The particles experience a force towards the positively charged plate indicating that they have a negative electric charge.

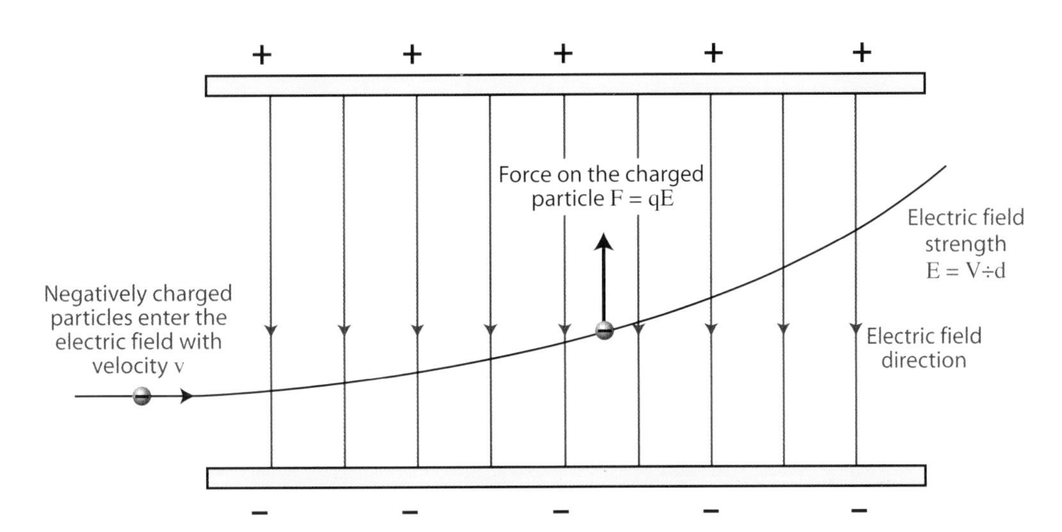

The charged particle experiences a force in the vertical direction only. Horizontally it does not experience any force. This means that we treat its motion in the following way:

Horizontally – constant velocity

Vertically – uniform acceleration from rest

The vertical force F is given by:

$$F = qE = q\frac{V}{d}$$

where q is the charge of the particle (coulombs)

E is the electric field strength = $\frac{V}{d}$

V is the potential difference between the plates (volts)

d is the separation of the plates (metres)

The acceleration of the particle is obtained from Newton's second law of motion:

$$a = \frac{F}{m}$$

$$= \frac{qV}{dm}$$

where a is the acceleration (ms^{-2})
m is the mass of the particle (kg)

The charged particles enter the electric field with a velocity v and travel a horizontal distance x in a time t. The vertical deflection of the charged particle in the same time is y.

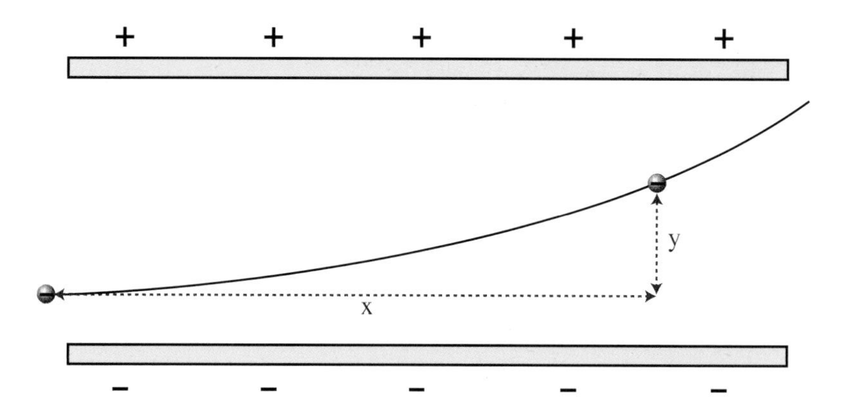

At any time, t, the horizontal displacement x = vt.

Since the vertical motion is one of uniform acceleration, **from rest**, it is necessary to use the equation of motion to determine the vertical displacement, y.

The vertical, y, displacement:

$$y = \tfrac{1}{2}at^2 = \tfrac{1}{2}a\frac{x^2}{v^2} \quad (\text{since } t = \frac{x}{v})$$

Since the acceleration and the initial horizontal velocity are constant we have that $y = kx^2$.

This is equation of a parabola, showing that the path of the charged particles when they move within the electric field is a parabola. When they exit the field they move in a straight line.

The time t to cross the electric field is $t = \frac{L}{v}$.

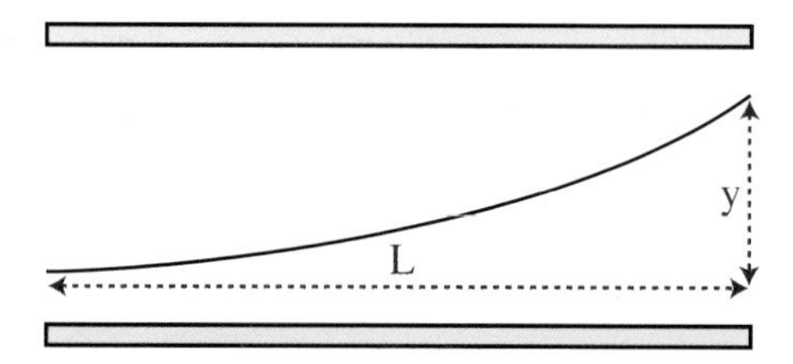

$y = \frac{1}{2}at^2$ The initial vertical velocity of the charged particles is 0.

$= \frac{1}{2}a\frac{L^2}{v^2}$ The time to move vertically is equal to the time to move horizontally.

$= \frac{1}{2}\frac{qV}{dm}\frac{L^2}{v^2}$ The vertical acceleration is $\frac{qV}{dm}$.

Worked Example

Example 1

The diagram shows two horizontal metal plates. The plates are 120 mm long and are separated by 40 mm. The upper plate is maintained at a potential of –20 V with respect to the lower one.

The region between the plates is a vacuum. An electron beam enters the electric field along a line midway between the plates. The electrons in the beam have a velocity of 6×10^6 ms^{-1}.

(a) Calculate the electric field strength at the point P.
State the direction of the electric field at P.

Solution $E = \frac{V}{d} = \frac{20}{40 \times 10^{-3}} = 500\ \text{Vm}^{-1}$ (or NC^{-1})

Direction upwards towards the upper plate.

(b) Calculate the force on an electron as it enters the magnetic field.

Solution $F = eE = 1.6 \times 10^{-19} \times 500 = 8.0 \times 10^{-17}$ N

(c) Sketch the path of the electron beam as it passes between the plates.

Solution

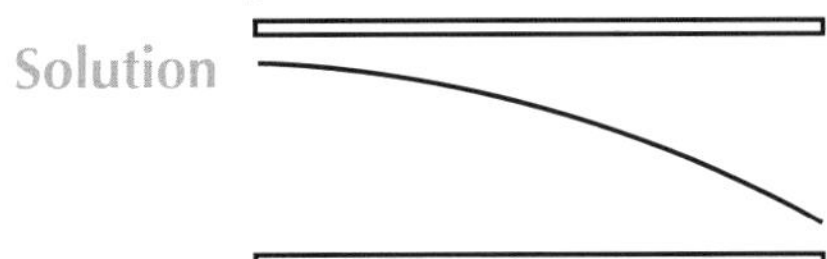

The electrons in the beam experience a force towards the lower plate, they are being repelled by the upper negative plate, the path is a parabola.

(d) Calculate the acceleration of an electron in the beam as it passes between the plates.
State the direction of the acceleration.

Solution $a = \frac{F}{m} = \frac{8.0 \times 10^{-17}}{9.1 \times 10^{-31}} = 8.8 \times 10^{13}\ \text{ms}^{-2}$

(e) Calculate the time taken for an electron in the beam to pass from one end of the plates to the other.

Solution $t = \frac{120 \times 10^{-3}}{6 \times 10^{6}} = 2.0 \times 10^{-8}$ s

(f) Calculate the distance from the mid-line at which the electron beam leaves the region between the plates.

Solution The vertical motion is uniform acceleration from rest

$$y = ut + \tfrac{1}{2}at^2 = 0 + \tfrac{1}{2}\left[8.8 \times 10^{13} \times \left(2 \times 10^{-8}\right)^2\right]$$

$$= 1.76 \times 10^{-2} \text{ m} = 17.6 \text{ mm}$$

(g) Calculate the final velocity of an electron in the beam as it leaves the electric field.

Solution The final velocity is the resultant of the constant horizontal velocity v_x and the final vertical velocity v_y

$$v_x = 6 \times 10^6 \text{ ms}^{-1}$$

$$v_y = u + at = 0 + 8.8 \times 10^{13} \times 2 \times 10^{-8} = 1.76 \times 10^6 \text{ ms}^{-1}$$

$$v_{resultant} = \sqrt{\left(6 \times 10^6\right)^2 + \left(1.76 \times 10^6\right)^2} = 6.25 \times 10^6 \text{ ms}^{-1}$$

The direction of the final velocity makes an angle θ with the horizontal.

$$\tan\theta = \frac{1.76 \times 10^6}{6 \times 10^6} = 0.2933, \text{ so } \theta \approx 16.3°$$

Example 2

A beam of singly charged positive ions is directed, at an angle, into a uniform electric field and follow the parabolic path shown in the diagram. The potential difference between the plates is 1200 V. The separation of the metal plates is 100 mm. The velocity of the ions as they enter the electric field is 5×10^5 ms^{-1} and they make an angle of 30° with the vertical as shown in the diagram.

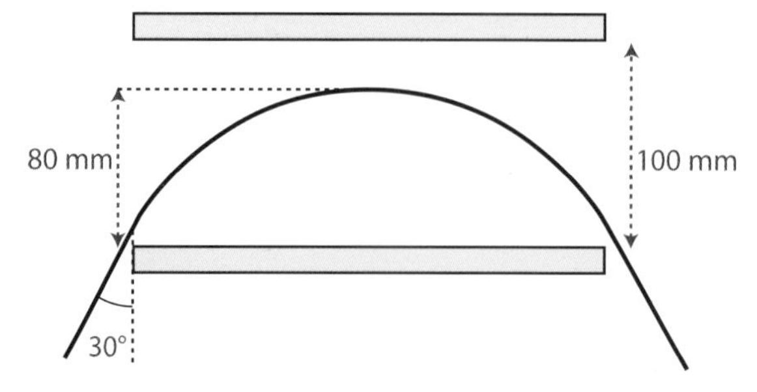

(a) Calculate the electric field strength between the plates and state its direction.

Solution $E = \frac{V}{d} = \frac{1200}{0.1} = 1.2 \times 10^4 \ Vm^{-1}$

The direction is down, towards the lower plate. The positive ions are being repelled by the positively charged upper plate.

(b) Calculate the acceleration of the ions in the beam.

Solution The vertical motion of the ions is one of uniform acceleration. The initial vertical velocity u_y of the ions is:

$(5 \times 10^5) \sin(90° - 30°) = (5 \times 10^5) \cos 30° = 4.33 \times 10^5 \ ms^{-1}$

When the vertical displacement of the ions in the electric field is 80 mm, the final vertical velocity v_y is zero.

$0 = u_y^2 + 2ay$

$0 = (4.33 \times 10^5)^2 + 2 \times a \times 80 \times 10^{-3}$ giving $a = -1.17 \times 10^{12} \ ms^{-2}$

(c) Calculate the electric force acting on the ions in the beam.

Solution The electric force on an ion in the beam is given by:

$F = qE = 1.6 \times 10^{-19} \times 1.2 \times 10^4 = 1.92 \times 10^{-15} \ N$

(d) Calculate the mass of an ion in the beam.

Solution The mass of an ion is found using Newton's 2nd law of motion.

$F = ma$ so $m = \frac{F}{a} = \frac{1.92 \times 10^{-15}}{1.17 \times 10^{12}} = 1.64 \times 10^{-27} \ kg$

Exercise 36

1 In an experiment a narrow beam of electrons enters a uniform electric field between two horizontal charged metal plates. The beam eventually emerges from the electric field.

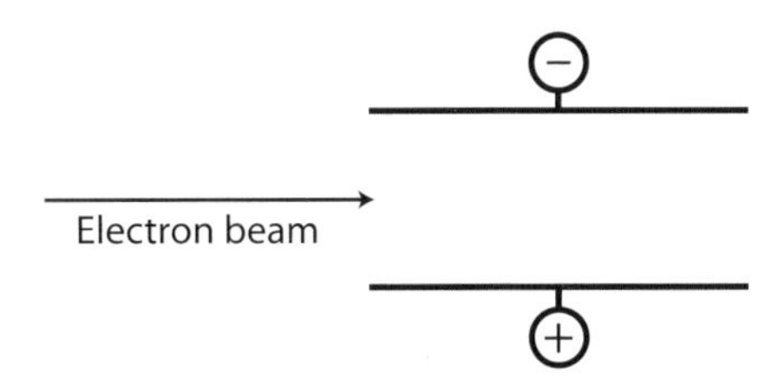

The experiment is repeated for a number of beams of negatively charged ions of different masses. All the ions have the same charge as the electron and enter the electric field with the same uniform velocity as the electron beam.

(a) Sketch a graph to show how the time spent in the electric field (y–axis) depends on the mass of the ions (x–axis).

(b) Sketch a graph to show how the kinetic energy of an ion of given mass (y–axis) depends on the time spent in the electric field (x–axis).

[CCEA 1999]

2 An ion has a mass of 6.64×10^{-26} kg and charge of 3.20×10^{-19} C. When in a certain uniform electric field it experiences an acceleration of 1.15×10^{11} ms^{-2} due to the field.

(a) Calculate the force the ion experiences due to the electric field.

(b) Calculate the force the ion experiences due to the Earth's gravitational field.

(c) Calculate the field strength.

[CCEA 2000]

3 A potential difference of 50 V is applied between two plane, parallel metal plates. The plates are separated by a distance of 20 mm in a vacuum. An electron is emitted from the negative plate with negligible speed.

(a) Calculate the speed with which the electron reaches the positive plate.

(b) How long does it take the electron to travel to the positive plate?

[CCEA 2008]

4 Ions each of charge – 3.20×10^{-19} C and mass 8.35×10^{-27} kg are accelerated from rest, in a vacuum, to a final speed of 5.75×10^{4} ms^{-1}.

Calculate the potential difference through which the ions are accelerated to this speed.

[CCEA 2001]

5 An electron beam enters a region of uniform electric field of strength 2000 V m^{-1}. The electrons are deflected towards plate A.

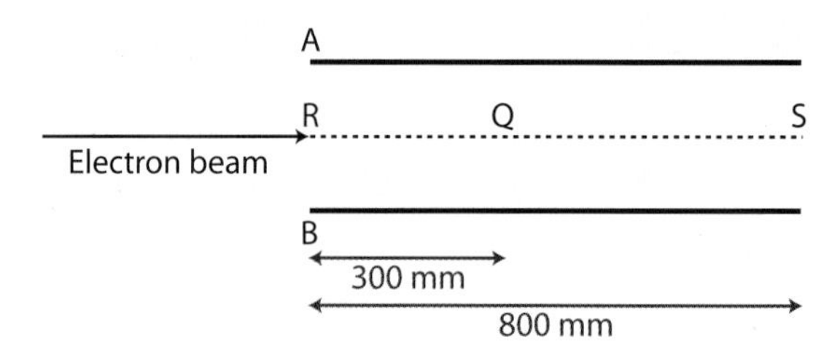

(a) State the polarity of plates A and B.

(b) The point Q on the line RS is 300 mm from the left-hand edge of the plates. What is the electric field strength at this point?

(c) Calculate the vertical force on an electron at Q due to the electric field.

(d) The electrons emerge from the plates at a point 16.0 mm above the line RS. Calculate the deflection of the electrons above the line RS at the point Q.

[CCEA 1995]

Deflection of Charged Particles in Magnetic Field

A moving charge in a magnetic field experiences a force which is perpendicular to both the velocity of the particle and the direction of the magnetic field. Fleming's left hand rule can be used to find the direction of this force.

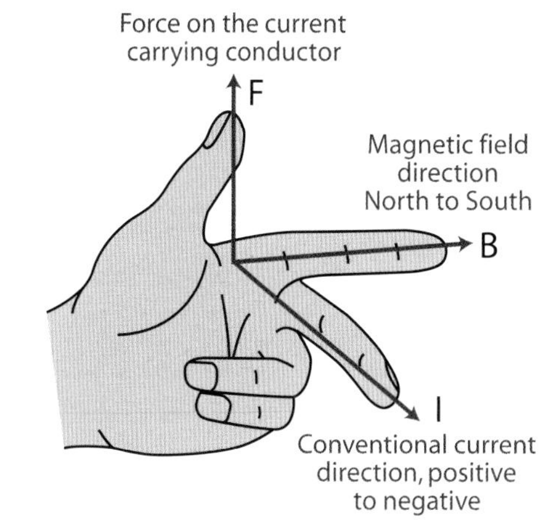

First finger is the magnetic Field direction (North to South).

SeCond finger is the Current direction (direction of movement of positive charge).

ThuMb is the direction of the Motion (Force).

To determine the direction of the force on an electron moving in a magnetic field you must remember that the movement of the electron is opposite to that of conventional current.

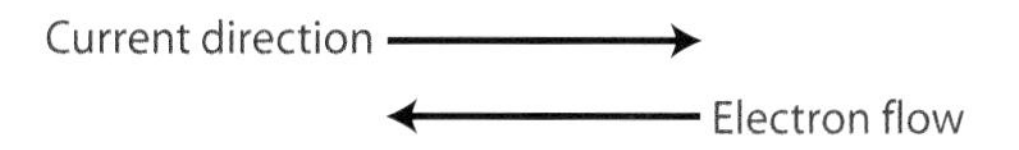

Since the force always acts at right angles to the velocity of the charged particles it causes the particles (electrons, protons, ions) to move in circular paths. The diagram below shows the paths taken by beams of negatively charged and positively charged particles when they enter a magnetic field.

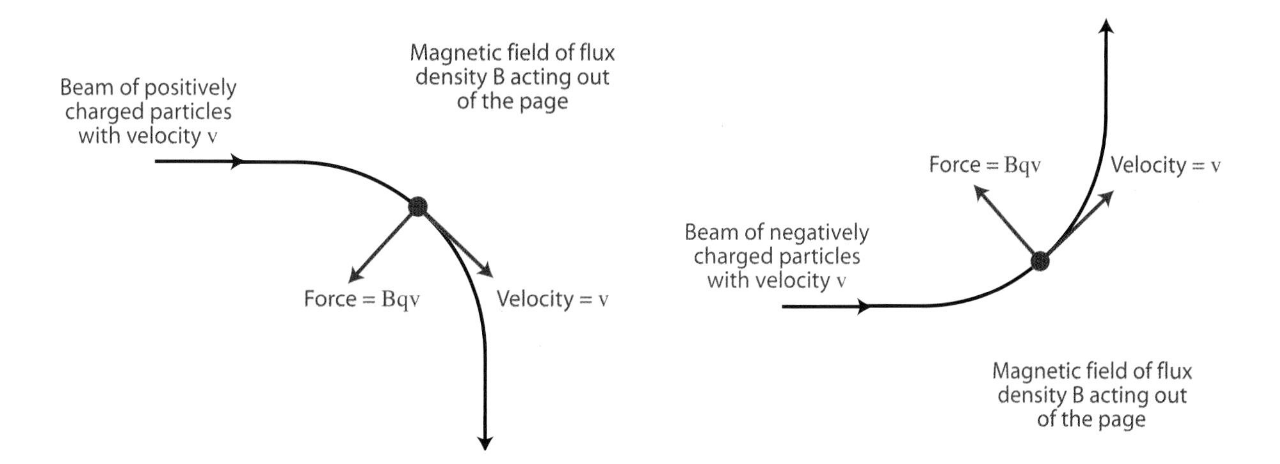

The magnitude of this force is given by

$$F = Bqv$$

where B is the magnetic field strength or flux density (Tesla)
q is the charge
v is the velocity of the charged particle

Since the charged particles are moving in a circle it is this magnetic force that provides the centripetal force.

$$Bqv = \frac{mv^2}{r}$$

where m is the mass of the charged particle
v is the velocity of the particles
r is the radius of the circle

A charged particle that is moving parallel to or anti-parallel to the line of magnetic flux does not experience a force due to the magnetic field.

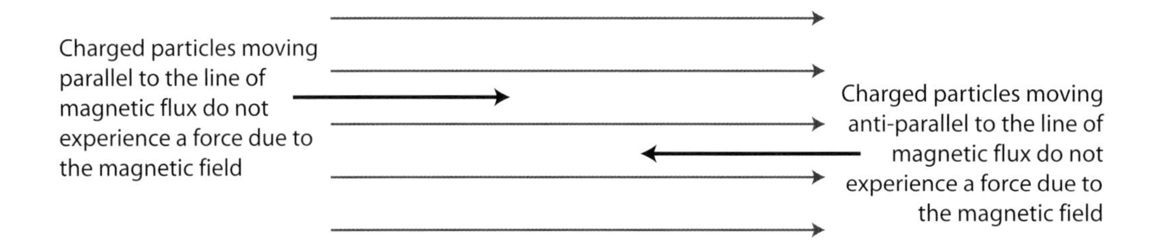

Worked Example

A fine beam of electrons is travelling in a vacuum with a velocity of 7.4×10^6 ms^{-1} from left to right as shown. The beam enters a region of uniform magnetic field of flux density 6.5×10^{-2} T. The direction of the magnetic field is at right angles to the direction of the electron beam, out of the plane of the paper.

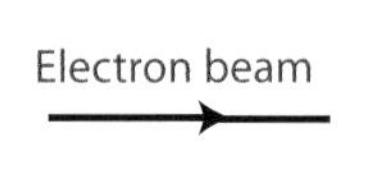

Region of uniform magnetic flux density with direction out of the page

When the beam emerges from the magnetic field, it is travelling at the same speed as it entered, but in the direction from right to left. The path of the emergent beam is parallel to, but displaced from, the direction of the incident beam.

(a) Draw the path of the electron beam within the magnetic field.

Solution

Application of Fleming's Left Hand Rule tells us that the force on the electrons is towards the top of the page as the electrons enter the magnetic field. This causes them to move in a semicircle emerging above the incident beam as shown opposite.

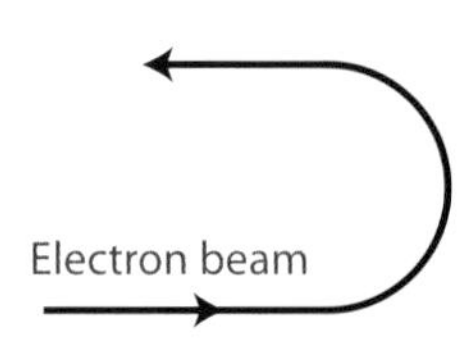

(b) Calculate the displacement between the emergent beam and the incident beam.

Solution

The displacement of the electron beam is the diameter of the semicircle, twice the radius of the circle.

$$Bev = \frac{mv^2}{r} \quad \text{Rearranging leads to} \quad r = \frac{mv}{Be}$$

$$r = \frac{9.1\times10^{-31} \times 7.4\times10^{6}}{6.5\times10^{-2} \times 1.6\times10^{-19}}$$

$$r = 6.5\times10^{-4}$$

$$\text{Displacement} = 2r = 1.3\times10^{-3}\ \text{m (1.3 mm)}$$

Exercise 37

1 The mass spectrograph is a device that uses the principle of the deflection of charged particles in a magnetic field. A plan view of a mass spectrograph is shown below. A beam of ions of charge +q and mass m, travelling at a constant speed v, enters a uniform magnetic field of flux density B applied throughout the shaded region which is evacuated. When the beam emerges from the slit, it moves in a semicircular path and leaves a trace on a photographic plate.

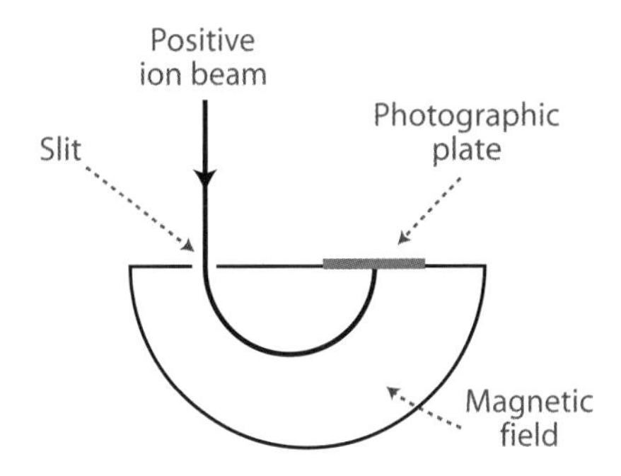

(a) State the name of the rule which you must use to determine the direction in which the magnetic field must be applied for the ions to move as shown.

(b) In what direction must the magnetic field be applied to give the deflection shown.

(c) Explain why the path of the ions in the magnetic field is semicircular.

(d) Find an expression in terms of B, q, m and v for the radius, r, of the semicircle followed by the ions.

(e) The specific charge of an ion is the ratio of the charge of the ion to the mass of the ion. Calculate the radius of the semicircular path if the ions have a specific charge of $+ 5.0\times10^{8}$ C kg^{-1} and are travelling at a speed of 4.5×10^{5} ms^{-1}. The value of B is 2.3×10^{-3} T.

(f) Sketch the path you would expect if the speed of the ions was increased. For comparison include the original path. Label the new path 2.

[CCEA 2008]

2 An electron beam travelling at a speed of 2.5×10^6 ms^{-1} enters a uniform magnetic field of flux density 150 mT at an angle of 90° to the direction of the field.

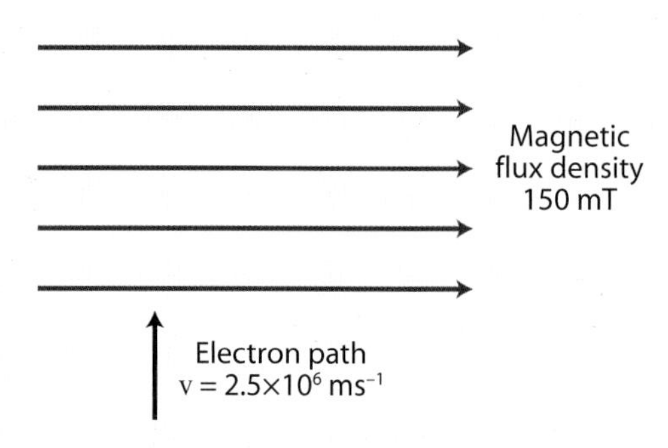

(a) Calculate the magnitude of the force on the electron as it enters the magnetic field.

(b) State the direction of the force on the electron.

[CCEA 2006]

3 A narrow beam of identical negatively charged ions moving with a constant velocity enters a uniform magnetic field. The direction of the magnetic field is perpendicular to the path of the ions and out of the plane of the paper. The ions follow the path shown in the diagram.

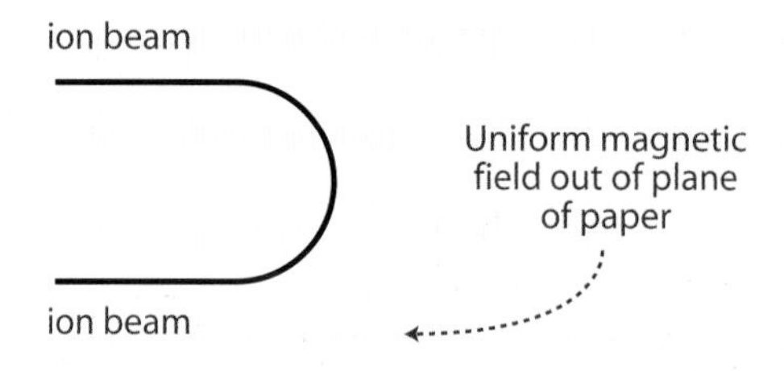

(a) Copy the diagram and mark the direction of travel of the ion beam. Explain how you arrived at your answer.

(b) While the ions are in the magnetic field region, does their momentum change? Explain your answer.

(c) Derive an expression for the momentum of the ions in terms of the magnetic flux density B, the charge of the ions q and the radius r of the semicircle.

[CCEA 1999]

Cathode Ray Oscilloscope (CRO)

An **oscilloscope** is an instrument that allows voltages to be viewed as a graph with voltage on the vertical axis (y–axis) plotted as a function of time on the horizontal axis (x–axis). The oscilloscope is one of the most versatile and widely-used electronic instruments. The oscilloscope can be used as a voltmeter to measure the size of a d.c. voltage and the amplitude of an a.c. voltage.

In addition to the amplitude of an a.c. voltage, an oscilloscope can be used measure the frequency of the a.c. voltage. Oscilloscopes are widely used in science and engineering. In medicine special purpose oscilloscopes are used to display the waveform of the heartbeat.

The main component of the cathode ray oscilloscope is a cathode ray tube. The pressure of gas inside the tube is less than 0.01 Pa, one ten millionth of atmospheric pressure. Inside the cathode ray tube is a hot cathode which provides electrons and a series of electrodes which accelerate and focus the electrons into a narrow fast moving beam. This assembly is called the **electron gun.**

Along the path of the beam are **deflecting plates.** The vertical plates deflect the beam in the Y direction and the horizontal plates deflect the beam in the X direction. The layout of these components is shown in the diagram below.

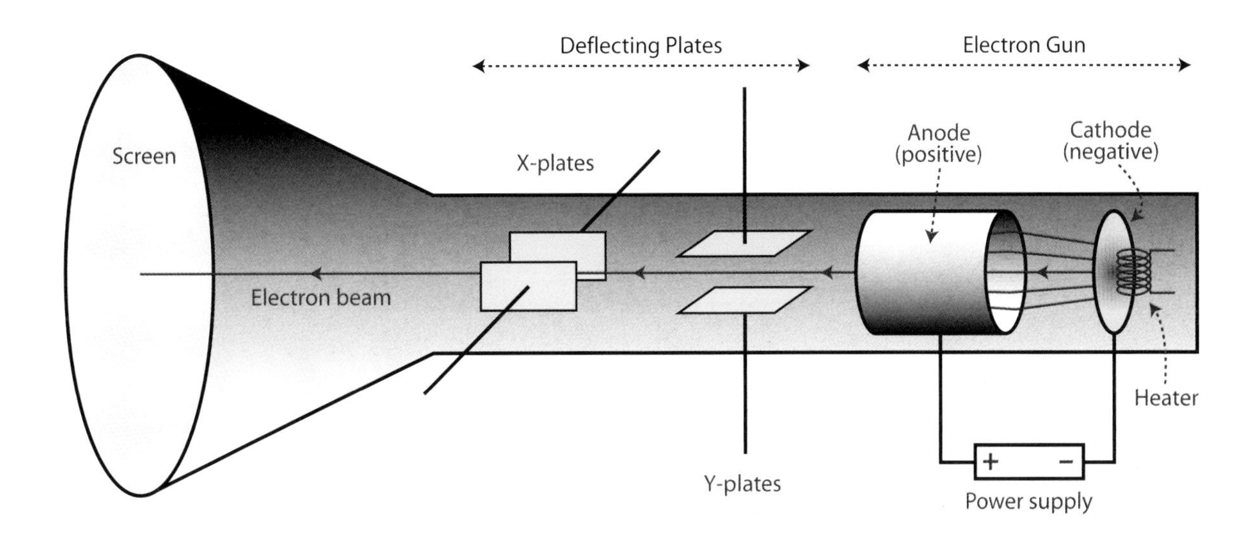

The inside of the cathode ray tube at the screen end is coated with phosphor. When the electrons strike the screen, this phosphor is excited and light is emitted from that point. This conversion of the electron's kinetic energy into light allows us to see various traces and waveforms as points or lines of light on an otherwise darkened screen.

The voltage to be examined is usually applied to the Y plates. An internally generated voltage is applied to the X plates. Its function is to sweep the electron beam across the screen and then to quickly return it to the start and begin all over again. This is known as the time base and uses a sawtooth voltage as shown in the diagram overleaf.

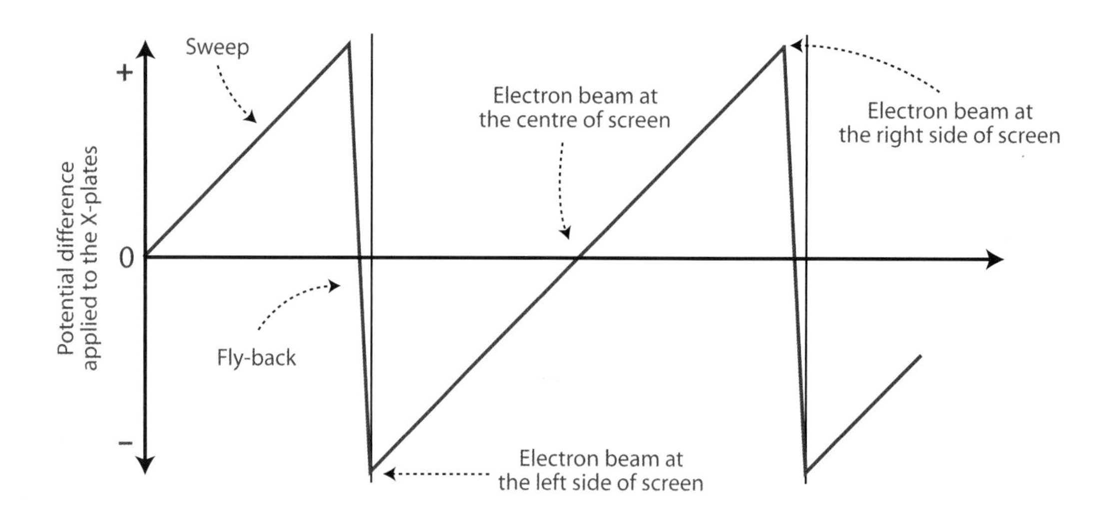

The oscilloscope is essentially a voltmeter with a very high resistance. It can be used to measure voltages both alternating and direct, display waveforms and measure time intervals. The display seen on the CRO screen depends on the type of voltage (a.c. or d.c) and on the whether the timebase is on or off.

You can regard the screen of the CRO as a graph, the vertical scale (y–axis) is voltage and the horizontal scale (x–axis) time. The voltage scale can be changed using a control known as the y sensitivity and marked in V/cm or mV/cm. The time scale can also be changed by the timebase control, this is marked in ms/cm or μs/cm.

A modern oscilloscope may have a lot more functions but the most important ones are shown in the diagram below.

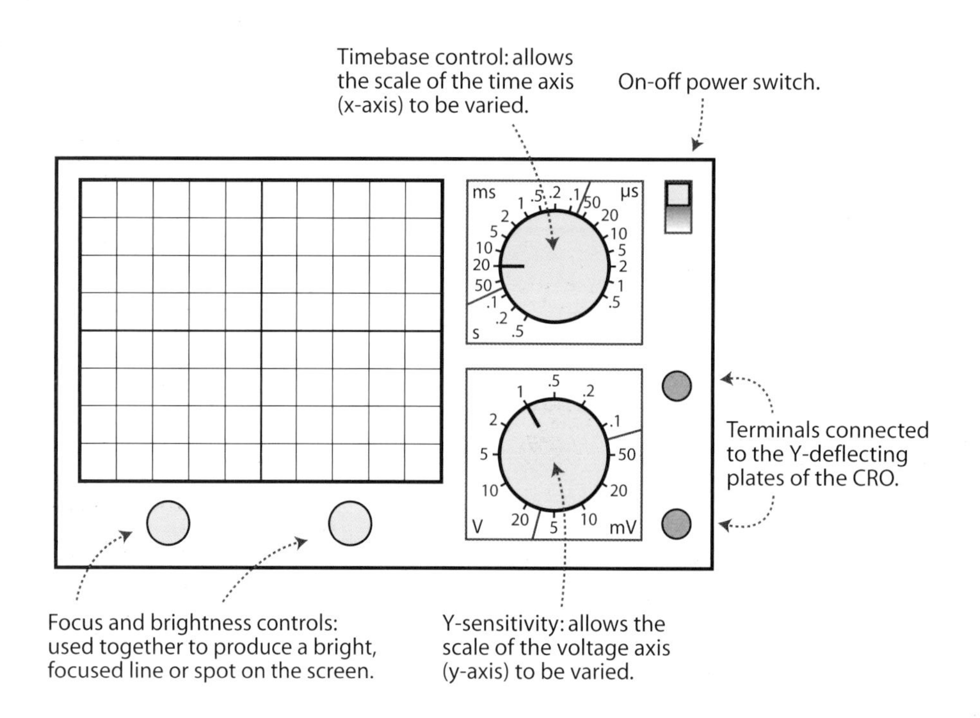

Measuring Voltage

The diagrams below show the appearance of the screen for a.c. and d.c. voltages with the timebase on and off.

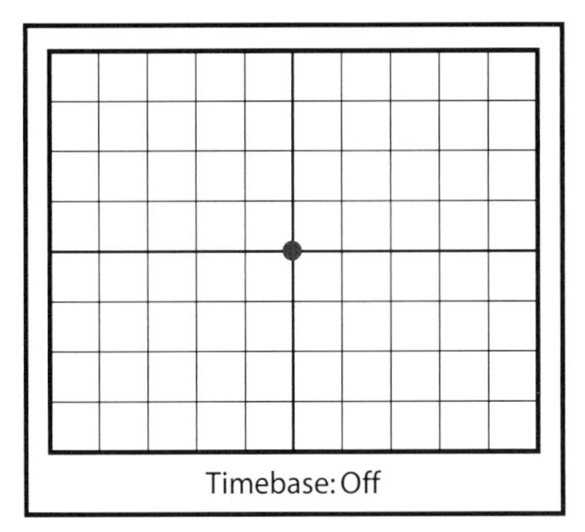

No voltage is applied to the y plates. The CRO is adjusted until the beam is at the centre of the screen. When the timebase is OFF this appears as spot, with the timebase ON a line is seen.

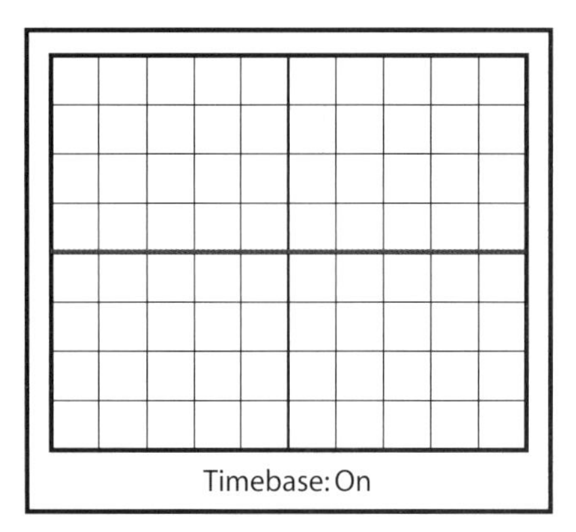

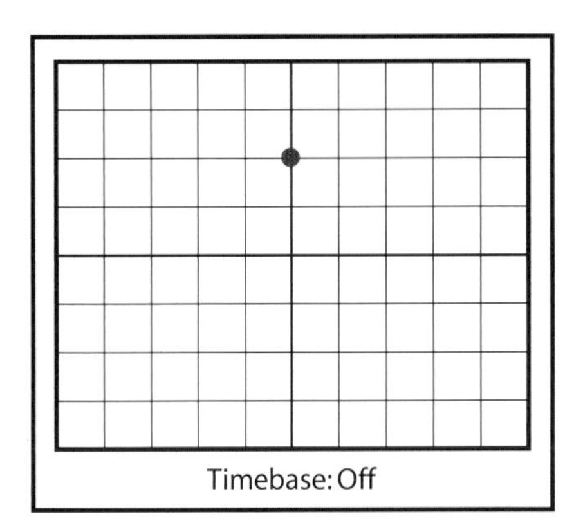

A + d.c. voltage is applied to the y plates.

With the timebase OFF the beam is deflected upwards by 2 cm.

If the y-sensitivity is set at 2V/cm then d.c. voltage is 2 cm × 2 V/cm = 4 V.

With the timebase ON the line is seen.

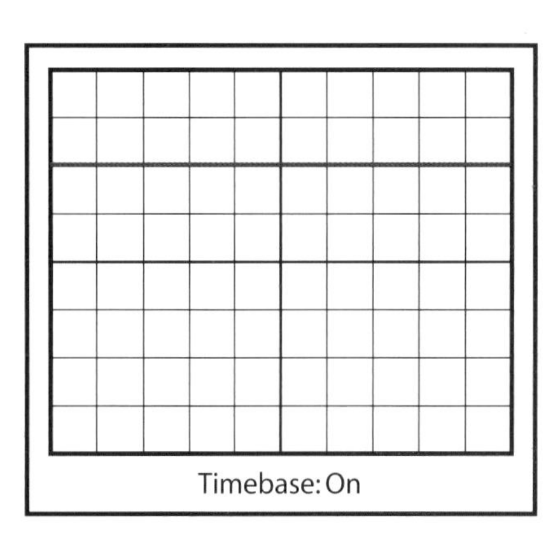

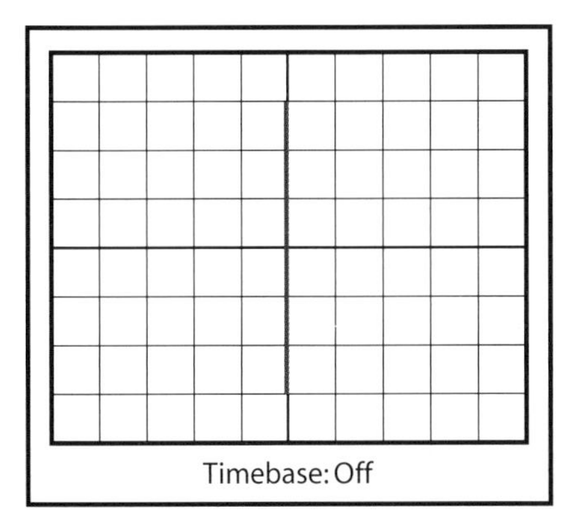

An a.c. voltage is applied to the y plates.

With the timebase OFF the beam is deflected upwards and downwards.

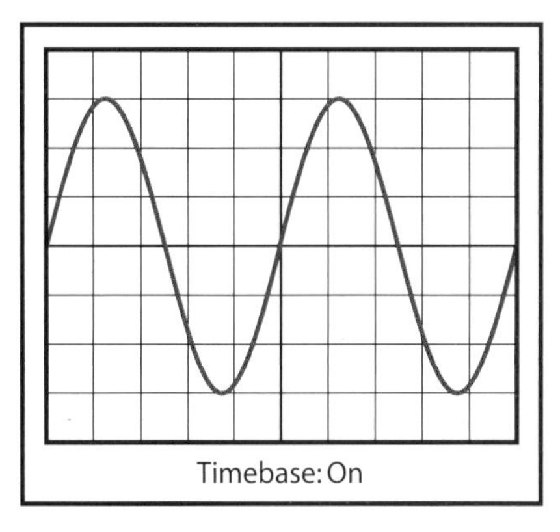

If the y-sensitivity is set at 100 mV/cm this represent a voltage that varies from +300 mV to –300 mV. The peak value of this voltage is 300 mV. If this voltage were measured using a meter a value less than this would be displayed since the meter gives an average value of the voltage.

With the timebase ON the CRO displays the waveform of the alternating voltage.

Measurement of Time and Frequency

Since the screen of the CRO shows a graph of voltage against time the instrument can be used measure the time interval between two events. Since time can be measured the CRO also provides us with a method of measuring the frequency of an alternating voltage.

The waveform on the left was displayed with the CRO settings below:

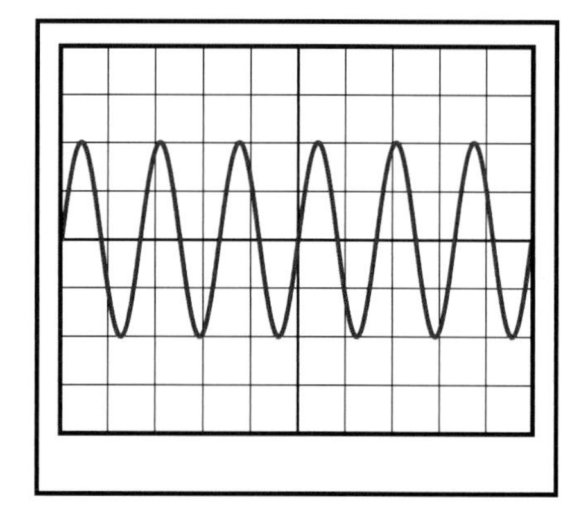

y-sensitivity = 2 V cm^{-1} and timebase setting = 5 ms cm^{-1}

There are 6 complete waves shown and width of the screen shows a total time of 10 cm × 5 ms cm^{-1} = 50 ms

$$\text{The period T of 1 wave} = \frac{50}{6} = 8.33 \text{ ms}$$

$$\text{Frequency} = \frac{1}{T} = \frac{1}{8.33 \times 10^{-3}} = 120 \text{ Hz}$$

Exercise 38

1 The screen of a cathode ray oscilloscope (CRO) measures 6 cm by 6 cm. The timebase is set at 250 μs cm^{-1} and y-sensitivity at 2.0 V cm^{-1}.

(a) A sinusoidal alternating voltage of frequency 1000 Hz and peak voltage (amplitude) 3.0 V is applied to the y-plates. Sketch the trace that would be displayed on the screen. Label this trace A

(b) The alternating voltage is now removed and replaced with a steady voltage of 3.0 V. Sketch the new trace that would be observed on the screen. Label this trace B.

(c) The alternating voltage is re-connected so that now it and the steady voltage are applied at the same time to the y-plates of the CRO. Sketch the trace that would be displayed on the screen. Label this trace C.

[CCEA 2000]

2 A voltage signal was applied to the y-plates of an oscilloscope and the trace below was obtained. Each square is 1 cm. The y–sensitivity was set as 5 mV cm^{-1} and the timebase was set at 50 μs cm^{-1}.

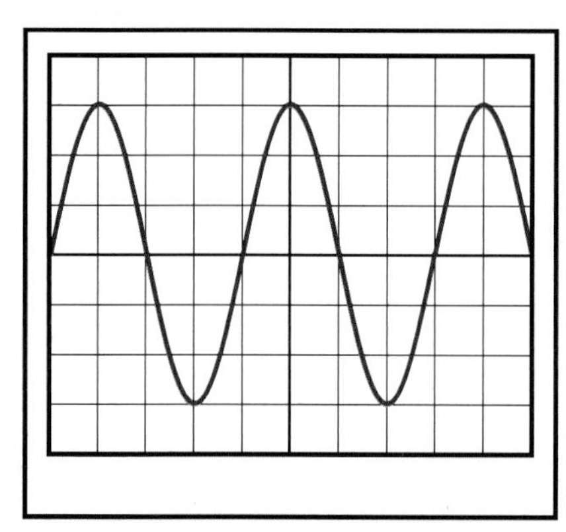

(a) Calculate the maximum (peak) voltage applied to the oscilloscope.

(b) Calculate the frequency of the signal.

(c) Calculate the timebase setting needed to show only **one** complete wave over the complete screen.

[CCEA 2007]

5.7 Particle Accelerators

5.7.1 Describe the basic principles of operation of a linear accelerator, cyclotron and synchrotron;

5.7.2 Compare and contrast the three types of accelerator;

5.7.3 Understand the concept of antimatter and that it can be produced and observed using high energy particle accelerators;

5.7.4 Describe the process of annihilation in terms of photon emission and conservation of energy and momentum;

Particle accelerators are a very important tool available to the physicist. By accelerating a particle to a very high velocity, one gives it a very high kinetic energy. This high energy particle can then be used for many purposes, for example:

- Studying the very smallest things we know of, for example the quarks inside a proton.
- Recreating conditions similar to those found at the birth of the universe though in a much smaller volume. This allows particles present in the very early universe to be produced and studied.

The first investigation into the uses of accelerated particles was carried out by John D.Cockcroft and Ernest Walton at the Cavendish Laboratory in Cambridge, England. This research was based on an idea that if you could give a proton enough energy, you would be able to overcome the repulsion from the nucleus of a target particle to penetrate and split it.

Particle accelerators accelerate subatomic particles to speeds almost equal to the speed of light, and then crash them into one another to see what happens. In these collisions some of the kinetic energy of the colliding particles turns into matter, and new particles are made. Matter and energy are interchangeable, you will recall the equation,

$$E = \Delta mc^2$$

where Δm is the change of mass (kg)
c is the velocity of light (ms^{-1})
E is the energy released (J)

Newton's laws of mechanics begin to break down when we apply them to objects moving at extremely high velocities. Einstein's theory of relativity predicts that the speed of light is an unattainable limit for any mass, even if we keep supplying it with energy.

Einstein also showed that energy itself has mass, so that a moving object has a greater mass than an object at rest. At low speeds this effect is unimportant, but close to the speed of light it becomes extremely important, and an object may have a total mass many times greater than its mass when it is stationary. The mass of an object that is at rest is known as its **rest** mass.

The prime purpose of an accelerator is to **not** to increase the velocity of particles, but to increase the energy of the particles. Once a particle is travelling at, say, 99% of the speed of light it is not going to increase its velocity very much, no matter how much more energy is supplied.

Its mass increases as it gains energy. When the speed of a particle increases from 0.99c to 0.999c the mass increases by a factor of 3. The velocity increase is about 1% but the mass increase is nearer 300%.

Extension material – not required by the specification

You can satisfy yourself that the increase in mass described previously is true by applying the equation below. As the velocity increases the mass increases according to the relationship below.

$$m = \frac{m_0}{\sqrt{1 - \frac{v^2}{c^2}}}$$

In this equation m_0 is the rest mass of a particle, m is the mass when it is moving with a velocity v and c is the speed of light. Similar expressions, know as a Lorentzian transformations also exist for time and length.

Simple Linear Accelerators

A cathode ray tube or electron tube commonly found in television sets is the simplest linear accelerator. A potential difference V volts is maintained between the anode and cathode and electrons accelerate in the field between them, converting electrical potential energy to kinetic energy.

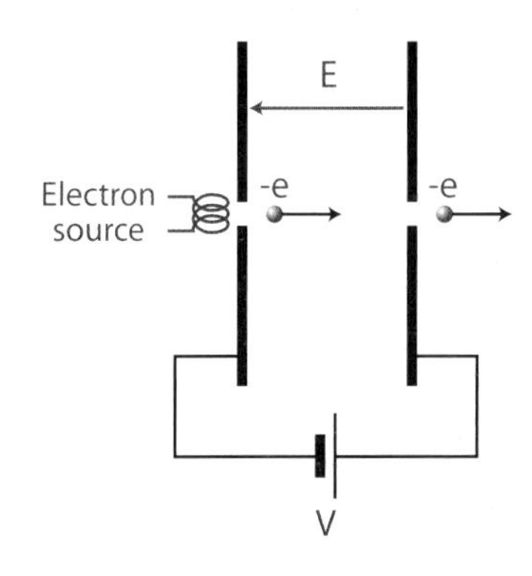

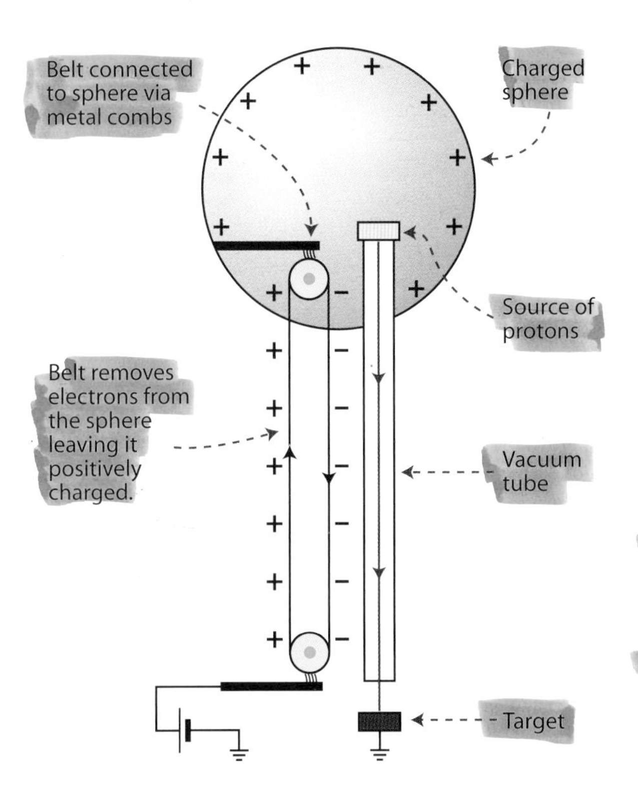

The Van de Graaff generator produces voltages of several million volts using a moving belt to remove electrons from a metal dome. The dome acquires a large positive charge and can be used to accelerate protons and positive ions.

The accelerated particles strike a target as shown in the diagram. This type of accelerator provided information on the structure of nuclei.

If we want a very high energy of around 1 GeV it would require a single potential difference of 1×10^9 volts. This is not a practical proposition since voltages of even a few hundred thousand volts are difficult to insulate.

Linear Accelerator (LINAC)

However, high energies can be achieved by accelerating the particles in a number of steps using lower voltages rather than supplying all the energy in one go. The linear accelerator is a series of metal cylinders, called drift tubes, arranged in a straight line.

A **high frequency alternating voltage** is used to accelerate the charged particles in a series of steps until they reach very high energy. Using an alternating voltage rather than a direct voltage has the advantage that transformers can be used to produce high voltages for acceleration in each stage.

However, the apparatus must be designed and operated in such a way that the particles are always accelerated in the same direction as they move from one drift tube to the next. The principle of **synchronous acceleration** is explained and illustrated below.

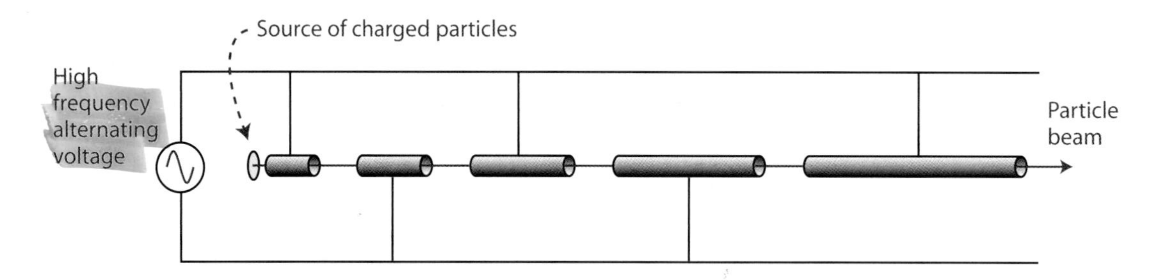

The electrons are injected into the linear accelerator using an electron gun arrangement similar to the simple design shown on page 177. If it is required that positively charged particles such as protons are to be accelerated then a suitable method is used to inject them into the accelerator.

There is no electric field inside the metal drift tubes so the electrons are not accelerated, they move with constant speed. The electric field exists in the gaps between the drift tubes. When the charged particles arrive at this gap they are accelerated across the gap and gain kinetic energy.

The drift tubes gradually become longer as the charged particles move down the linear accelerator. This increase in length ensures that the charged particles arrive at the gap between the electrodes at the correct time to receive maximum acceleration across the gap. Gaining energy in this way is known as synchronous acceleration.

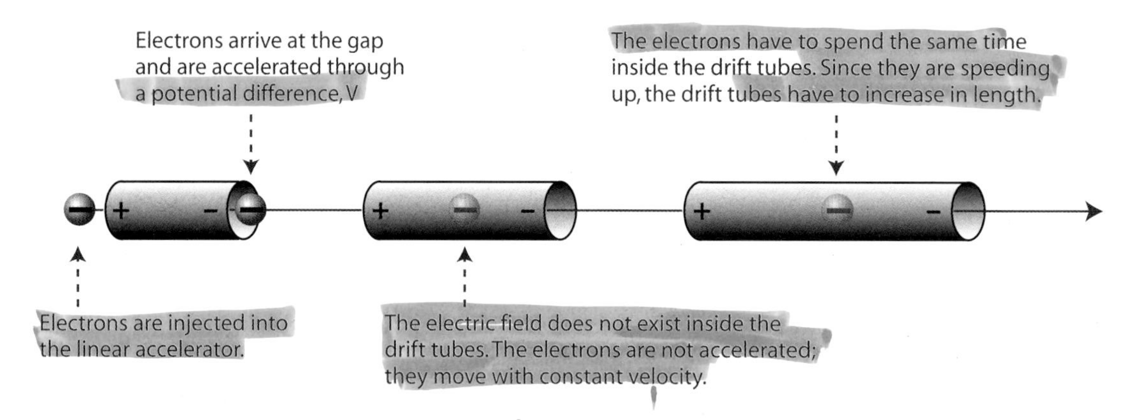

The charged particles should spend the correct period of time inside the drift tubes so that they gain the maximum increase in energy as they move from one drift tube to the next. This means the particles should arrive at the gap between the drift tubes just as the potential difference between the tubes reaches a maximum. This maximum is the peak value of the applied alternating voltage.

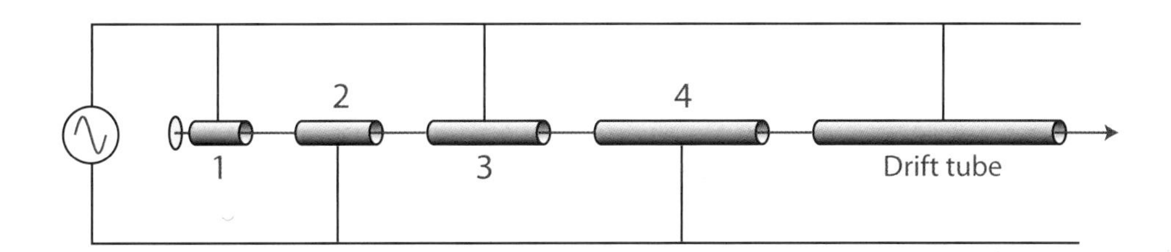

When drift tube 1 is at a maximum positive potential drift tube 2 is at a maximum negative potential. Similarly when tube 3 is positive tube 4 is negative.

In the case of electrons being accelerated they should emerge from drift tube 1 just as drift tube 2 is at a maximum positive potential and tubes 1 and 3 will be at their maximum negative potential. The time they spend inside tube 2 should be long enough so that when they emerge from tube 2, tube 3 will have changed from maximum negative to maximum positive. The time for this is **half the period** of the applied alternating voltage.

If we assume that the charged particles arrive at the gaps between the drift tubes when the potential difference between the tubes is at its peak voltage V. The increase in energy as the charged particles are accelerated across the gap is **qV**, where q is the charge on the particle.

After being accelerated across N gaps the kinetic energy will be **NqV**.

The Stanford accelerator in California is 3 km long and can accelerate electrons to energies of around 50 GeV. The scale of this accelerator can be gauged from the aerial photograph below.

A line of drift tubes inside a typical linear accelerator. *(Source: Fermilab)*

Advantages of the Linear Accelerator

- The particle beam has a small cross section area, this means Linacs produce high intensity particle beams. In other words a large number of particles per second per unit area.
- Linacs can be used to accelerate heavy ions to energies greater than those available in the cyclotron and synchrotron. These ring type accelerators, are limited by the strength of the magnetic fields required to make the ions move in a curved path.
- Linacs can produce a continuous stream of particles, whereas the synchrotron produces bursts of accelerated particles.
- Energy losses are small, no synchrotron radiation emitted by the accelerated particles.
- This makes the linear accelerator more practical for the production of antimatter.
- The particles are accelerated in a straight line so magnets are not required to bend the particle beam, this reduces the cost of building a linear accelerator.

Disadvantages of the Linear Accelerator.

- Linear accelerators can be very long when high energies are needed.
- The electrodes (drift tube) are used only once per acceleration.

The Cyclotron

When a charged particle moves with a velocity v at right angles to a magnetic field of flux density B it experiences a force at right angles to its velocity. Its path is then circular as shown in the diagram. This is the path of a positively charged particle in a magnetic field directed into the paper, you can use Fleming's Left Hand Rule to verify this.

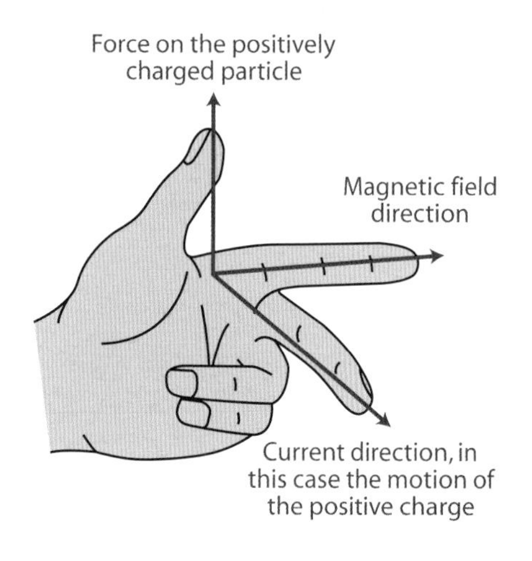

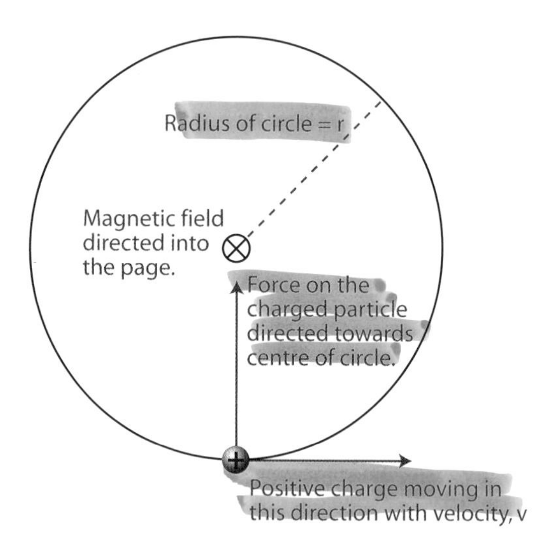

The principle of the cyclotron involves the technique of synchronous acceleration but with an added magnetic field which results in the accelerating particles moving in a spiral path as they cross and re-cross the same gap between two electrodes.

The force on the charged particle due to the magnetic field is $F = Bqv$

This provides the centripetal force to make the particle move in a circle.

$$Bqv = \frac{mv^2}{r}$$

The radius of the circular path is then:

$$r = \frac{mv}{Bq}$$

The time taken for the particle to make one complete revolution is T.

$$T = \frac{2\pi r}{v} = \frac{2\pi m}{Bq}$$

The above expression for T tells us that the period of the particle in the magnetic field is independent of both its velocity and the radius of its path, it depends only on the mass and charge of the particle and the value of the magnetic field flux density. This is an important property of charged particles moving in a circular path in a magnetic field. It tells us that when the velocity of a particle is increased whilst the particle moves in a magnetic field, the radius of its orbit will increase but the time taken for one revolution will remain the same.

In a cyclotron, a uniform magnetic field is maintained in the gap between the two cylindrical poles of an electromagnet. In the gap, two hollow D-shaped electrodes, known as **dees**, are arranged inside a highly evacuated box as shown in the diagram.

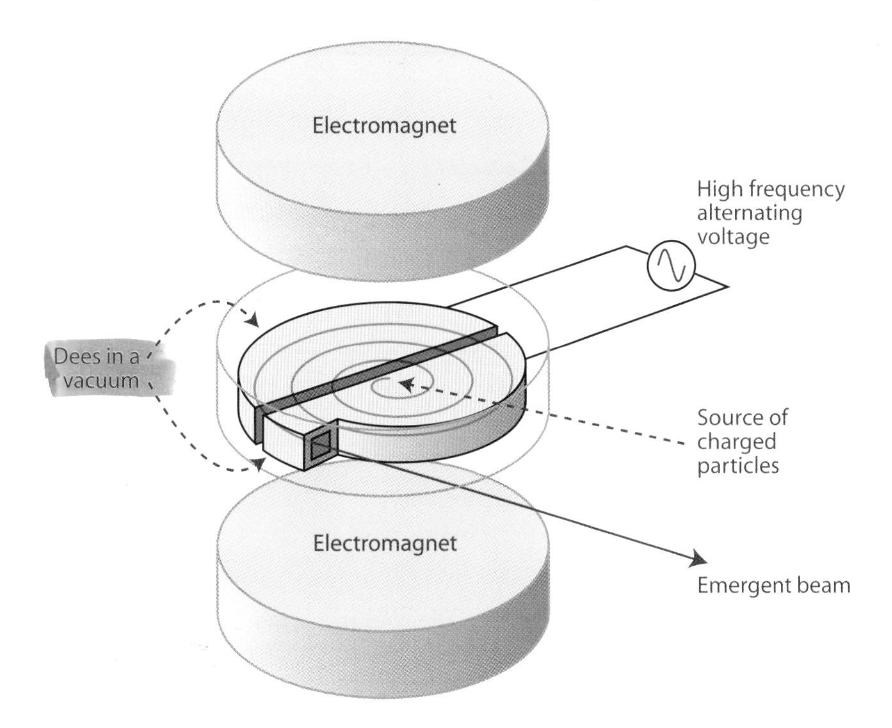

The lines of magnetic flux act at right angles to the dees. An alternating potential difference is maintained between the dees so that an alternating electric field is set up in the gap between them. At the same time, the region inside the dees is free of electric fields and so the particles do not experience an electric force when inside the dees, as was the case for charged particles inside the drift tubes of the linear accelerator.

Suppose that a positive ion leaves the ion source, which is at the centre of the machine, when dee1 is negative with respect to dee 2. The ion accelerates into dee 1 and travels through a half circle inside this dee since, in this region, it is under the action of the magnetic field alone. If the machine is properly adjusted, the ion returns to the gap between the dees when dee 2 is negative with respect to dee 1.

In this case, the ion accelerates again and enters dee 2, where it once more executes a half circle due to the magnetic field.

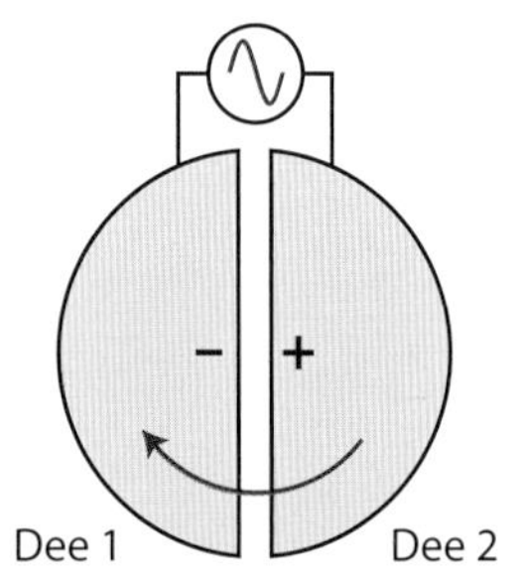

The positive ions are accelerated across the gap from dee 2 to dee 1.

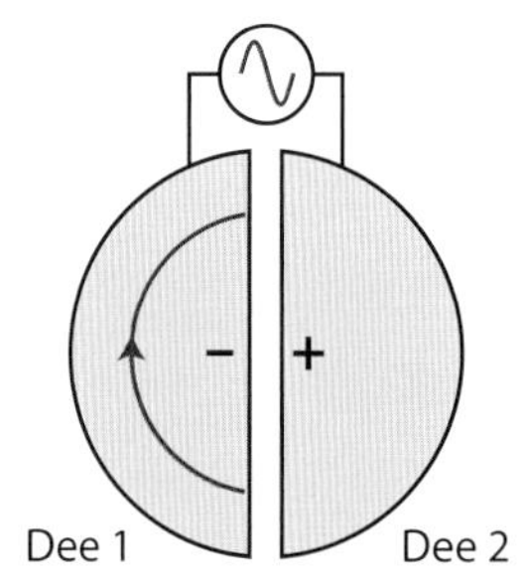

During the time when dee 2 is positive the positive ions move inside dee 1 with constant speed. The time spent inside dee 1 equals ½ the period of the a.c. voltage.

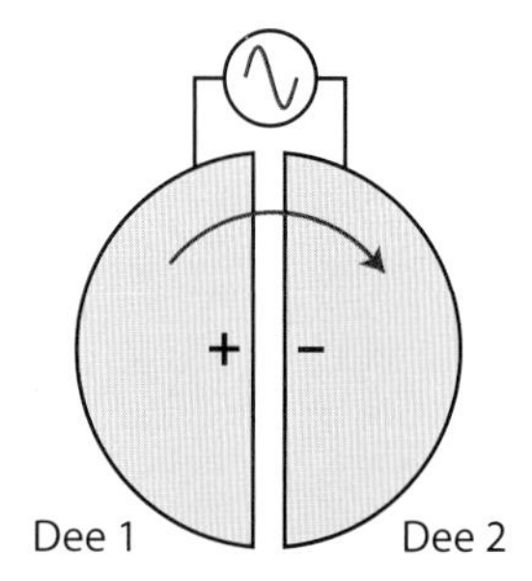

The positive ions are accelerated across the gap from dee 1 to dee 2.

In this way the particle may be accelerated repeatedly provided it always arrives at the gap when the field is such as to produce a force which accelerates it towards the dee which it is entering.

This condition will be fulfilled if the alternating voltage applied to the dees executes exactly one half-cycle in the time taken for the particle to travel through half a revolution. This requires the period of the alternating voltage to equal the period of one complete orbit of the particle.

Thus, if **f** is the frequency of the alternating voltage, the condition for synchronous acceleration is,

$$f = \frac{1}{T} = \frac{qB}{2\pi m}$$

As the charged particle gains energy at each successive crossing of the gap, the radius of its path in the magnetic field increases. The maximum energy which can be obtained depends on the radius available in the magnetic field and the maximum value of the field which can be achieved.

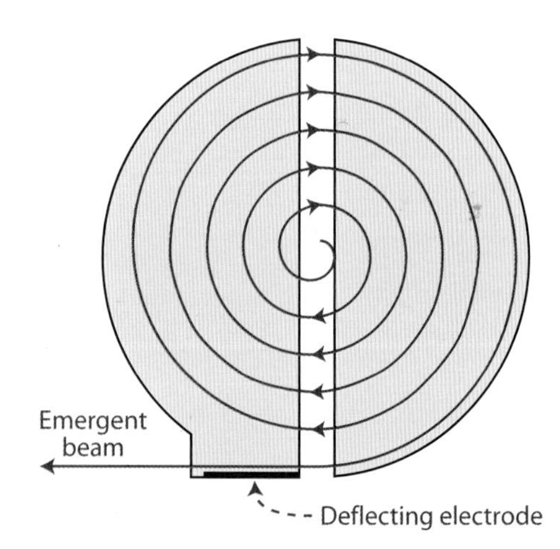

When the beam of particles reaches the largest radius possible in the machine, it is extracted from the accelerator by a deflecting electrode so that they leave the machine before striking the target under investigation.

One of the problems with the cyclotron results from the

increase in mass of the particles as they move close to the speed of light. This relativistic increase in mass was mentioned at the start of this chapter.

In the cyclotron when the energy of the particle is above 25 MeV the relativistic mass increase becomes significant and the time for particle orbits becomes larger than that calculated on the assumption of constant mass. Eventually particles arrive too late to be accelerated by the voltage between the dees and end up completely out of step with the a.c. supply.

Because of this relativistic increase in the mass the frequency of the alternating voltage required for synchronous acceleration depends on the mass. Since the mass is increasing the frequency required for synchronous acceleration will decrease. This means that the frequency and the time that the particles spend within the dees get out of step so limiting the maximum energy of the particles. One way to avoid this is to use an enormous voltage so that large energies are reached in just a few orbits.

Another factor limiting the energy of the particles is the energy lost by electromagnetic radiation when particles are accelerated. In the cyclotron since the particles are made to move in a circle centripetal acceleration is always present. Synchrotron radiation is the name given to the electromagnetic radiation which occurs when charged particles are accelerated in a curved path or orbit.

Advantages of the Cyclotron

- Cyclotrons are more compact than linear accelerators.
- The electrodes are used repeatedly to accelerate the particles.
- The Cyclotron produces a stream of particles as opposed to the bursts produced by the synchrotron.

Disadvantages of the Cyclotron

- The relativistic increase in the mass causes the frequency of the accelerating voltage and the time the particles spend inside the dees become out of step, this limits the maximum energy of the accelerated particles
- Energy is lost due to synchrotron radiation.

The Synchrotron

To overcome the problems associated with the cyclotron a more effective approach is to change the accelerating frequency. A machine designed to do this is the synchrotron. The synchrotron is similar to the cyclotron in that it accelerates particles in a circle.

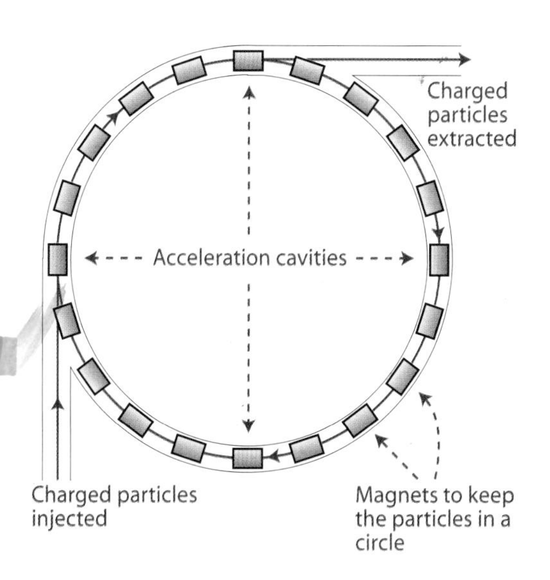

In the synchrotron, particles are accelerated and held in a circle by means of a magnetic field. However the difference between the synchrotron and cyclotron is that the synchrotron varies the strength of this magnetic field. This means particles do not move in a spiral but in a perfect circle of fixed radius.

The acceleration of the particles is achieved by the application of a high frequency alternating voltage at various cavities along the circumference of the ring.

Instead of reducing the frequency and allowing the orbital radius to increase, the technique used is to increase the magnetic field strength and keep the charged particles in an orbit of fixed radius. The advantage of this is that a magnetic field is only applied at the circumference of the orbit and not over all of a large circular area. This allows larger higher energy machines to be built.

The path followed by a particle of mass m and charge q, moving through a magnetic field of strength B at speed v has a radius of curvature given by:

$$r = \frac{mv}{Bq}$$

See page 187 for the derivation of this expression.

As the particles accelerate the magnetic field is increased to keep them on the same orbit. The particles take less and less time to complete their orbit so the frequency of the accelerating alternating voltage must increase as well.

The magnets perform two functions in the synchrotron – they bend the beam into a circular path and focus it to keep as many particles as possible on the ideal orbit.

Effective focusing of particle beams is very important as it:

- increases the beam intensity, i.e. more particles per unit area per second
- reduces the area of cross section of the particle beam, so that smaller evacuated beam tubes and smaller gaps between magnetic poles can be used.

Several accelerating cavities are positioned around the ring. These contain the alternating electric fields synchronised with the beam's orbital period and accelerate the charges as they reach and pass through each cavity.

In a proton synchrotron the frequency of the applied voltage must be increased as the particle velocity increases. In an electron synchrotron the electrons are already travelling at very close to the speed of light when they are injected. Their orbital frequency remains more or less constant – so a constant alternating frequency is used.

A linear accelerator is often used to inject particles into the synchrotron. This linear accelerator intersects the main ring of the synchrotron at a tangent. The incoming beam is deflected onto its orbital path by a magnet and the accelerated beam is extracted by deflecting it out of the ring in a similar way.

A synchrotron has several extraction points which allows the same machine to be used by several different research groups at the same time. The diagram on the next page shows the structure common to many of the synchrotrons located around the world.

The booster and storage rings are the large circular sections where the electrons orbit. In the booster ring, the electrons are accelerated repeatedly until they are at a sufficient energy to be injected into the storage ring. In the storage ring, the electrons orbit continuously and are accessed by the various experiments located around this ring.

The largest synchrotron is the large electron-positron collider (LEP) at CERN (*European Organisation for Nuclear Research*). This has a circumference of 27 km and operates in an underground tunnel which crosses the Swiss-French border near Geneva.

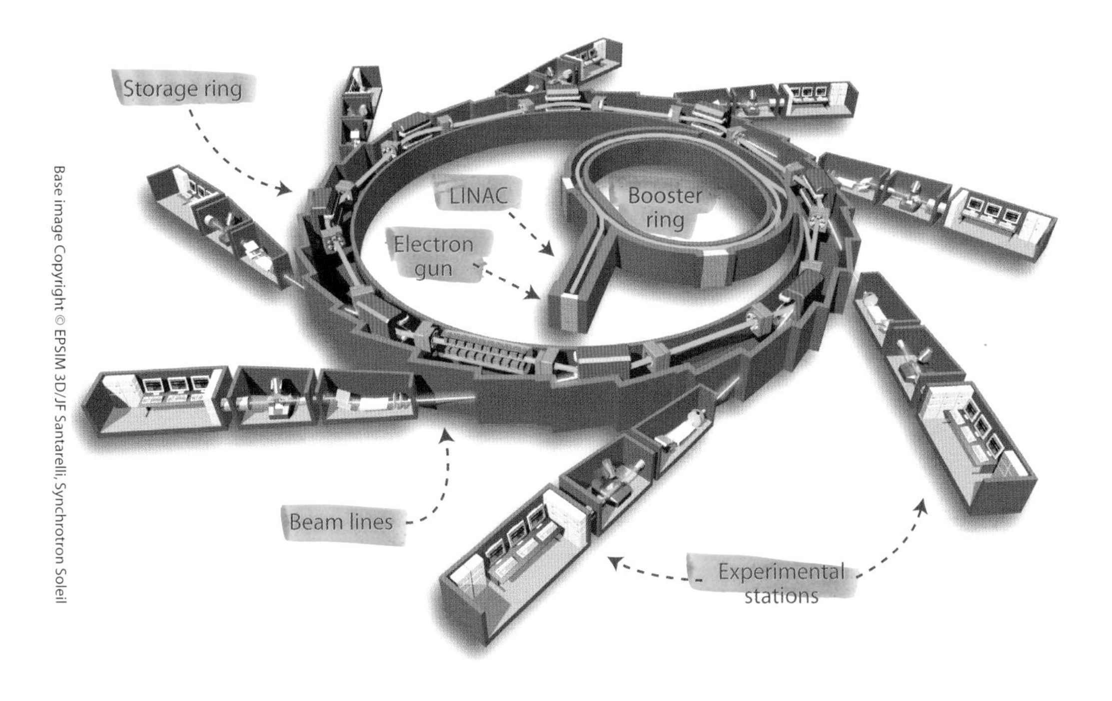

The LEP at CERN seen from the air. *(Copyright CERN)*

Up until 1996 it accelerated electrons to energies of 50 GeV and collided them with positrons (anti-electrons) of the same energy moving in the opposite direction. The 1996 upgrade, almost doubled the collision energy. At close to the speed of light electrons and positrons complete about 11 000 orbits of LEP per second.

The Tevatron at Fermilab near Chicago accelerates protons and antiprotons to 1000 GeV or 1 TeV. It has a circumference of 6 km.

Advantages of the Synchrotron

- Particles can be extracted at various points along the path allowing a number of different experiments to be carried out.
- The energy losses due to synchrotron radiation are reduced since the charged particles move in a circle of much greater radius than the cyclotron.

Disadvantages of the Synchrotron

- Although the effect is less than in the Cyclotron, the accelerated particles still lose energy by emitting electromagnetic radiation (synchrotron radiation) when they move in a circle.
- Cost is a major concern when building large accelerators.

Large Hadron Collider

The Large Hadron Collider (LHC) is a gigantic particle accelerator. It spans the border between Switzerland and France about 100 m underground.

The LHC is the world's largest and most powerful particle accelerator, it is the latest addition to CERN's accelerator complex. It mainly consists of a 27 km ring of superconducting magnets with a number of accelerating structures to boost the energy of the particles along the way.

Two beams of subatomic particles such as protons and antiprotons travel in opposite directions inside the circular accelerator, gaining energy with every lap. Physicists try to recreate the conditions just after the Big Bang, by colliding the two beams head-on at very high energy. Teams of physicists from around the world analyse the particles created in the collisions using special detectors in a number of experiments.

The proton-antiproton collision energies will be comparable to the typical energy of particles 10^{-12} s after the Big Bang and will be used to look for the hypothetical particles believed to present at this time.

The particles are guided around the accelerator ring by a strong magnetic field, achieved using superconducting electromagnets. These are built from coils of special electric cable that operates in a superconducting state, efficiently conducting electricity without resistance or loss of energy. This requires chilling the magnets to about –271°C. Liquid helium is used to cool the superconducting coils.

A worker inside one of the detectors of the Large Hadron Collider.
(Image by Maximilien Brice. Copyright CERN)

Two physicists carrying out tests on part of the LHC.
(Image by Maximilien Brice, Claudia Marcelloni. Copyright CERN)

Comparison of the Three Particle Accelerators

Type	Path taken by the particles	How are the particles accelerated	Advantages	Disadvantages
Linear	Straight line.	Accelerated across the gap between neighbouring drift tubes using a high frequency alternating potential difference.	Beam cross section area is small so Linacs produce particle beams with a large number of particles per second per unit area – high beam intensities. Linacs can be used to accelerate heavy ions to energies greater than those available in the cyclotron and synchrotron. These ring type accelerators, are limited by the strength of the magnetic fields required to make the ions move in a curved path. Linacs can produce a continuous stream of particles, whereas the synchrotron produces bursts of accelerated particles. Energy losses are small, i.e. no synchrotron radiation is emitted by the accelerated particles. This makes the linear accelerator more practical for the production of antimatter. As there are no bending magnets, thus the cost of an accelerator is reduced.	Machines are very long when high energies are needed. The electrodes (drift tube) are used only once per acceleration.
Cyclotron	Particles move outwards in a circular path of increasing radius (spiral). A magnetic field exists over the total area of the accelerator to achieve this.	Particles are accelerated across the gap between two semi cylindrical electrodes (dees) using a high frequency alternating potential difference.	Are more compact than the linear accelerator. The electrodes are used repeatedly. The Cyclotron produces a stream of particles as opposed to the bursts produced by the synchrotron.	Due to the relativistic increase in the mass the frequency of the accelerating voltage and the time the particles spend inside the dees become out of step. Energy is lost due to synchrotron radiation.
Synchrotron	Particles move in a circle of fixed radius. The strength of the magnetic field is gradually increased to achieve this.	Acceleration takes place at several points along the path. A high frequency alternating potential difference is used. To overcome the relativistic mass increase the frequency is altered to maintain synchronous acceleration.	Particles can be extracted at various points along the path allowing a number of different experiments to be carried out. The energy losses due to synchrotron radiation are reduced since the charged particles move in a circle of much greater radius than the cyclotron.	Although the effect is less than in the Cyclotron the accelerated particles still lose energy by emitting electromagnetic radiation (synchrotron radiation) when they move in a circle. Cost is a major concern when building large accelerators.

Antiparticles and Antimatter

Every particle has an associated antiparticle with the same mass and in some cases opposite electric charge. For example, the antiparticle of the electron is the positively charged antielectron, or positron. The antiparticle of the proton has the same mass as the proton but has a negative charge. In a few cases a particle is its own antiparticle.

Antiparticles are produced naturally in beta decay, and in particle accelerators.

The equation describes the beta decay of actinium to thorium with the emission of an electron (β–) and antineutrino.

$$^{228}_{89}\text{Ac} \rightarrow {}^{228}_{90}\text{Th} + {}^{0}_{-1}\text{e} + {}^{0}_{0}\bar{\nu}_{e}$$

This equation describes the beta plus decay in which a positron (β+) and a neutrino are emitted.

$$^{22}_{11}\text{Na} \rightarrow {}^{22}_{10}\text{Ne} + {}^{0}_{+1}\text{e} + {}^{0}_{0}\nu_{e}$$

Particle-antiparticle pairs can annihilate each other, producing photons. Since the charges of the particle and antiparticle are opposite, charge is conserved. The antielectrons produced in natural radioactivity meet electrons resulting in annihilation and producing pairs of gamma rays.

When a positron and an electron meet they annihilate each other. Their energy is converted to two gamma ray photons. Two gamma ray photons are emitted in opposite directions so that momentum is conserved.

The energy released is obtained from Einstein's mass-energy equation $E = \Delta mc^2$.

$$\text{Mass of electron + positron} = 2 \times 9.1\times10^{-31}\ \text{kg}$$

$$E = 2 \times 9.1\times10^{-31} \times \left(3\times10^{8}\right)^{2}$$

$$= 1.64\times10^{-13}\ \text{J}$$

$$= 1.022\ \text{MeV}$$

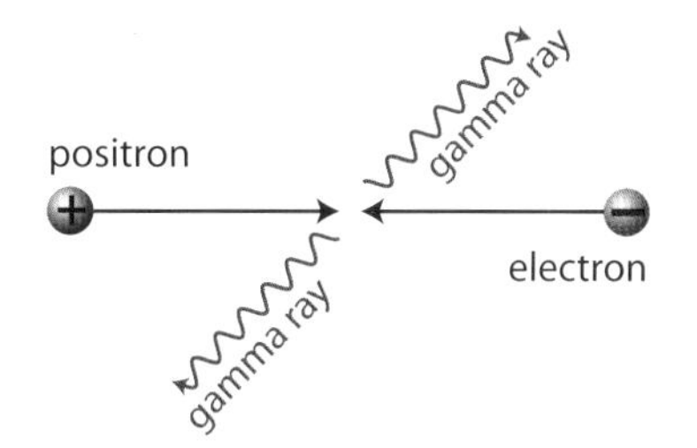

Each gamma ray has an energy of just over 0.51 MeV. Using the Planck relationship $E = hc/\lambda$ the wavelength of the gamma rays is 2.44×10^{-12} m.

The reverse process is also possible, a gamma ray can, provided it has enough energy, produce a fundamental particle and its antiparticle. This normally happens in the vicinity of a nucleus. The diagram shows the paths taken by the electron and the positron in a magnetic field. The opposite charges on the particles is revealed by the opposite circular paths of each they move in the magnetic field.

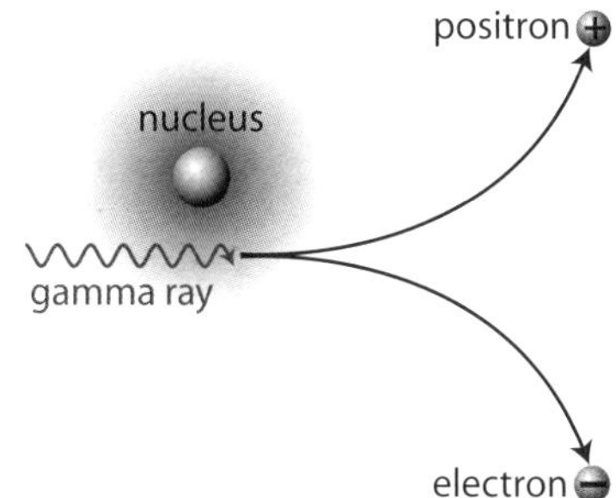

Antimatter is the extension of the concept of the antiparticle to matter, where antimatter is composed of antiparticles in the same way that normal matter is composed of particles. An antiproton and a positron can form an antihydrogen atom, which has almost exactly the same properties as a hydrogen atom. Antimatter in very short-lived. Antihydrogen atoms survive for only 4×10^{-11} s before annihilation with ordinary matter takes place. Ordinary matter is the dominant type of matter in the Universe.

Although there is an antiparticle for every particle it does not mean they occur in equal numbers in the universe. If they did we would have equal amounts of matter and antimatter. The formation of matter after the Big Bang resulted in a universe consisting almost entirely of matter, rather than being a half-and-half mixture of matter and antimatter. Why this has happened is a question that physicists are yet to completely answer.

Antimatter is difficult to produce. If all the antiprotons produced at CERN during one year were to meet and annihilate an equal number of protons the energy produced would light a 100 watt electric bulb for three seconds.

Particle physicists regularly use collisions between electrons and their antiparticles, positrons, to investigate matter and fundamental forces at high energies. Particles and their antiparticles turn into energy when they annihilate which, at high energies, can produce new particles and antiparticles.

At low energies the electron-positron annihilations can be used to reveal the workings of the brain in the technique called **Positron Emission Tomography (PET)**. Positron emission tomography (PET) is a nuclear medicine imaging technique which produces a three-dimensional image or picture of functional processes in the body.

In PET, the positrons come from the decay of radioactive nuclei in a special fluid injected into the patient. The positrons then annihilate with electrons in nearby atoms. As the electron and positron are almost at rest when they annihilate they have only enough annihilation energy to make two gamma-rays which shoot off in opposite directions to conserve momentum. The system detects these pairs of gamma rays emitted from the body. Images of structures within the body are then reconstructed by computer analysis. PET scanner are used mostly in the medical imaging of tumours and for clinical diagnosis of certain diffuse brain diseases such as those causing various types of dementia. PET is also an important research tool to map normal human brain and heart function.

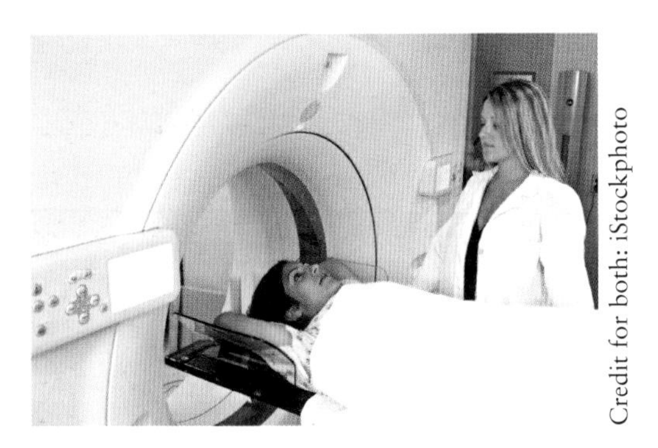

Credit for both: iStockphoto

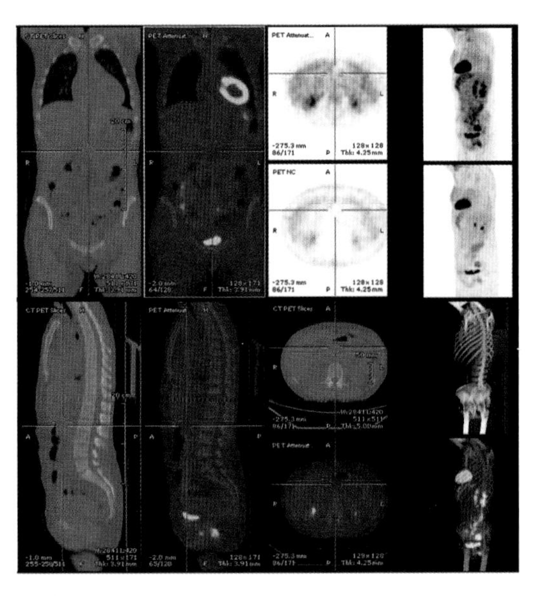

Cancerous growths absorb more of the injected solution than other regions so that the emission of gamma rays is more intense from these regions making them easier to spot.

Exercise 39

1 Beams of very high energy protons used in many studies of nuclei are produced using a proton synchrotron.

(a) Sketch a simple diagram to show the essential features of such a machine.

(b) (i) Briefly, distinguish between the magnet arrangement required for a synchrotron and a cyclotron.

(ii) In the synchrotron how does the strength of the field change as the protons are accelerated.

(iii) In the synchrotron how does the frequency of the applied voltage change as the protons are accelerated? Explain your answer.

(iv) Calculate the flux density of the magnetic field required to keep protons of speed 1.5×10^7 ms^{-1} on a circular path of diameter 100 m.

2 The diagram below shows the first five drift tubes of a linear accelerator used to accelerate protons.

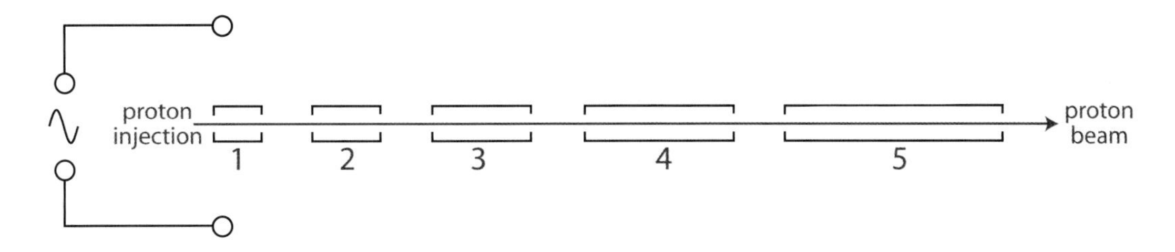

(a) Copy and complete the diagram by drawing the connections from the high frequency alternating voltage supply to the five drift tubes.

(b) How, if at all, does the frequency of the applied voltage change during the passes of a pulse of protons through the accelerator?

(c) Describe the motion of the protons inside the drift tubes.

(d) What relationship exists between the time the protons spend within each drift tube and the period of the applied voltage? How is this relationship maintained as the protons are accelerated?

(e) Derive an expression for the length **L** of a drift tube, in terms of the velocity **v** of the proton and the frequency **f** of the applied voltage.

3 In a linear accelerator, each proton is injected into the first drift tube with a kinetic energy of 500 keV. The second drift tube is 20% longer than the first tube.

(a) Calculate the kinetic energy of a proton in the second drift tube.

(b) Calculate the potential difference between the two drift tubes as the protons cross the gap between them.

(c) This accelerator has 43 drift tubes. Calculate the kinetic energy of the protons leaving the accelerator. Give your answer in MeV.

Exercise 40

Past paper questions

1 The cyclotron is a particle accelerator often used for accelerating protons. The particles are accelerated as they pass between two electrodes called **dees**. These electrodes and the proton source are shown in the diagram below. An alternating potential difference of constant frequency is applied between the dees. A uniform magnetic field is applied normal to the plane of the dees.

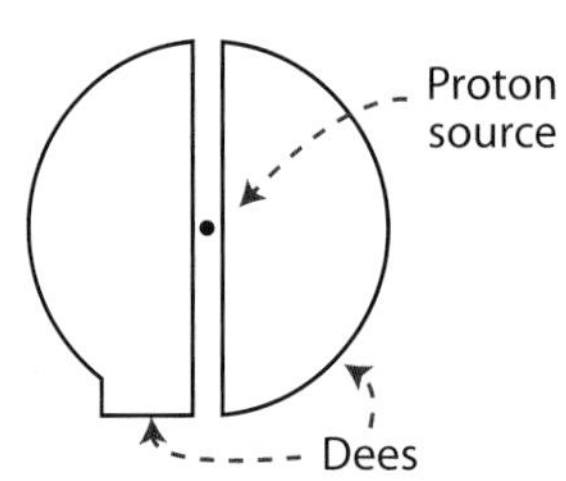

(a) Copy the diagram and sketch the path the protons take after they have been emitted from the source.

(b) Is the direction of the magnetic field into or out of the plane of the paper?
Explain how you arrived at your answer.

(c) Explain how the alternating potential difference between the dees accelerates the protons during one revolution of their path.

(d) Show that the magnetic flux density B of the applied magnetic field is given by the expression:

$$B = \frac{2\pi fm}{e}$$

where f is the frequency of the alternating potential difference
m is the proton mass
e is the elementary charge

(e) When the cyclotron is used to accelerate protons the magnetic flux density is 1.5 mT. Calculate the frequency of the applied potential difference used in the accelerator.

[CCEA 2006]

2 (a) State two features of the synchrotron accelerator that distinguishes it from the cyclotron.

(b) Show that, in a synchrotron, the path followed by a particle of mass m and charge q at a speed v has a radius of curvature given by

$$r = \frac{mv}{Bq}$$

(c) The large electron positron collider (LEP) at CERN accelerates electrons and positrons to kinetic energies of 50 GeV and has a radius of 4.25 km. The super proton synchrotron (SPS) at CERN accelerates protons and antiprotons to kinetic energies of 450 GeV in a 1.1 km radius circle.

(i) Both the LEP and SPS involves particles and their antiparticles. What is an antiparticle?

(ii) In the LEP, what happens when an electron collides with a positron?

(iii) Assume that the magnetic field used in both machines has the same flux density. Using the expression in part (b) explain why the velocity of the particles in the LEP collider is greater than that in the SPS accelerator.

[CCEA 2007]

3 When an electron and a positron meet, two photons are produced.

(a) (i) State the name given to this type of interaction.

(ii) Why is this type of interaction rare?

(b) Collisions of the type described above are carried out in particle accelerators. Describe the principle of operation of the linear accelerator. You should use a sketch to help explain your answer.

[CCEA 2008]

5.8 Fundamental Particles

Students should be able to:

5.8.1 Explain the concept of a fundamental particle;

5.8.2 Identify the four fundamental forces and their associated exchange particles;

5.8.3 Classify particles as gauge bosons, leptons and hadrons (mesons and baryons);

5.8.4 State examples of each class of particle;

5.8.5 Describe the structure of hadrons in terms of quarks;

5.8.6 Understand the concept of conservation of charge, lepton number and baryon number;

5.8.7 Describe β– decay in terms of the basic quark model.

What is a Fundamental Particle?

An elementary particle or fundamental particle is a particle not known to be made up of smaller particles. A fundamental particle has no substructure, it is one of the basic building blocks of the universe from which all other particles are made.

In the 1930s our understanding of the structure of matter seemed almost complete. Our picture of the atom was one of a relatively tiny but massive nucleus orbited by electrons. We had made sense of atomic spectra, electron orbits and the link between the two. The neutron had been detected and its discovery explained nuclear isotopes. Some seventy years ago, the fundamental particles, the building blocks of all matter, were considered to the proton, neutron and electron.

However some questions had yet to be answered. Firstly, what force holds the protons and neutrons together to form the nucleus, how does it prevent the electrical repulsion of the positively charged protons splitting the nucleus apart? Secondly, what are the forces involved in the radioactive decays of nuclei that make alpha particles, beta particles, and gamma rays?

Particle accelerators allowed physicists to study the nucleus and the interactions of neutrons and protons that form it. Particle experiments study collisions of high energy particles produced by accelerators and sophisticated detectors surrounding the collision point are used to identify each of the many particles that may be produced in a single collision.

Accelerator experiments revealed many more particles of a type similar to protons and neutrons. A whole new family of particles called mesons was also discovered. By the 1960s hundreds of different particles had been identified, and physicists had no complete understanding of how they were related to each other or of the fundamental forces that are involved with them.

These new particles had a wide range of properties. Some were very massive, some had no mass, some seemed to consist of smaller parts while others were apparently point-like, they had various charges, some were stable and some decayed almost as soon as they were created. Patterns began to emerge from the many discoveries and experiments, and a simpler picture gradually formed.

One classification of these particles that emerged from all the observations was:

- Hadrons (pronounced haedrons)
- Leptons
- Gauge Bosons

Hadrons

These are particles that are affected by the strong interaction, this is the force that acts between neutrons and protons within the nucleus. As more hadrons were discovered they formed two sub-groups within the hadron family, baryons and mesons.

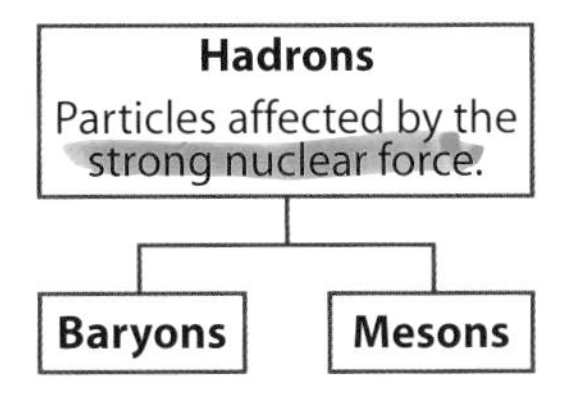

The neutron and proton are the most familiar members of the baryon family. The proton has slightly less mass than the neutron.

The neutron and proton are given baryon numbers of 1.
Their antiparticles have a baryon number of –1.

Table of Baryons

Particle	Symbol	Relative mass	Relative charge	Baryon number B
neutron	n^0	1	0	1
proton	p^+	0.999	+1	1

Antiparticle	Symbol	Relative mass	Relative charge	Baryon number B
antineutron	$\overline{n^0}$	1	0	–1
antiproton	p^-	0.999	–1	–1

Mesons have a mass less than the proton but greater than the electron. The pi-meson family consisting of two charged particles π^+, π^- and the neutral π^0 play a role in the strong nuclear force. The mesons have a baryon number of 0 as do their antiparticles, as they are not baryons. The baryon number B is a quantity that is conserved during interactions.

Table of Mesons

Particle	Symbol	Relative mass	Relative charge	Baryon number B
pion	π^+	0.149	+1	0
pion	π^0	0.144	0	0

Antiparticle	Symbol	Relative mass	Relative charge	Baryon number B
pion	π^-	0.149	–1	0
pion	$\overline{\pi^0}$	0.144	0	0

Leptons

Leptons are particles that are not affected by the strong interaction.

Leptons appear to be indivisible fundamental particles, i.e. they cannot be broken into smaller particles. There are six leptons in the present structure, the electron (e), the muon (μ), and the tau (τ) particle and their associated neutrinos.

The different varieties of the elementary particles are commonly called **flavours.**

The six flavours of leptons form three generations, each of increasing mass.

Each lepton is given a lepton number of 1 and their antiparticles –1. The lepton number, L, is conserved during interactions.

Table of Leptons

Generation	Particle	Symbol	Relative mass	Relative charge	Lepton number L	Antiparticle	Symbol	Relative mass	Relative charge	Lepton number L
1	electron	e^0	1	–1	1	positron	e^+	1	+1	–1
1	electron-neutrino	v_e	0	0	1	antielectron-neutrino	$\overline{v_e}$	0	0	–1
2	muon	μ^-	207	–1	1	antimuon	μ^+	207	+1	–1
2	muon-neutrino	v_μ	0	0	1	antimuon-neutrino	$\overline{v_\mu}$	0	0	–1
3	tau	τ^-	3490	–1	1	antitau	τ^+	3490	+1	–1
3	tau-neutrino	v_τ	0	0	1	antitau-neutrino	$\overline{v_\tau}$	0	0	–1

Fundamental Forces of Nature

Before we examine the gauge bosons in more detail we need to find out about the four fundamental forces that occur in nature.

Gravity, which affects particles with mass. It is very weak, it is only noticeable when large masses are present. Our weight is due to the gravitational attraction between ourselves and the earth which has a mass of 6×10^{24} kg. Gravity is always attractive and has an infinite range. Gravity is the force that determines the structure of large scale matter such as stars and galaxies.

The **Electromagnetic** force, which affects particles with charge. It is much stronger than gravity. Matter is electrically neutral because the opposite charge on electrons and protons cancel each other. It also has an infinite range. Electromagnetic forces determine the structure of atoms as well as determining the properties of materials and the results of chemical processes.

The **Strong nuclear** force exists between neutrons and protons in the nucleus. It is clearly strong enough to overcome the electrical repulsion of the protons. It is a very short range force and only exists when hadrons are within a distance of around 10^{-15} m of each other. The strong nuclear force determines the structure of the nucleus.

The **Weak interaction** is the name given to the force that induces beta decay. Beta decay occurs when a neutron decays to a proton and creates an electron and antineutrino in the process. This nuclear event creates a particle, the antineutrino, that is not affected by the electromagnetic force or the strong nuclear force. The weak interaction is the short range force needed to explain this effect.

Exchange Particles and the Fundamental Forces

The modern understanding of the four fundamental forces comes from what is known as Quantum Field Theory. All the fundamental forces can be treated as the exchange of particles. These exchange particles are the **gauge bosons**. Each fundamental force is attributed to the exchange of at least one gauge boson. Physicists have found that, using the idea of exchange particles, they can explain very precisely the force of one particle acting on another.

The gauge bosons play a role in the fundamental forces found in nature. There are four kinds of gauge bosons: photons, W and Z bosons, gluons and gravitons.

Photons are the gauge bosons of the electromagnetic interaction, such as the repulsion between two electrons. The W and Z bosons are the exchange particles of the weak interaction which governs beta decay. Gluons play a role in the strong interaction, i.e. the force that exists between neutrons and protons. Gravitons are believed to play a similar role in gravity, however the graviton, unlike the other exchange particles, has yet to be detected. Bosons are named after the Indian physicist Satyendra Nath Bose and 'gauge' is a theory found in quantum physics.

Table of the Fundamental Forces and their Exchange Particles

Force	What it does	Strength (Comparative)	Range	Exchange particle (gauge boson)
Strong nuclear	Holds the nucleus together	1	1×10^{-15} ~ diameter of a nucleus	Gluons
Electromagnetic	Attractive and repulsive force between charged particles	~ 1/150	Infinite	Photon
Weak interaction	Induces beta decay	1×10^{-6}	1×10^{-18} m ~ diameter of a proton	W and Z bosons
Gravity	Attractive force between masses	$\sim 1 \times 10^{-39}$	Infinite	Graviton (by analogy only)

This is a very new way to regard force, an explanation using a simple analogy might help. The repulsion between two electrons uses the concept of exchange particles that transfer back and forth between elementary particles. In the case of the electron-electron interaction the repulsion between the two electrons is visualized as the exchange of electromagnetic energy in the form of a photon.

You can think about this force and the exchange particles as being analogous to the following situation. Two ice skaters are standing on ice. One skater appears to throw an object towards the other one, who appears to grab the object and is pushed backwards. The object thrown could be a football but, although you cannot see it, you certainly notice its effect. The object thrown is analogous to the exchange of photons. If the football where thrown more often this would be equivalent to a greater rate of exchange of photons, in other words, the repulsive force is stronger.

The above analogy only explains repulsive forces: there is no analogy based on everyday actions that can be used to describe the purely quantum mechanical effect of attraction using exchange particles.

Extension Material – not on the syllabus

Richard Feynman devised a way to represent particle interactions using diagrams. The diagrams are not meant to show the paths that the particle take. They were used when quantum field theory was being developed to allow equations to be devised that were helpful in predicting the outcome of experiments.

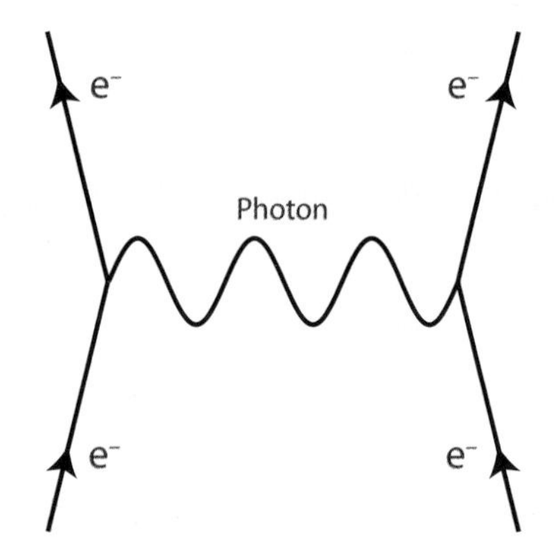

This represents two electrons repelling each other. The exchange particle or gauge boson in this interaction is the photon.

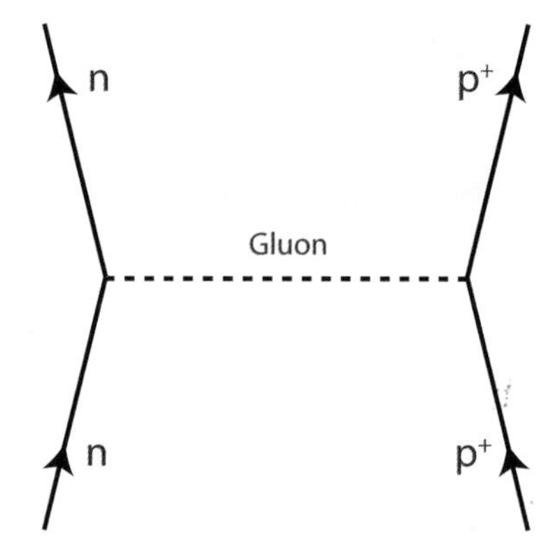

The strong nuclear force between the neutron and proton is represented in this diagram. The gauge boson is the gluon. The actual gluon in this case being the π^o meson.

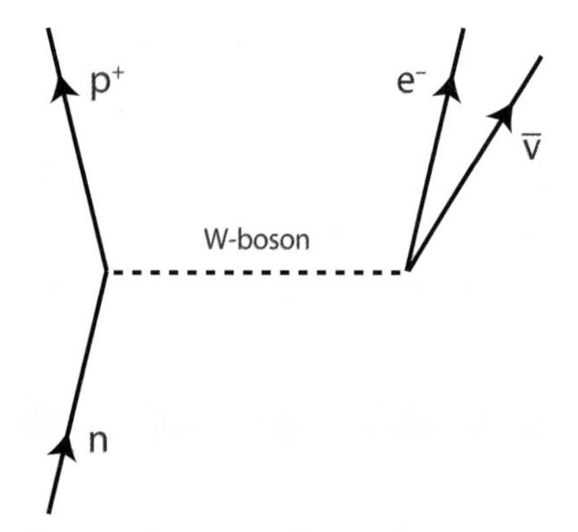

This represents beta decay. A neutron transforms to a proton, with an electron and an antineutrino being emitted as a result, the gauge boson in the weak interaction being the W^- particle.

The Quark Model of the Hadrons

In 1964 physicists came up with the idea that neutrons and protons and all those new particles could be explained by a considering them to be made up of smaller particles, now known as **quarks**.

This quark model could explain the neutron, proton and most of the mesons using just two types of quarks and their corresponding antiparticles, antiquarks. These are now called up, down. To explain some of the mesons with more unusual or strange properties the strange quark was added.

Up to then the charge of the electron was considered to be the smallest electric charge possible.

In the quark model electric charges of $+\frac{2}{3}e$ and $-\frac{1}{3}e$ are possible.

Antiquarks are the antimatter partners of quarks, they have the same masses as, but the opposite charge from, the corresponding quarks. When a quark meets an antiquark, they may annihilate.

The quark model has been confirmed by many experiments. It is now part of the Standard Model of Fundamental Particles and Interactions. Observations from particle accelerator experiments reveal that there are six types of quarks. A free quark cannot exist, they are always combined in twos (mesons) or in threes (hadrons)

The current model has six quarks and six antiquarks. They have been given the names up, down, strange, charm, bottom, and top. It is a difficult task to adequately describe the curious properties of such particles, physicists say they have different flavours, such as charm and strangeness.

The table below shows the properties of these quarks and their corresponding antiquark. Notice that they also have three generations, just as the leptons have, each generation having an increasing mass.

Generation	Quark	Symbol	Charge Q	Baryon number B
1	up	u	$+\frac{2}{3}e$	$\frac{1}{3}$
1	down	d	$-\frac{1}{3}e$	$\frac{1}{3}$
2	strange	s	$-\frac{1}{3}e$	$\frac{1}{3}$
2	charm	c	$+\frac{2}{3}e$	$\frac{1}{3}$
3	top	t	$+\frac{2}{3}e$	$\frac{1}{3}$
3	bottom	b	$-\frac{1}{3}e$	$\frac{1}{3}$

Antiquark	Symbol	Charge Q	Baryon number B
anti-up	$\bar{u}$	$-\frac{2}{3}e$	$-\frac{1}{3}$
anti-down	$\bar{d}$	$+\frac{1}{3}e$	$-\frac{1}{3}$
anti-strange	$\bar{s}$	$+\frac{1}{3}e$	$-\frac{1}{3}$
anti-charm	$\bar{c}$	$-\frac{2}{3}e$	$-\frac{1}{3}$
anti-top	$\bar{t}$	$-\frac{2}{3}e$	$-\frac{1}{3}$
anti-bottom	$\bar{b}$	$-\frac{1}{3}e$	$-\frac{1}{3}$

In the quark model baryons, neutrons and protons, **consist of three quarks**, known as a triplet.

The proton consists of 2 up quarks and 1 down quark

Quark	u	u	d	
Charge	$= \frac{2}{3}e +$	$\frac{2}{3}e +$	$\left(-\frac{1}{3}e\right)$	$= 1e$
Baryon number	$= \frac{1}{3} +$	$\frac{1}{3} +$	$\frac{1}{3}$	$= 1$

The neutron consists of 2 down quarks and 1 up quark

Quark	u	d	d	
Charge	$= \frac{2}{3}e +$	$\left(-\frac{1}{3}e\right) +$	$\left(-\frac{1}{3}e\right)$	$= 0$
Baryon number	$= \frac{1}{3} +$	$\frac{1}{3} +$	$\frac{1}{3}$	$= 1$

In the quark model **mesons consist of two quarks**, a quark and an anti-quark, known as a doublet.

The π^o consists of 1 up quark and 1 anti-up quark

Quark	u	$\bar{u}$	
Charge	$= \frac{2}{3}e +$	$\left(-\frac{2}{3}e\right)$	$= 0$
Baryon number	$= \frac{1}{3} +$	$\left(-\frac{1}{3}\right)$	$= 0$

Exercise 41

1 Determine the quark composition of the antiproton p^-.

2 Determine the quark composition of the mesons π^+ and π^-.

3 A muon can decay by the weak interaction in the way shown below.

$$\mu^- = e^- + X + v_\mu$$

(a) Name the exchange particle of the weak interaction.

(b) Identify the particle marked X in the above equation. Remember the lepton number L is conserved in this type of interaction.

4 In a collision between a proton and an antiproton the following reaction resulted.

$$p^+ + p^- = ? \rightarrow 4\pi^+ + 4\pi^-$$

(a) Show that the rule governing baryon numbers was not broken in this interaction.

(b) What name is given to the type of process in which a particle and its antiparticle interact?

(c) Immediately after the proton and the antiproton interact what was formed before the eight pi-mesons were created?

5 A student writes down the equation below to show positron emission. However it is wrong because it violates two conservation laws.

$$p^+ \rightarrow \bar{n} + e^+ + \bar{\nu}_e$$

(a) State and explain the two conservation laws that are violated.

(b) Give the correct equation.

6 From the following list of particles,

$$p^+ \quad n \quad e^+ \quad \nu_e \quad \mu^- \quad \pi^0$$

identify all those that are examples of:

(a) hadrons

(b) leptons

(c) baryons

(d) mesons

Beta Decay and the Quark Model

The weak interaction force induces beta decay. Inside the nucleus a neutron changes to a proton plus an electron (β^-) and an antineutrino. The equation below illustrates this process.

$${}^{1}_{0}n \rightarrow {}^{1}_{1}p + {}^{0}_{-1}e + {}^{0}_{0}\bar{\nu}$$

In terms of the quarks that make up the neutrons the process involves one of the down quarks that make up the neutron changing to an up quark.

$${}^{\frac{1}{3}}_{-\frac{1}{3}}d \rightarrow {}^{\frac{1}{3}}_{\frac{2}{3}}u + {}^{0}_{-1}e + {}^{0}_{0}\bar{\nu}$$

This process is represented in the diagram below.

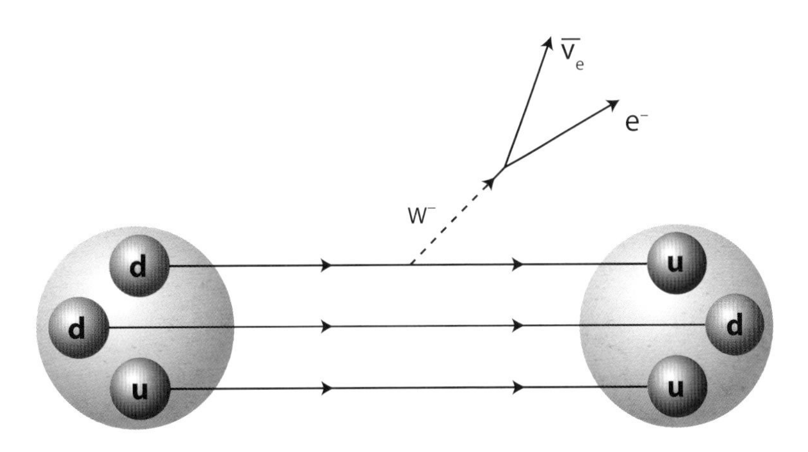

Exercise 42

Past paper questions

1 (a) What is meant by a fundamental particle?

(b) From the following list of subatomic particles select those which are not fundamental:

neutron neutrino pi-meson electron tau proton

(c) State the difference between leptons and hadrons.

(d) Hadrons are subdivided into two classes of particle.

(i) Name these two classes.

(ii) Describe the difference between these two classes in terms of the quark structure.

(iii) For two of the particles in the above list that have a quark structure give details of this structure.

[CCEA 2008]

2 (a) Consider the neutral atom of beryllium, ${}^{9}_{4}\text{Be}$. State, with an explanation the number of leptons, baryons and mesons that this neutral atom contains.

(b) The following equation represents the reaction between particle X and a neutron.

$$X + n \rightarrow p^{+} + e^{-1}$$

By applying the conservation of charge and the conservation of lepton number, identify particle X.

[CCEA 2007]

3 (a) Copy and complete the table below by filling each blank space with a suitable example or particle type.

Particle type	Example
Baryon	
Lepton	
	Pion
antiparticle	

(b) State the quark structure of the proton and use it to show that the charge on the proton is +1e.

(c) (i) Write down two equations representing a reaction which describes (β^{-})decay in terms of quarks. Each equation should include the exchange particle involved in the process.

(ii) Name the force responsible for this process.

[CCEA 2006]

Unit 6 (A2 3) Practical Techniques

6.1 Measuring

The correct use of common school laboratory apparatus is described in the pages that follow. Physics relies on accurate measurements of physical quantities such as mass, length, time and temperature. To improve the accuracy and precision of such measurements instruments such as metre rules, vernier callipers, stop clocks and thermometers are used.

It is important that you know how to use these devices properly. In measuring any quantity there is always some degree of uncertainty. Appreciation of the uncertainty associated with each measuring instrument is equally important.

Measuring Length

Using a Metre Rule

Although this may be one of the simplest length measuring instruments to be found in a school laboratory, care must be taken with its use to avoid errors.

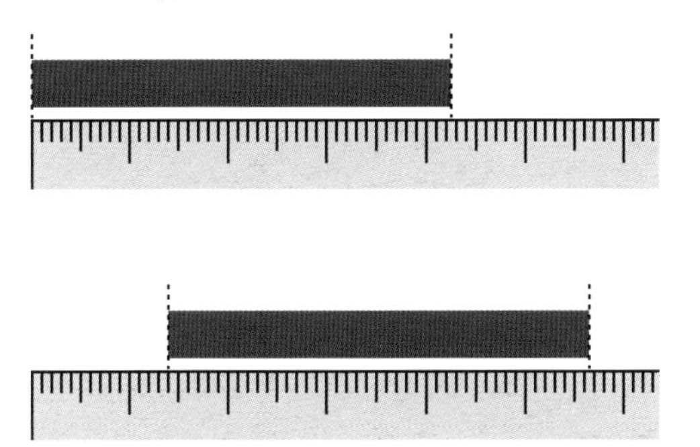

This is bad practice. The end of the metre rule may be worn giving rise to a zero error and an inaccurate measurement of the length.

It is good practice to place the metre rule against the object so that you have two readings to take and subtracting them will give you the length of the object. It avoids a zero error in the measurement. Of course the measurement of length still has an uncertainty associated with it.

The smallest division on the metre is usually 1 mm. If we say that each reading of the metre rule has an uncertainty of ± 0.5 mm then subtracting the two readings to obtain the length has an associated uncertainty of ± 1 mm.

For example, if the two readings are 14.0 cm and 56.5 cm the length is 42.5 cm and if we quote the length with the associated uncertainty then we would write this as (42.5 ± 0.1) cm i.e. an uncertainty in the length of about 0.25%.

For lengths greater than 1 m it is better to use a tape measure, with 1 mm divisions. Tape measures can be used to measure distances up to several hundred metres with good accuracy.

Parallax Error

Parallax error occurs when any scale is not viewed at right angles as shown. Failure to view the scale at right angles will give a reading which is either too high or too low.

Having the scale of the metre rule as close as possible to the object will reduce the possibility of a parallax error.

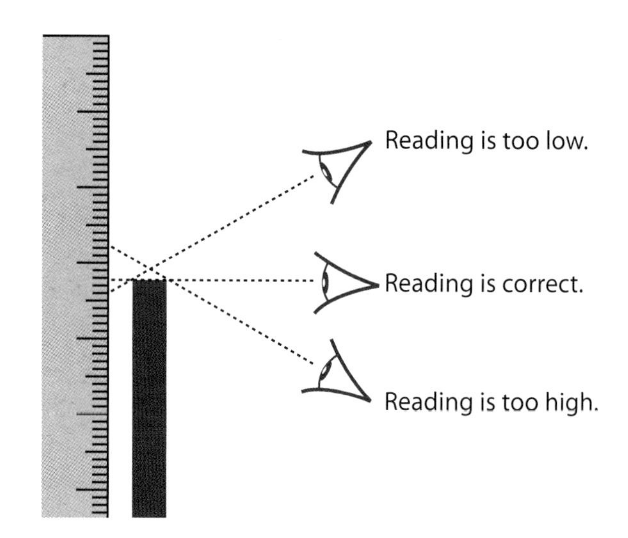

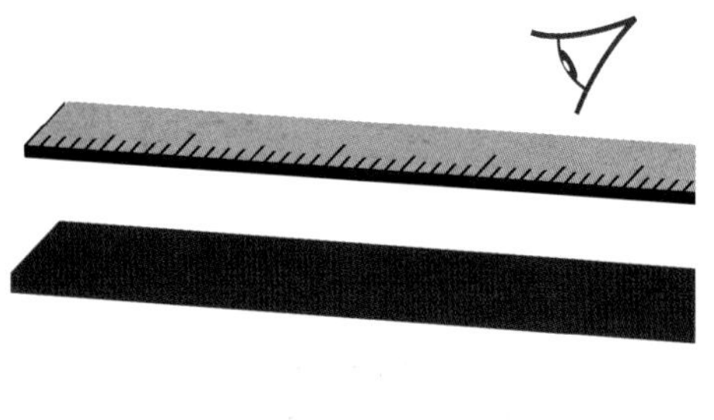

The object to be measured is too far from the scale increasing the possibility of parallax error.

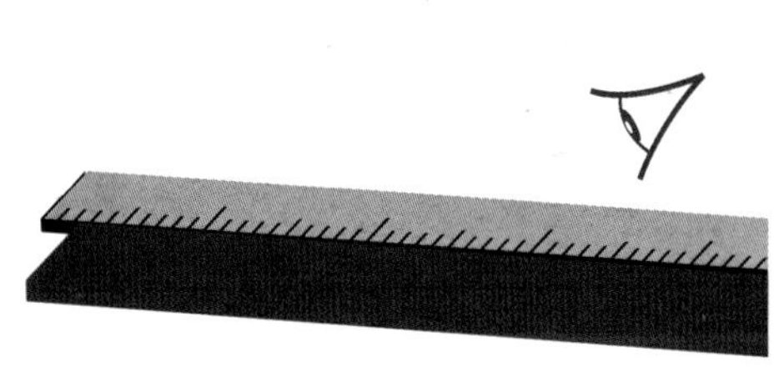

Moving the object closer to the scale as shown is good practice since it reduces the possibility of parallax error.

Vernier Calliper

The Vernier Calliper is a precision instrument that can be used to measure internal and external distances extremely accurately. The example shown below is a manual calliper. Measurements are interpreted from the scale by the user.

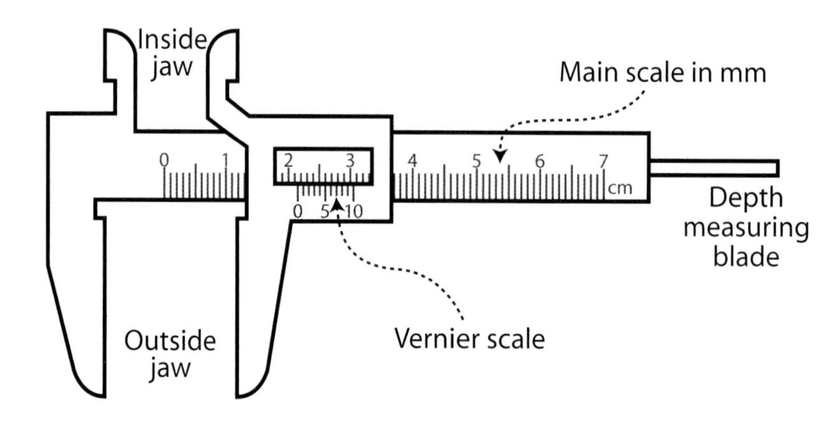

The internal jaws can be used measure the internal diameter of a tube, the external jaws can be used to measure the external diameter of the tube or the width of a block. The depth measuring blade can be used to measure the depth of, say, a hole drilled in a metal bar.

The calliper in the diagram below can read to ± **0.1 mm**. To take the reading you should follow the two steps shown.

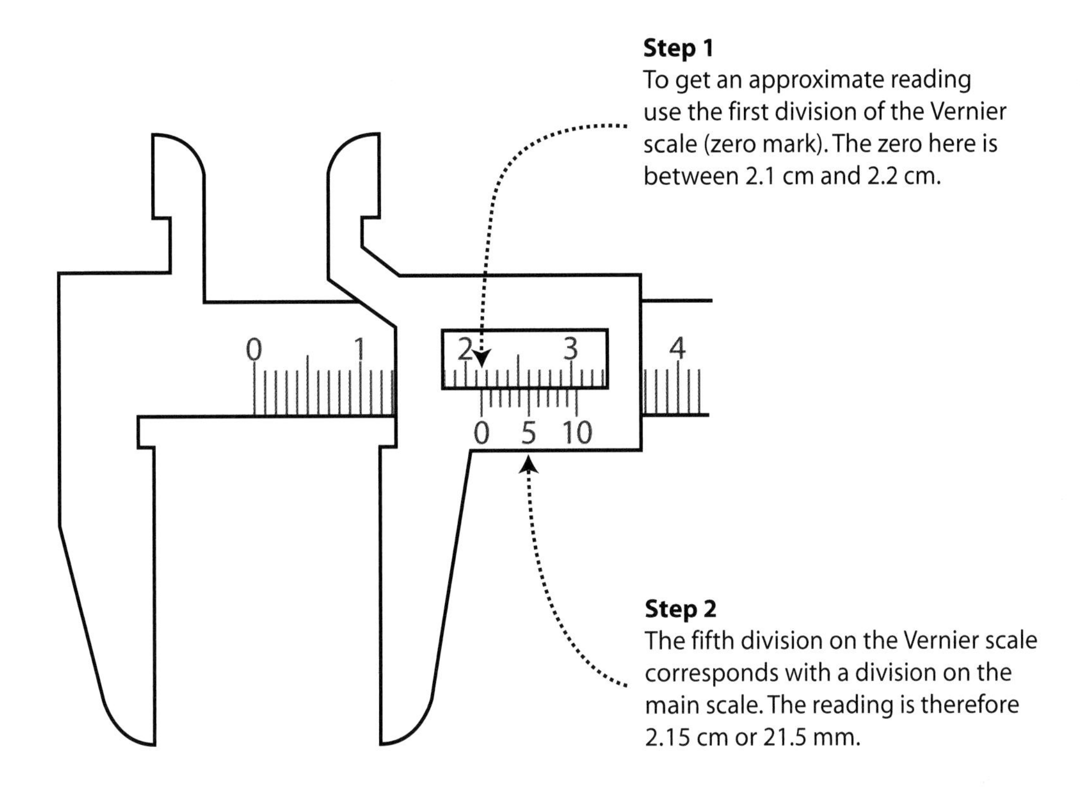

Exercise 43

Work out these readings.

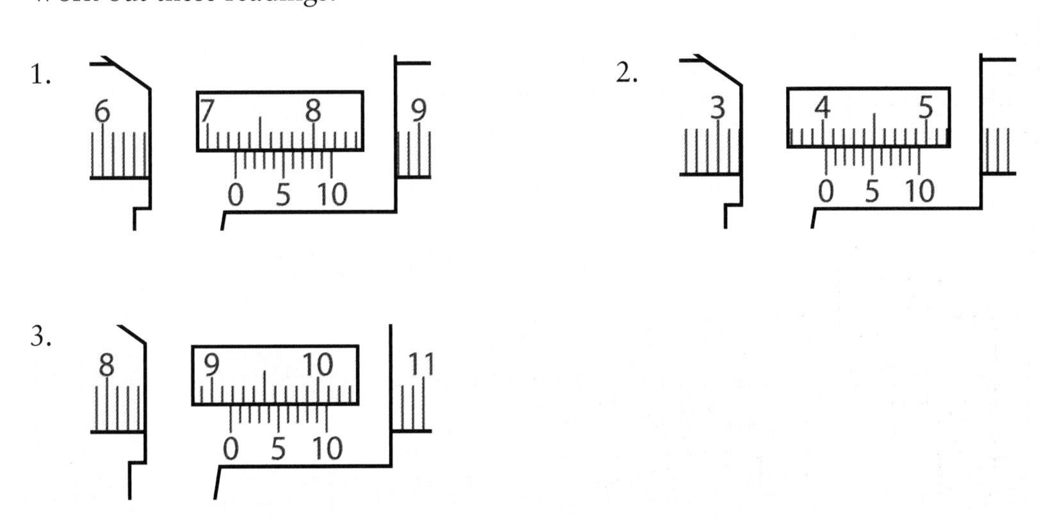

Using a Micrometer Gauge

A micrometer gauge can measure distance to an accuracy of ± 0.01 mm. It is particularly useful for measuring the diameter of a wire or the thickness of a glass microscope slide. The area of cross section of a wire can be calculated when the diameter of the wire is measured using a micrometer gauge.

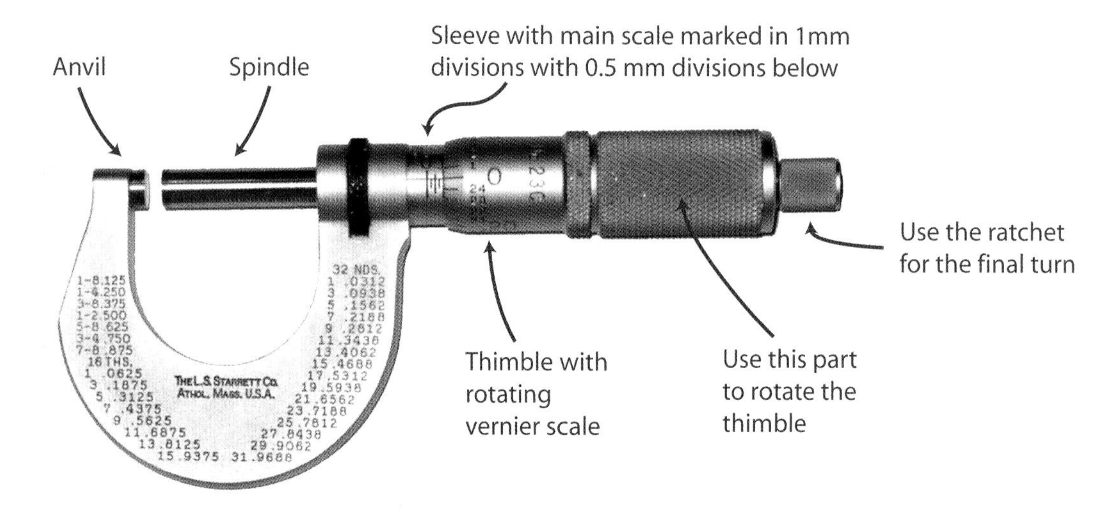

The divisions along the micrometer are 1 mm.

There are (in the micrometer normally used in school) 50 divisions around the barrel.

To move the thimble 1 mm along the barrel requires the thimble to moved through 2 complete turns. There are 50 divisions around the thimble, so to move 1 mm the barrel is turned 100 divisions. This means that 1 division around the barrel = 0.01 mm

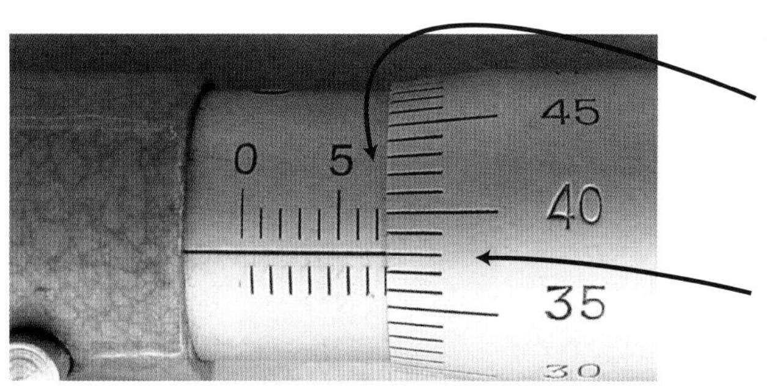

The top scale gives a reading between 7.0 and 7.5 mm. (Note that the 0.5 mm division on the lower scale is not all visible in this image.)

On the Vernier scale, division 38 lines up with the main scale. This is 0.38 mm. The complete reading is therefore 7.38 mm.

Exercise 44

Work out these readings.

1. 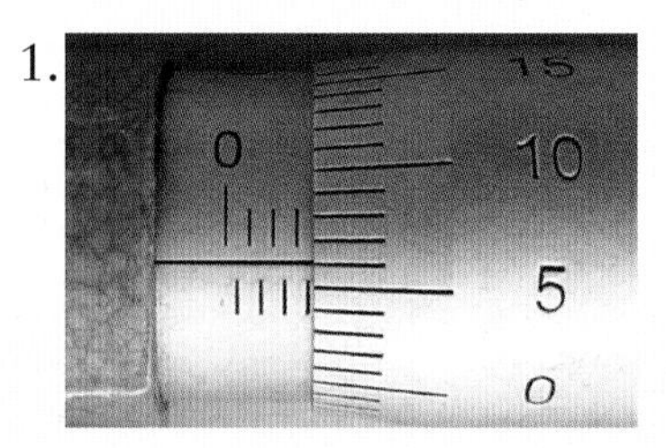

2. 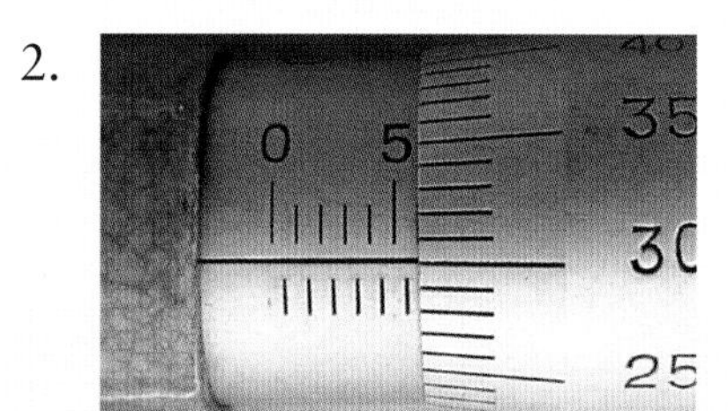

3. 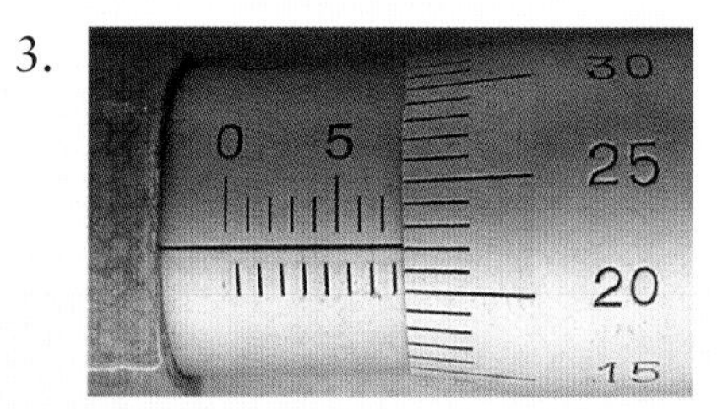

Measuring Volume

Using a Graduated Cylinder

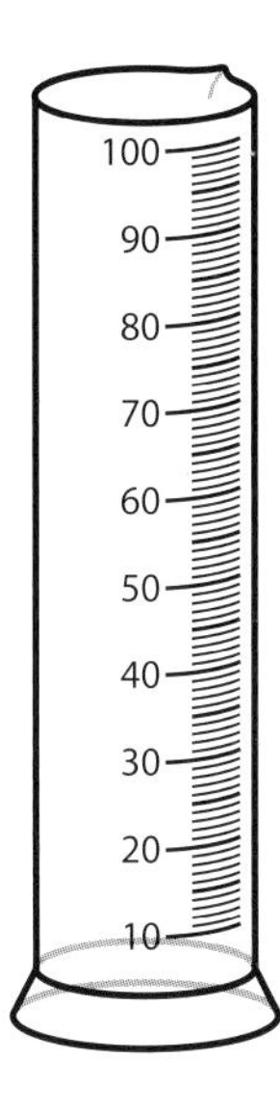

Graduated cylinders are used to measure the volume of a liquid. They come in a range of sizes from 10 cm^3 to 1000 cm^3. Liquids in glass containers curve at the edges; this curvature is called the **meniscus**. With water in glass, the meniscus will curve up at the edges and down in the centre so we say you read the bottom of the meniscus. When reading the volume you should have your eye level with the curved surface (meniscus) of the liquid to avoid parallax error. In some plastic cylinders water has a flat surface. However it still best to take the reading at the centre rather than at the edge.

The visibility of the meniscus can be improved by using a card with a dark stripe on it, placed behind the cylinder. Adjusting the position of the card you will either see a white meniscus against a black background or a black meniscus against a white background. There are some liquids where the curve goes the other way. In this case you would take the reading at the top of the meniscus.

Like most measuring instruments it is important to work out the volume represented by each of the marked divisions.

The volume of an irregular object such as a stone can be found using the displacement method. In this technique a graduated cylinder is partly filled with water and the volume measured. The stone is carefully lowered into the water and when completely covered with water the new volume is measured. The difference between the two readings gives the volume of the stone.

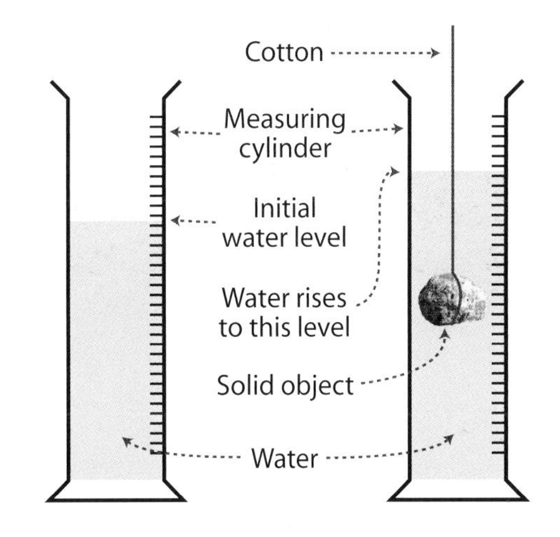

The method can be adapted to find the volume of an object that floats, like a cork. The stone is tied to the cork and, provided both are completely covered, the volume of the cork can be found.

Measuring Angles – Using a Protractor

Notice that numbers marked on the protractor run in both directions so be careful which you use when taking measurements.

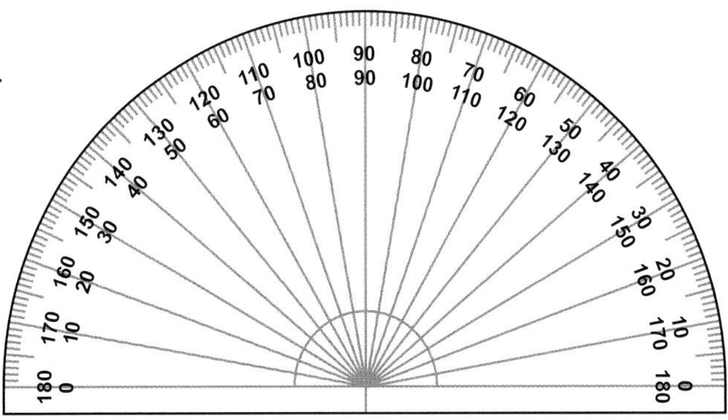

To measure the angle between two lines follow the steps below.

Find the centre of the straight edge of the protractor. This is the cross as shown below.

Place the cross over the point of the angle you wish to measure or draw.

Line up the zero on the straight edge of the protractor with one of the sides of the angle or the line already drawn.

Find the point where the second side of the angle intersects the curved edge of the protractor (you may need to extend the lines). The value at the intersection is the measure of the angle in degrees.

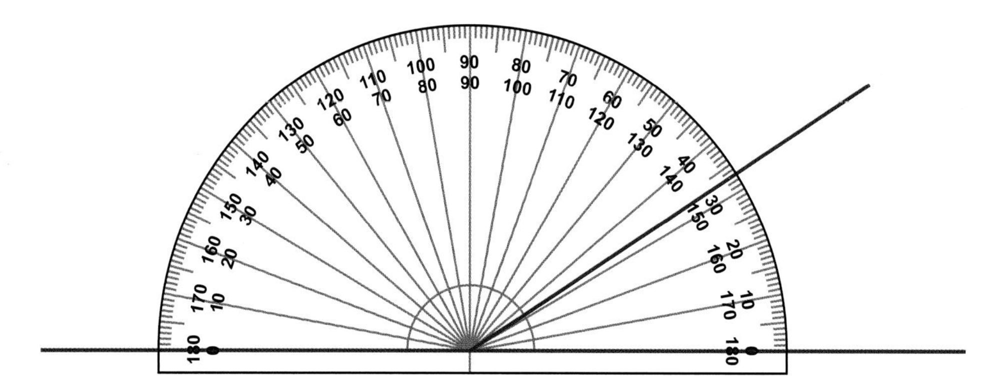

There is of course uncertainty associated with the measurement of this angle. The reading on the scale can be read to ± ½ of one division. This corresponds to ± 0.5°. The positioning of the line along the zero of the protractor also involves an uncertainty of ± ½ of one division, again this is ± 0.5°.

The overall uncertainty in the measurement of the angle is ± 1°.

The angle shown above should be quoted as (33 ± 1)°.

Measuring Weight – Using a Spring Balance or Newton Meter

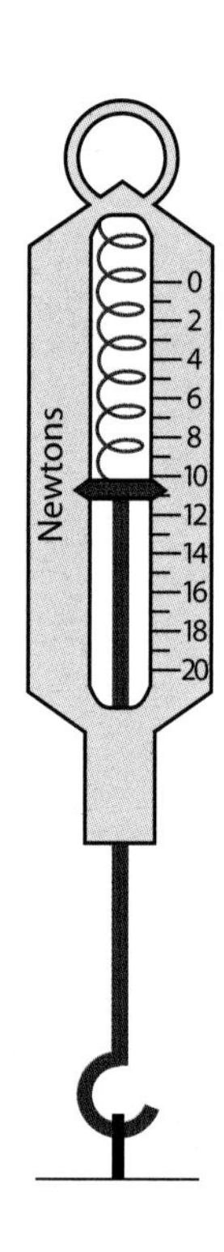

This is used to measure force. It consists of a spring which extends when a force is applied to one end. The spring obeys Hooke's law, so the extension produced is proportional to the force. This allows a scale calibrated in newtons to be placed alongside the spring.

The maximum force that can be measured depends on the strength of the spring. Spring balances with ranges of 0 – 10 N, 0 – 20 N and 0 – 50 N are common.

Measuring Mass – Using an Electronic Top Pan Balance

This is used to measure the mass of an object. The object is placed on the pan and the force it exerts is detected by a sensor which converts this to an electrical signal. A conversion factor is applied and the display will show the mass of the object in grammes or kilogrammes.

Most electronic balances have a 'tare' facility which when pressed sets the reading to zero. This is useful when measuring a required mass of a solid or liquid in a beaker or other container.

The container is placed on the pan, the tare button is pressed and the reading goes to zero. The reading then shown is the mass of material added to the container.

The final decimal place indicates the uncertainty. For the electronic balance shown above this is ± 0.1 g.

Measuring Temperature

In Physics we use the Kelvin and the Celsius temperature scale. On the Celsius scale water freezes at 0 °C, on the Kelvin scale this is 273 K. Water boils at 100 °C or 373 K.

The thermometers found in school laboratories are normally calibrated from –10 °C to 110 °C. It is possible to get thermometers capable of measuring higher temperatures than 110 °C.

It is important to determine the temperature difference indicated by the smallest division shown on the thermometer. This is normally 1 °C, so it is possible to read the scale to ± 0.5 °C. Therefore a temperature change would measured to ± 1 °C.

Thermometers are made of glass and therefore fragile. Most are round and they can easily roll off a bench. However some are triangular in cross section or have a small plastic triangle around them to reduce the likelihood of rolling.

When recording the temperature of a liquid it is important to stir the liquid to ensure thorough mixing. When heating a liquid do not use the thermometer to stir the liquid unless it is a robust type clearly intended for the purpose. When recording the temperature of a liquid ensure than the bulb of the thermometer is completely covered by the liquid or is as close as possible to the position at which the temperature is to be measured.

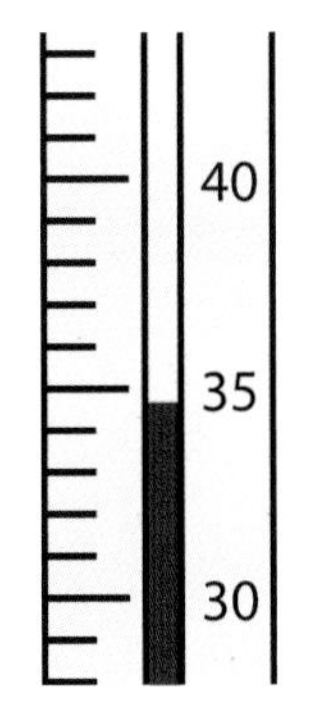

Smallest division: 1.0 °C

Uncertainty: ±0.5 °C

Reading: 34.5 °C

The example on the left shows how to measure temperature accurately using a thermometer.

Measuring Current – using an Ammeter

An ammeter measures electric current. An ammeter is connected in **series** with the other components in a circuit. The positive terminal of the ammeter is connected to the positive side of the cell. If there are other components in the circuit you should trace the connections from the positive terminal of the ammeter to the positive terminal of the cell. The circuit below shows an ammeter in series with a bulb.

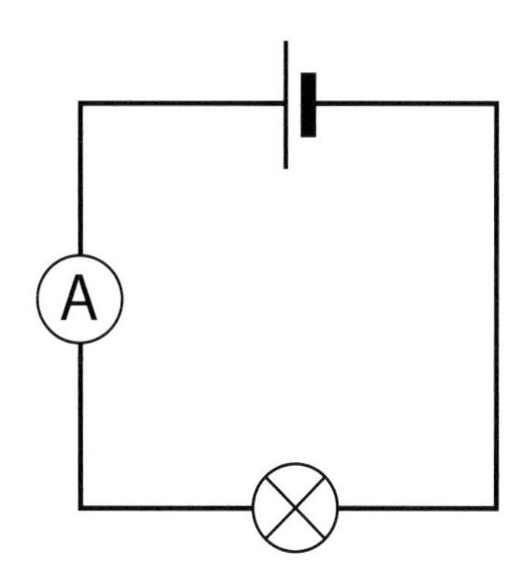

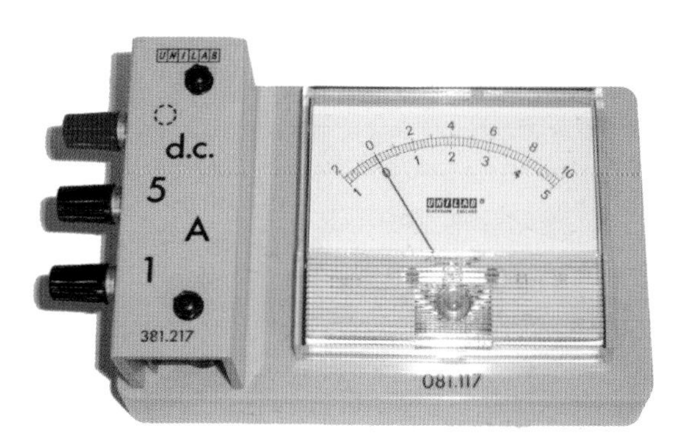

The scale of an ammeter can measure amperes (A), milliamps ($1 \text{ mA} = 1 \times 10^{-3}\text{A}$) or possibly microamps ($1 \text{ μA} = 1 \times 10^{-6}\text{A}$). You should also determine what the smallest division of the scale represents. This is important when it comes to considering the uncertainty associated with a measurement of current. The uncertainty is normally taken as ± ½ the smallest division on the scale.

The ammeter shown above is of a type found in many school laboratories. The top scale can measure currents up to a maximum of 1.0 A, the lower scale to a maximum of 5.0 A. The scale to be used is determined by which of the terminals 1A or 5A is used to connect the meter into the circuit.

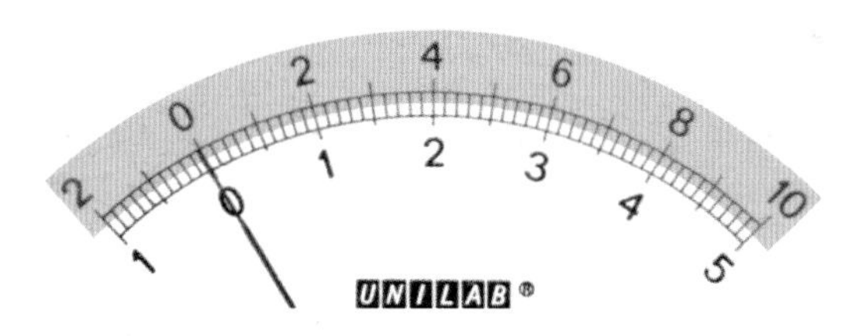

Using the ammeter shown above, the top scale has the following features:
Maximum current is 1.0 A
Smallest division is 0.02 A
Uncertainty is ± 0.01 A (± ½ division)

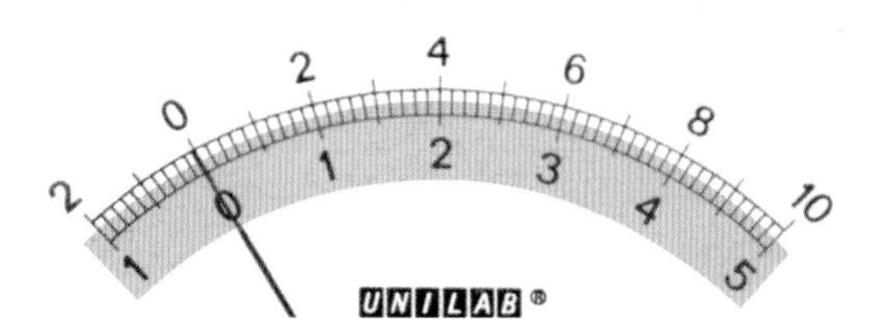

Using the ammeter shown above, the bottom scale has the following features:
Maximum current is 5.0 A
Smallest division is 0.1 A
Uncertainty is ± 0.05 A (± ½ division)

The type of meter shown can easily be converted to a milliameter by changing the shunt. The procedure outlined above can be used to determine which scale to use and the uncertainty associated with that scale.

Digital ammeters are also found in school laboratories. The one shown below can measure current as large as 10 amperes. The final decimal place indicates that the uncertainty associated with the use of this ammeter is ± 0.01 A.

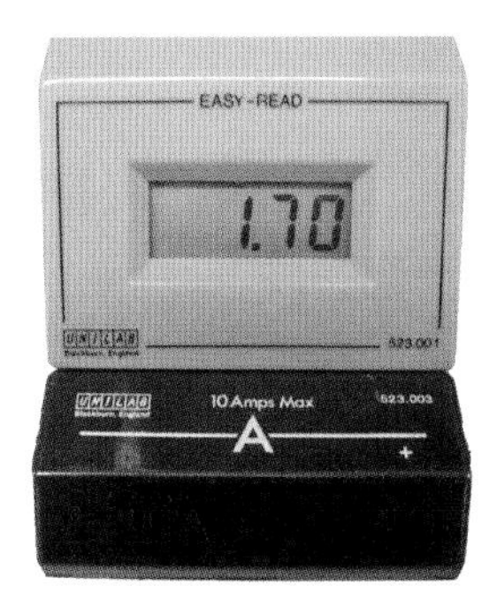

Measuring Potential Difference – Using a Voltmeter

A voltmeter measures potential difference. It is connected in **parallel** with the component across which the potential difference is to be measured. The positive terminal of the voltmeter is connected to the end of the component which is nearest to the positive terminal of the cell, battery or power supply and the negative terminal to the other end of the component.

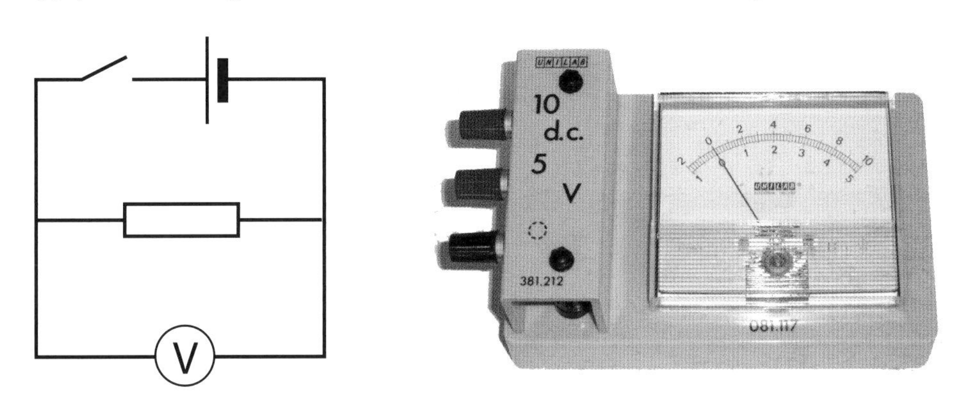

When building a circuit the voltmeter can be the last item attached.

The voltmeter shown has a multiplier which allows the top scale to have a maximum reading of 10V and the lower scale a maximum of 5.0V.

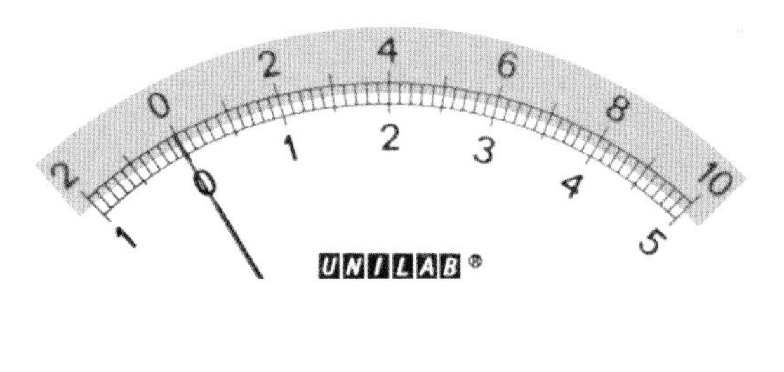

Using the voltmeter shown above the top scale has the following features:
Maximum current is 10.0 V
Smallest division is 0.02 V
Uncertainty is ± 0.01 V (± ½ division)

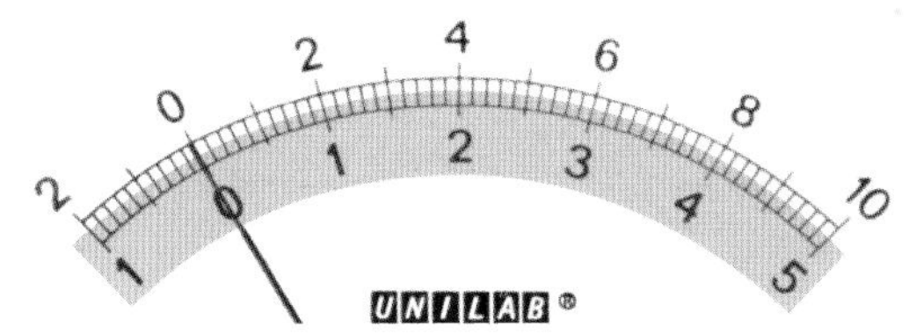

Using the voltmeter shown above the bottom scale has the following features:
Maximum current is 5.0 V
Smallest division is 0.1 V
Uncertainty is ± 0.05 V (± ½ division)

Digital voltmeters are also found in school laboratories. The one shown below can measure a potential difference as large as 20 V.

The final decimal place indicates that the uncertainty associated with the use of this voltmeter is ± 0.01 V.

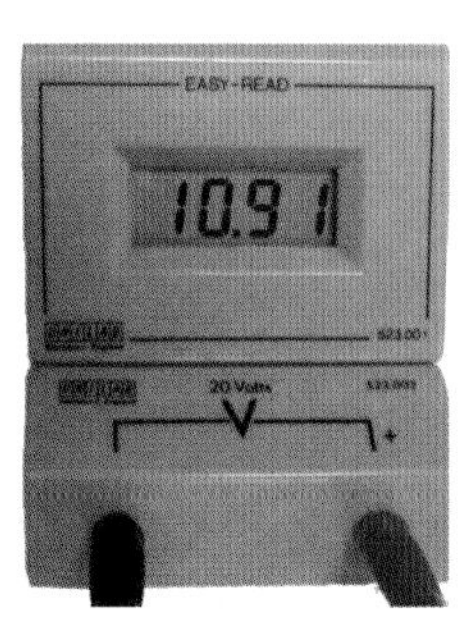

Zero Error

Before you use a meter it is important that it reads zero before any current passes through it or a potential difference is applied to it. The one shown below has a zero error. It may be possible to set the pointer to read zero. If this cannot be done then the zero error value must be subtracted from all your readings.

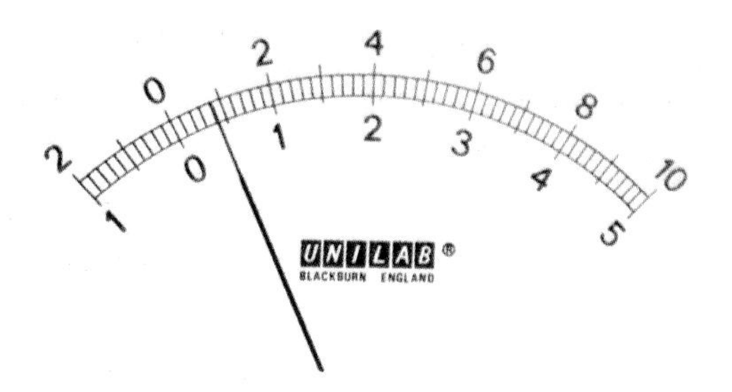

6.2 Precision, Accuracy and Errors

It requires skill to carry out an experiment well. Skill involves being able to manipulate equipment, identifying variables that can be measured, varied and controlled, and being able to take readings from a range of measuring instruments. To become a skilful experimenter requires practice but what follows is a number of simple techniques that can improve your experimental technique and lead to more precise measurements.

The Difference Between an Accurate Measurement and a Precise Measurement

An accurate measurement is one that is close to the true value of a physical quantity. A precise measurement is one taken with a measuring device that can give an exact value when used with skill. For example using a vernier scale that reads to 0.1 mm will give a more precise value than a metre rule that reads to 1 mm. Of course it requires more skill to take a reading with a vernier scale than with a metre rule.

Precise and **accurate** The measurements that are close to the true value and the measurements are very similar ie random and systematic uncertainties are small.	 True 2.45 Readings taken 2.42, 2.46, 2.44, 2.45 Average 2.44
Precise but **inaccurate** The measurements show very small differences but their average value is far from the true value.	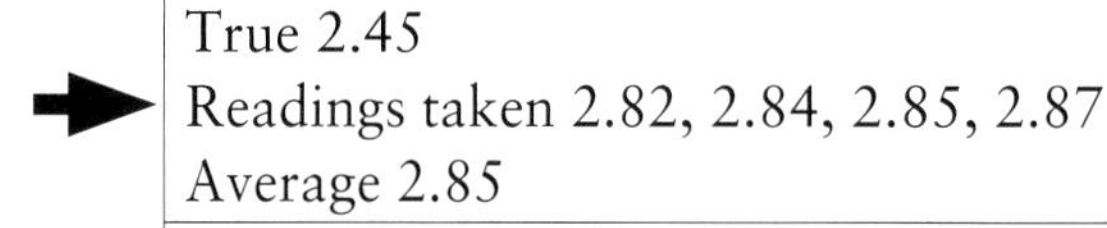 True 2.45 Readings taken 2.82, 2.84, 2.85, 2.87 Average 2.85
Imprecise but **accurate** The measurements show large variation but an average value that is close to the true value.	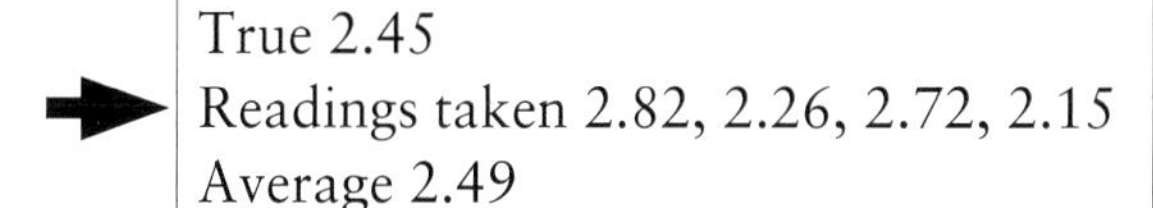 True 2.45 Readings taken 2.82, 2.26, 2.72, 2.15 Average 2.49
Imprecise and **inaccurate** The measurements show large variations and an average value far from the true value.	True 2.45 Readings taken 2.62, 2.76, 2.14, 2.95 Average 2.62

What are Systematic Errors?

A systematic uncertainty will result in all readings being either above or below the accepted value. In other words it leads to inaccuracy of the measurement although the measurements taken may well be precise (very small differences between them). This uncertainty **cannot** be eliminated by repeating readings and then averaging.

Examples of systematic uncertainty are:

Zero Error

Zero error on an instrument (the scale reading is not zero before measurements are taken).

Where appropriate, instruments should be checked for any zero error. Where there is a zero error, the meter should be adjusted to zero or, if this not possible, the zero error should be noted and all recorded readings should then be adjusted.

Note: Remember to record all readings as they are taken. Do not allow for zero error 'in your head' and then write down the adjusted value.

Parallax Error

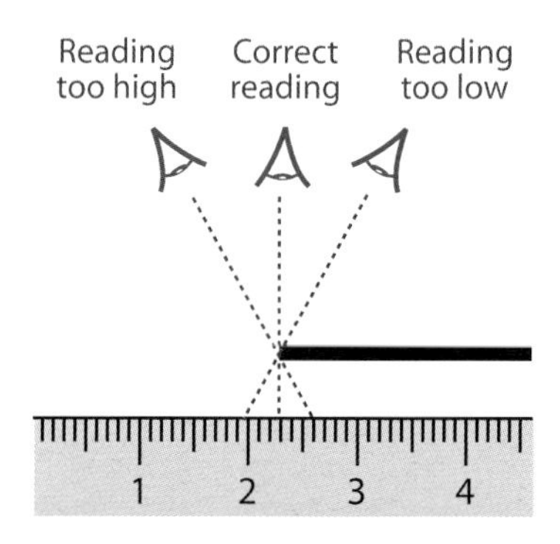

Parallax error occurs when the scale is not viewed normally when taking a reading. To reduce parallax errors, always:

- have the scale as close as possible to the pointer;
- view the scale normally.

Over-tightening a Micrometer

Over-tightening a micrometer gauge when taking a measurement will lead to a systematic error that will always give a smaller value than the true value.

Always use the ratchet, because this will slip when the jaws meet any resistance. This is particularly important if you are measuring the diameter of a wire.

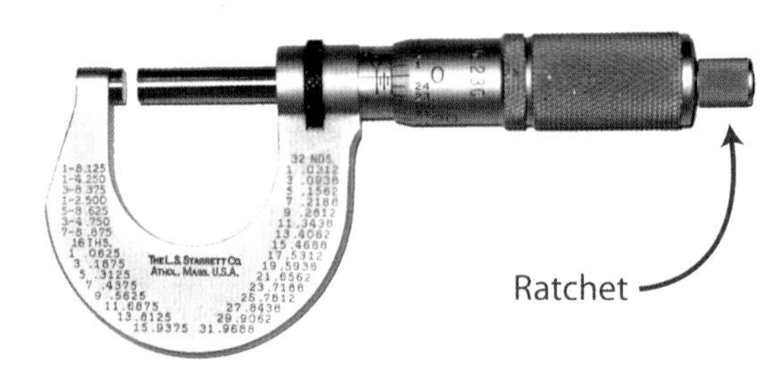

Techniques Designed to Improve the Accuracy of Your Measurements

Timing Oscillations

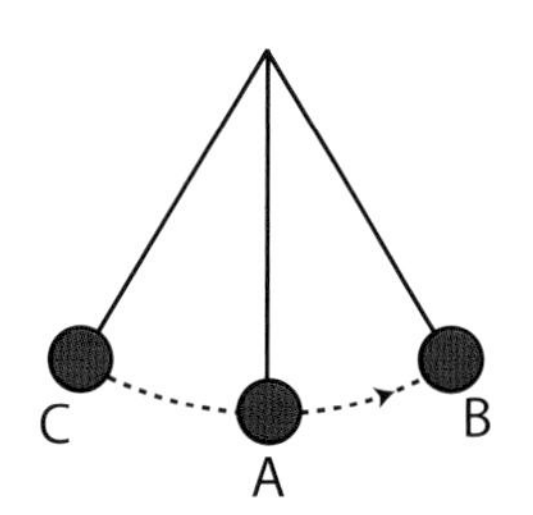

Start your timing when the oscillation is at one extreme, ie when the vibrating object is momentarily at rest. In the case of the simple pendulum, one oscillation would be from C to A to B and back to C again. Start the object oscillating before you start timing and watch the object until the vibrations are no longer noticeable. This will determine how many oscillations are noticeable.

It is poor practice to measure the period of oscillation by timing just one. The error in human timing is likely to be around 0.2 s. It is better to time 20 such oscillations. However in some circumstances the oscillations may die out quickly and only 3 or 5 complete oscillations may be noticeable.

When you decide to start timing, begin by saying zero as you start. Timing 20 oscillations will reduce the uncertainty in determining the time for 1 oscillation, the period. A simple pendulum with a period of just 1 second would have an error of 0.2 s, i.e. 20%. However, if you time 20 such oscillations the error is only 0.2 s in 20 seconds, ie 1%.

When you take measurements, vary the quantity in a logical manner, e.g. increase the length of the pendulum in equal sized steps. This will allow trends to be more noticeable.

Measuring Current and Potential Difference

Always draw the circuit diagram before you start building the circuit.

Start at the positive terminal and insert components as you follow the circuit from positive to negative.

Voltmeters should be left until all the series components have been connected. Remember voltmeters are connected in parallel.

Ammeters and voltmeters are always connected as positive to positive or red to red.

When using analogue meters establish what each division on the scale represents.

When taking a reading look vertically down on the scale. This reduces the possibility of parallax error in your reading.

If you are using a digital meter do not change the scale in the middle of the experiment.

In many cases try to change the potential difference in equal steps. However when dealing with light emitting diodes (LEDs) it may be necessary to change the potential difference in very small steps when the current is beginning to increase.

6.3 Analysis and Interpretation of Results

The first step in the analysis of your data is the recording of measurements in a suitable table. The table should have sufficient columns for all the measurements and possible calculations you need to make.

Columns need headings. These should be the quantity in the appropriate units for that quantity.

As an example, consider the table that could be used for the investigation of the period of a simple pendulum and the length of the pendulum.

Measurement repeated 3 times to give an average

Units shown on **all** column headings

Systematic and a good range

Length of the pendulum/m	Time for 20 oscillations/s			Average time for 20 oscillations/s	Period/s
	1st	2nd	3rd		
1.2					
1.0					
0.8					
0.6					
0.4					
0.2					

The length is varied in a systematic manner: it is gradually increased in length using steps of 0.2 m.

Increasing the length from 0.2 to 1.2 m covers a good range of values. For each length the time for 20 oscillations is measured. To improve the accuracy and ensure reliability, this is done 3 times and the average taken. The final step is to calculate the period of the pendulum by dividing the average time by 20.

To reduce the uncertainty in the measurement of the periodic time of any vibrating system, it is advisable to time sufficient oscillations so that a total time of around 20 seconds or better is to be measured. In the case above, a pendulum with length of 0.2 m would yield around 18 seconds and the length of 1.2 m would yield a total time of around 40 seconds.

However it is not always possible to obtain sufficient oscillations to achieve a total time of at least 20 seconds. In this case you need to determine the maximum number of oscillations that you can detect before they cease to be noticed.

AS Practical 2002

In this a bifilar pendulum was set up for you. The pendulum was made to vibrate about its centre. The length of the vertical cords was to be varied and the effect this had on the periodic time of oscillation was to be investigated.

The timing was to be carried out using a stopwatch or stopclock. The length of the supporting cords was to be decreased from 400 mm to about 200 mm and 5 sets of readings were to be taken.

The results were to be recorded in a table that was partly completed with the first value of L and the period T column shown with the appropriate unit.

When the bifilar pendulum was set swinging it was found that 5 oscillations were easily observable. More than this and they became very difficult to see. As you can see from the table, it was decided that 5 oscillations should be timed.

L/mm	Time for 5 oscillations/s				T/s
	1st	2nd	3rd	Average	
400	9.55	9.40	9.51	9.48	1.90
350	8.90	8.85	8.82	8.85	1.77
300	8.21	8.30	8.25	8.25	1.65
250	7.35	7.51	7.53	7.46	1.49
200	6.52	6.75	6.62	6.63	1.33

The relationship between T and L is given by one of the following equations. Which one?

1. $T = A\sqrt{L}$ 2. $T = \dfrac{A}{\sqrt{L}}$ 3. $T = \dfrac{A}{L^2}$

From the trend shown by the results it is clear that 1 is the correct relationship. As the length L increases the period T also increases. Equations 2 and 3 indicate that as L increases T would decrease.

To draw a suitable straight line graph from the results to find the constant A then $\sqrt{L}$ should be plotted on the x–axis and T on the y–axis. A new table containing the appropriate values is then produced.

$\sqrt{L}$ / $mm^{1/2}$	14.14	15.81	17.32	18.71	20.0
T/s	1.33	1.49	1.65	1.77	1.90

The graph obtained using these values is shown below.

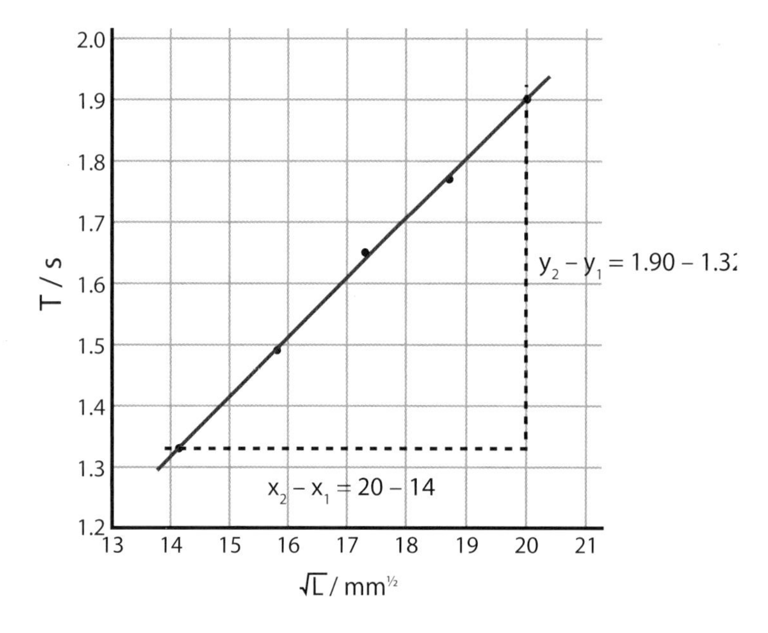

To find the value of A we need to take the gradient of this line. Why?

The relationship between L and T is given by $T = A\sqrt{L}$. This corresponds to $y = mx$, the equation for a straight line that passes through the origin (0,0). Note that in this case there is no requirement to find the intercept so there is no need to plot the graph from the origin.

By comparing the two equations we see that $y \equiv T \quad x \equiv \sqrt{L}$ and the gradient $m \equiv A$.

$$\text{The gradient} = \frac{y_2 - y_1}{x_2 - x_1} = \frac{1.90 - 1.32}{20.0 - 14.0} = \frac{0.58}{6.0} = 0.097$$

The gradient may have units and in this case it has the units of s mm$^{-½}$.

Graphs

Graphs are commonly used to show the results of experiments. Graphs allow you to deduce relationships much more quickly than using a table. They provide a visual picture of how two quantities depend on each other: they show up anomalous readings and, if straight lines, the gradient can be used to find an average value of the ratio of the two quantities.

Dependent and Independent Variables

Plot the independent variable (the one you have been changing) along the horizontal axis and the dependent variable along the vertical axis. The exception to this rule occurs where you need to plot a particular graph to find a required quantity. For example, in the case of stretching a spring, the equation $F = kx$ applies. F is the force, x is the extension and k is the spring constant. In this instance F is the independent variable but to find the spring constant, the force F is plotted along the y–axis and the extension along the x–axis, because the gradient is then the spring constant.

Labels and Units

Label both axes to show the quantity that is being plotted.

Indicate on the axes the unit of measurement used for the quantity. Sometimes the quantity may just be a number so a unit is not required.

Scales

Choose scales on the axes to make the plotting of values simple. Generally this means letting 10 small divisions on the graph paper equal 1, 2, 5, 10 or some multiple of these numbers. Do not make life difficult for yourself by letting small divisions equal 3 or 7. This will take you longer to plot the graph, increase your chances of mis-plotting points and makes it difficult for others to read the data.

Choose the range of the scales on the axes so that the points are spread out. As a general rule the graph you draw should fill at least three quarters of the graph paper grid in both the x and y directions.

Plotting Points

Plot the results clearly, and use a sharp pencil rather than a pen. Pencil is much easier to erase should you make a mistake in plotting. Use crosses or dots with circles around them.

Lines and Curves

The graphs that you will encounter during an A level course will generally represent a smooth variation of one quantity with another so a smooth curve or straight line will be appropriate. Draw a best fit line, which may be a smooth curve or a straight line that passes through or close to all your points as shown below. In general you should **not** join the points with short straight lines.

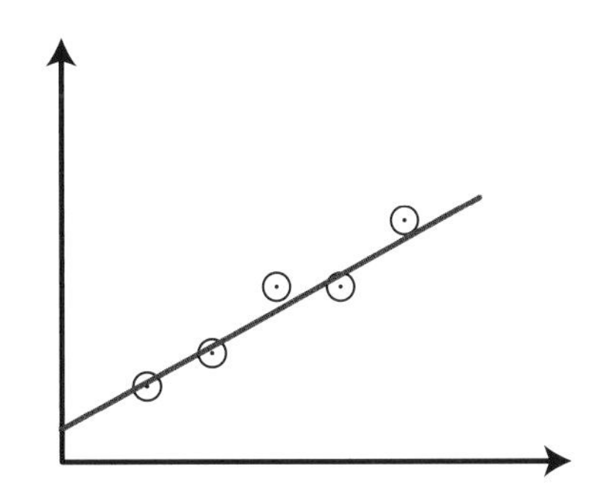

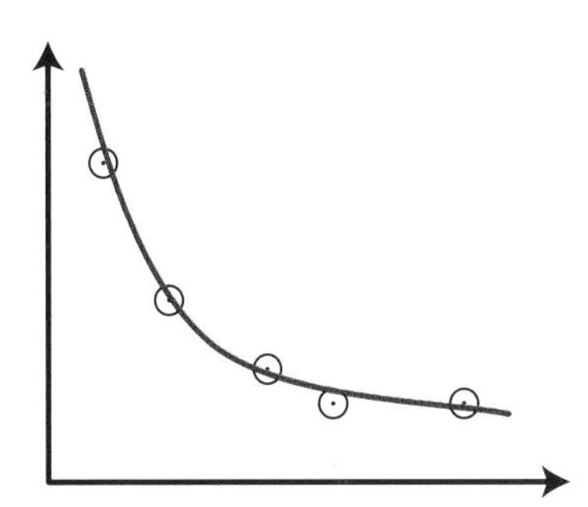

Worked Examples

1 A ball rolls from rest down a sloped plane. Measurements are made of the distance travelled along the slope, d, and the time taken, t.

The relationship between d and t is: $d = \frac{1}{2}at^2$ where 'a' is the acceleration of the ball.

(a) What **straight line** graph would you plot to display your results?

(b) How could you find the acceleration from the graph?

Solution

The equation of a straight line that passes through the origin (0,0) is $y = mx$.

By comparing this with the relationship above we see that d should be plotted on the y–axis and t^2 on the x–axis.

In the equation for the straight line m represents the gradient. The gradient of the graph of d against t^2 is equal to $\frac{1}{2}a$.

2 The unknown e.m.f. of a cell, E, is linked to the terminal voltage, V, and the current I by the equation $E = V + Ir$ where r is the unknown internal resistance of the cell.

In an experiment, corresponding values of V and I are recorded as the resistance in an external circuit is changed. The e.m.f. and the internal resistance are both constant.

(a) What **straight line** graph would you plot? Which variable would be on the vertical axis?

(b) Draw a sketch of the graph you would expect to obtain.

(c) How would you find the e.m.f. of the cell and the internal resistance from this graph?

Solution

First the equation has to be arranged so that V becomes the subject of the equation.

This gives $V = E - Ir$. The equation of a straight lines is $y = mx + c$. By comparing these two equation we see that V should be plotted on the y–axis and I on the x–axis. The intercept on the y–axis is c and in this case it will give us a value for E. The gradient of the graph is negative and will give a value for r.

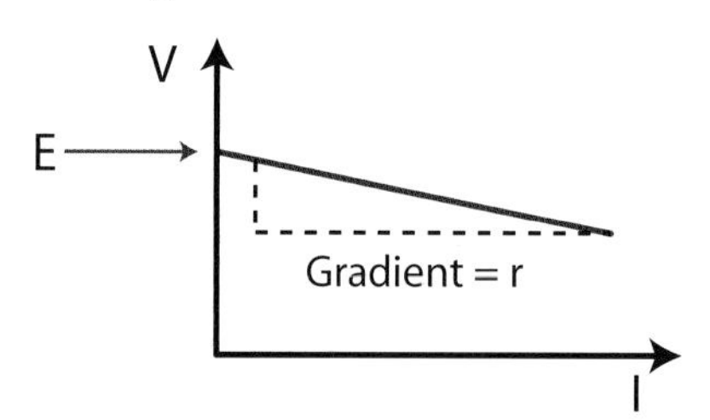

3 The focal length f, of an inaccessible lens (inside a cylinder) can be found by a technique called 'the displacement method'. The distance between an illuminated object and a screen is measured, this is s, as shown in the diagram. The cylinder containing the lens is moved until a sharp image is obtained on the screen. The position of the cylinder is noted. The cylinder

is moved again until a new image on the screen is obtained. The distance between the two positions of the cylinder containing the lens is found, this is d, as shown in the diagram.

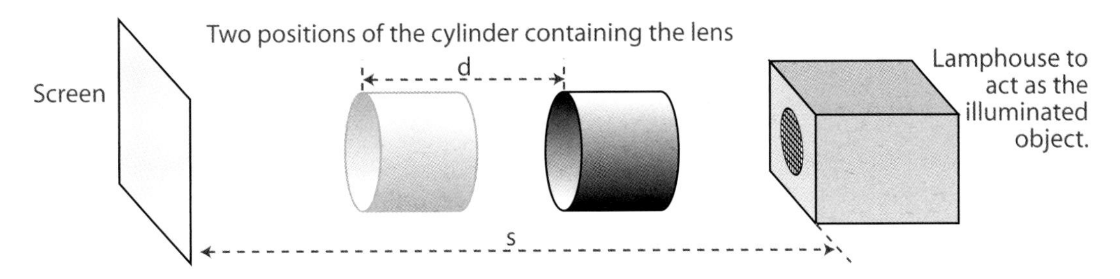

The mathematical relationship between the quantities is $f = \frac{s^2 - d^2}{4s}$

Below is a table of results of s and d.

s in cm	50.0	54.0	58.0	62.0	66.0	70.0
d in cm	22.4	27.5	32.3	36.9	41.4	45.8

(a) What **straight line** graph should you plot so that the gradient can be used to calculate the focal length of the lens?

(b) Copy the table above and calculate the values required to plot the graph. Insert the values in the appropriate spaces in your table and add appropriate labels.

(c) Plot the graph, labelling carefully the units on each axis and use it to find the focal length of the lens.

Solution

Rearrange the equation to give

$s^2 = d^2 + 4sf$

Divide both sides by s

$s = \frac{d^2}{s} + 4f$

Plot *s* on the y–axis and $\frac{d^2}{s}$ on the x–axis. The intercept is equal to 4f.

Alternatively, rearrange the equation to give

$4fs = s^2 - d^2$

$(s^2 - d^2) = 4fs$

Plot $(s^2 - d^2)$ on the y–axis and s on the x–axis. The slope is 4f.

Combining Uncertainties

Often the aim of an experiment is to find the value of a quantity that depends on the measurement of several quantities, each with its own associated uncertainty.

In an experiment to determine the resistivity of a material the diameter of a wire is measured. Then that value is used to calculate the area of cross section of the wire. What is the uncertainty in the calculation of the area of cross section? There are two methods of dealing with this.

Method 1: Maximum, Minimum and Range

In this method the first thing to do is to calculate the area using the best value of the diameter and using the maximum and minimum values as determined by the measuring instrument. The average diameter of the wire was measured as 0.32 mm using a micrometer gauge giving an uncertainty of ± 0.01 mm.

$$A_{\text{Best}} = \frac{3.142 \times (0.32 \times 10^{-3})^2}{4} = 8.0 \times 10^{-8} \text{m}^2$$

$$A_{\text{Min}} = \frac{3.142 \times (0.31 \times 10^{-3})^2}{4} = 7.55 \times 10^{-8} \text{m}^2$$

$$A_{\text{Max}} = \frac{3.142 \times (0.33 \times 10^{-3})^2}{4} = 8.55 \times 10^{-8} \text{m}^2$$

This gives a range of 1.0×10^{-8} in the values for the area. The calculated value for the area can be written as: $(8.0 \pm 0.5) \times 10^{-8}$ m^2

Method 2: Combining Percentage Uncertainties

Consider a measured quantity A and its associated uncertainty ΔA, and another measured quantity B with its associated uncertainty ΔB. X is the quantity we wish to measure and its uncertainty is ΔX. The value of X may be found by combining the values of A and B in various ways. For each way that A and B could be combined the final uncertainty ΔX is simply the sum of the uncertainties in A and B. This is a simplified version of a statistical method that is applicable to AS and A2 Physics.

How X is found from A and B	Final uncertainty in X
$X = A \times B$	$\Delta X = \Delta A + \Delta B$
$X = A \div B$	$\Delta X = \Delta A + \Delta B$
$X = A \times B^n$	$\Delta X = \Delta A + n\,\Delta B$
$X = A \div B^n$	$\Delta X = \Delta A + n\,\Delta B$
$X = kA^n$ where k is a constant	$\Delta X = n\,\Delta A$

The uncertainties, ΔA and ΔB, are best quoted as percentages of the measured values A and B. The uncertainty ΔX is then a percentage of the final measured value.

For example, if the period of an oscillating pendulum is 1.1 s with an uncertainty of ± 0.1 s, the percentage uncertainty in this measurement is $(0.1 \times 100) \div 1.1 = \pm 9\%$

The numerical examples below show how the rules shown in the table are applied.

Example 1 Measurement of Resistance

The resistance of a length of wire is found by measuring the current passing through it and the potential difference across it. In such an experiment the values of these quantities and their uncertainties were found to be V = 5.2 ± 0.2 and I = 1.2 ± 0.1.

$$\Delta V = \frac{0.2 \times 100}{5.2} = 3.8\% \quad \Delta I = \frac{0.1 \times 100}{1.2} = 8.3\%$$

$$R = \frac{V}{I} \quad \Delta R = \Delta V + \Delta I = 3.8\% + 8.3\% = 12.1\%$$

$$R = \frac{5.2}{1.2} = 4.3\ \Omega \pm 12.1\% = (4.3 \pm 0.5)\ \Omega$$

Example 2 Measurement of Density

The diameter of a steel ball was measured using a micrometer gauge and found to be 8.65 mm and the uncertainty was ± 0.01 mm. The mass of the steel ball was found using an electronic balance, the measured value being 2.82 g with an uncertainty of ± 0.01 g. What is the density of the steel?

The volume of the sphere is given by $\frac{\pi d^3}{6}$, d being the diameter of the sphere.

The uncertainty in the measurement of the diameter is $\Delta d = \frac{0.01 \times 100}{8.65} = \pm 0.12\%$

However, the uncertainty in the volume ΔV is three times this since diameter has to be cubed to find the volume.

The uncertainty ΔV = ± 0.36%

$$\text{Density} = \frac{\text{Mass}}{\text{Volume}} = \frac{2.82 \times 10^{-3}}{3.39 \times 10^{-7}} = 8318\ \text{kg m}^{-3}$$

$\Delta m = 0.01 \times 100\% \div 2.82 = 0.35\%$

The uncertainty in the density $\Delta D = \Delta m + \Delta V = 0.35\% + 0.36\% = \pm 0.71\% = \pm 59\ \text{kg m}^{-3}$

The final value for the density can be quoted as $(8318 \pm 59)\ \text{kg m}^{-3}$

Exercise 45

In an experiment to measure the resistance of a piece of wire a voltmeter capable of measuring a maximum potential difference of 10 V was used. The smallest division on the scale was 0.2 V. The ammeter used was capable of measuring current as large as 1.0 A and the smallest division on its scale was 0.1 A. The experiment yielded readings of 4.9 V and 0.3 A.

Calculate the resistance of the wire along with the uncertainty in its value. Use the percentage method to determine the uncertainty in the resistance.

Uncertainties from Graphs

The slope or gradient of a graph provides a means of determining an average value for a physical quantity. The intercept on either the x– or y–axis is dependent on the slope. A small change in the slope can produce a large change in the value of the intercept.

The points plotted may not all lie on a straight line. It may be necessary to judge the best fit line.

The slope of the best fit line will give you the best value for a physical quantity and the intercept on the appropriate will give you the best value for this quantity. The placing of this line of best fit can be aided by calculating the average x value and average y value and plotting this point. This is known as the **centroid** and the line of best fit is drawn so that it passes through this point.

To estimate the uncertainty in the slope and the intercept, follow the procedure outlined below:

Draw the line of best fit as outlined above.

Now draw two more lines, one of maximum slope and one of minimum slope through the plotted points. The gradients of these two lines will give you a maximum and minimum value for the slope. The difference between these two values gives you a range and the uncertainty can be taken as half the value of the range.

Similarly the line of best fit will give best value for the intercept. The range of the intercept values can be found from where the lines of maximum and minimum slope cut the appropriate axis. The uncertainty in the intercept value is again half the range.

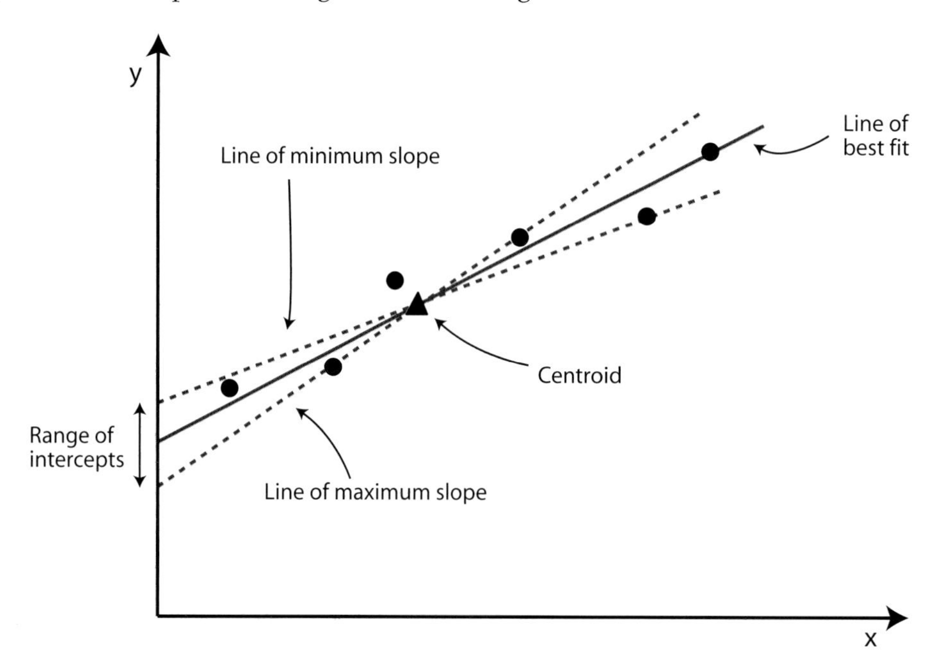

Unit 7
Synoptic Assessment

7.1 Synoptic Assessment

The A-level Physics assessment expects students to show that they have an overall understanding of the Physics they are studying. Synoptic assessment will test your grasp of the connections between the different topics at AS and A2. Synoptic assessment will be assessed in modules A2, 1 A2, 2 and in the planning design question which is part of the practical assessment. The details of the assessment of the A2 modules is shown below.

A2 1 Momentum, Thermal Physics, Circular Motion, Oscillations and Atomic and Nuclear Physics

A written examination, consisting of a number of compulsory short answer questions, some of which will afford the opportunity for extended writing.

Elements of synoptic assessment will be embedded within the questions, and in a data analysis question.

A2 2 Fields and their Applications

A written examination, consisting of a number of compulsory short answer questions, some of which will afford the opportunity for extended writing.

Elements of synoptic assessment will be embedded within the questions, and in a longer question drawing together different strands of the specification.

A2 3 Practical Techniques

A test of practical skills, consisting of two experimental tests (40 marks) and one question on planning and design (20 marks).

Synoptic assessment will be assessed through the planning and design question.

Synoptic assessment requires you to demonstrate that you can:

- build on the material you covered in the AS modules,
- make and use connections between different areas of physics,
- apply your knowledge and understanding of more than one area of physics to a particular situation or context,
- use your knowledge and understanding of principles and concepts of physics in planning experimental work and in analysing and evaluating data,
- use ideas and skills which are found through physics.

In this section you will cover data analysis, planning and design and look at some of the longer questions that have been asked in previous years.

Data Analysis

Significant Figures

If an experiment provides a numerical value for a physical quantity then it is good practice when reporting your measurement to give:

- the measured value for the quantity,
- the uncertainty in your measurement,
- the appropriate unit.

The numerical value for the quantity and the uncertainty should be quoted to an appropriate number of significant figures.

Let us review the rules concerning significant figures.

The number of significant figures in a number is found by counting all the digits from the first non-zero digit on the left. A zero between two non-zero digits is significant.

Here are some examples:

845.470 has six significant figures. You start counting from the 8 which is the first non-zero digit on the left. The last, or trailing, zero has been considered significant otherwise it would not have been necessary to include it.

0.0516 has three significant figures, counting from the 5 which is the first non-zero digit on the left. The leading zeroes are essential to give the magnitude of the number.

Rounding

Rounding involves reducing the number of significant digits in a number. The result of rounding is a number having fewer non-zero digits yet similar in magnitude. The result is less precise but easier to use.

For example: 23 rounded to the nearest ten is 20, because 23 is closer to 20 than to 30.

The procedure for rounding is:

- Decide how many significant figures you want,
- Decide which is the last digit to keep,
- Increase it by 1 if the next digit is 5 or more (this is called rounding up),
- Leave it the same if the next digit is 4 or less (this is called rounding down).

Example 1

Write 8.143 to 3 significant figures .... Answer **8.14**

Solution

8.143 ← This is less than 5 so 3rd significant figure is not changed.

↑ 3rd significant figure

This is an example of rounding down

Example 2

Write 6.245 to 3 significant figures .... Answer **6.25**

Solution

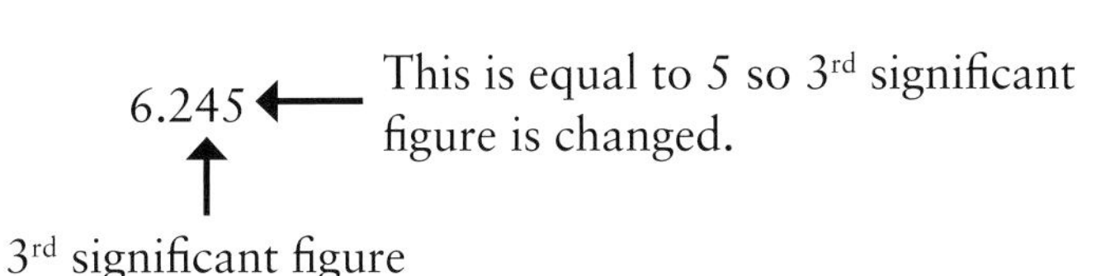

This is an example of rounding up

Note!

When you have to perform calculations on a set of measurements then the result should be given to the same number of significant figures as the initial values.

For example, $3.25^2 = 10.5625$, but this should be quoted as 10.6.

Exercise 46

1 To how many significant figures are each of the following numbers quoted to?

(a) 273.16

(b) 9.81

(c) 3.1412

(d) 0.003450

(e) 0.1001

2 Re-write each of the following values to the number of significant figures given in brackets alongside each.

(a) 645.5701 (4)

(b) 0.0125 (1)

(c) 1678 (2)

(d) 1.245 (2)

Scientific Notation

This **is** sometimes known as **standard form.** It provides us with a way of writing numbers that are too large or small to be conveniently written in standard decimal notation. In scientific notation, numbers are written in the form:

$$a \times 10^b$$

← Exponent – this is an integer

↑ Coefficient – this is a real number

In the **normalized** form, a, the coefficient has value between 1 and 10.

For example, 1,250,000 would be written as 1.25×10^6.

For numbers less than 1, the exponent has a negative value.

For example, 0.00035 would be written as 3.5×10^{-4}.

Significant Figures and Scientific Notation

The same rules for significant figures apply to numbers expressed in scientific notation.

Remember that in the normalized form of scientific notation, leading and trailing digits do not occur, so all digits are significant.

For example, 0.00011 (two significant figures) becomes 1.1×10^{-4}, and 0.000111500 (six significant figures) becomes 1.11500×10^{-4}.

For example, if we quote 1600 to four significant figures it is written as 1.600×10^{3}.

However, is we quote 1600 to just three significant figures it is written as 1.60×10^{3}.

Exercise 47

1 Write each of the following in normalized scientific notation. The number in brackets is the number of significant figures you should quote the value to.

(a) Density of mercury 13552 kg m^{-3} (3)

(b) Speed of light 299 792 458 ms^{-1} (2)

(c) Expansion coefficient of brass 0.0000193 K^{-1} (2)

Orders of Magnitude

The order of magnitude of a physical quantity is the nearest power of ten to the value of the quantity.

For example, the order of magnitude of 8.85×10^{5} is **10^{6}**.

Exercise 48

1 Give the order of magnitude of the following quantities.

(a) The mass of the earth = 6.0×10^{24} kg

(b) The universal gravitation constant G = 6.67×10^{-11} $m^3kg^{-1}s^{-2}$

(c) The speed of sound in air = 333 ms^{-1}

Significant Figures and Uncertainties

When reporting the value of some physical quantity, obtained through experimental work, you should be aware of the uncertainties associated with any measurements used. This will, in turn, determine the number of significant figures to which you should quote your final value.

For example, say you have measured a block of metal (shown on the right) and used the values to calculate its volume.

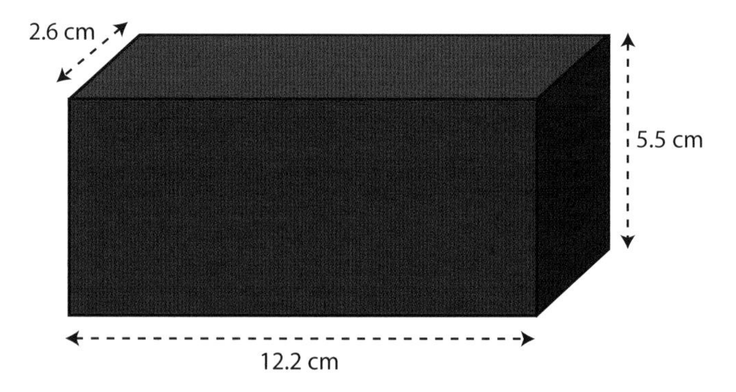

The calculated volume of the block = 12.2 × 5.5 × 2.6 = 174.46 cm^3. However, this ignores the uncertainty in each measurement of length. A ruler was used, so the uncertainty in each measurement could be 0.1 cm. This uncertainty is due the scale of the ruler and the skill of the person taking the measurement. The smallest division on the ruler is 1 mm and the person taking the measurements judges that they can read the scale to ± ½ the smallest division. Since two readings are taken and then subtracted to find the length the uncertainty becomes ± 1 mm.

Taking this into account the volume of the block would have the range of values shown below:

Minimum volume = 12.1 × 5.4 × 2.5= 163.35 cm^3

Measured volume = 12.2 × 5.5 × 2.6 = 174.46 cm^3

Maximum volume = 12.3 × 5.6 × 2.7 = 185.98 cm^3

The actual volume of the block is between 163 cm^3 and 186 cm^3.

Considering the uncertainties in this simple measurement it is inappropriate to quote the volume of the block to five significant figures. It is much more appropriate to quote the result to three significant figures, since the measurements of length are given to 3 significant figures, i.e. 174 cm^3.

Considering the difference between the maximum and minimum values for the volume the measurement of the block's volume should be given as (174 ± 12) cm^3.

Exercise 49

1 A capacitor initially uncharged is connected to a battery. The voltage across the capacitor is measured as it charges. The results are shown below.

Voltage/V	0.00	1.66	3.65	5.97	7.44	8.97
Time/s	0.00	2.00	5.00	10.00	15.00	25.00

State the number of significant figures to which the voltage readings are stated.

[CCEA 2007]

2 The energy E of alpha particles is often given in MeV. An experimenter requires to calculate the reciprocal of the square root of the energy, i.e. $E^{-\frac{1}{2}}$.

Copy and complete the table below by calculating the value of $E^{-\frac{1}{2}}$, giving your answer to three significant figures.

E/MeV	$E^{-\frac{1}{2}}$ / $MeV^{-\frac{1}{2}}$
2.00	
4.00	
7.50	
9.00	

[CCEA 2004]

3 In the study of atomic line spectra it is often convenient to use the wavenumber of the spectral line. The wavenumber is calculated as $1/\lambda$, where λ is the wavelength in mm. The wavenumbers for some of the spectral lines emitted by sodium are shown below.

Wavenumber/mm^{-1}	3500	3730	3850	3930

(a) To how many significant figures are the wavenumbers quoted?

(b) The wavelength of spectral line X is 251 nm. Calculate its wavenumber to the correct number of significant figures.

[CCEA 2004]

Obtaining Straight Line Graphs

In many of the experiments you undertake, a graph of one of the variables you measured against the other measured variable will not give a straight line. The graph obtained will often be a curve. Obtaining a straight line graph will require some manipulation of the measured quantities. If we have an equation relating the quantities the task is simpler. Let us begin with the example below.

The viscosity η of a liquid is a measure of the fluid friction an object experiences when moving through the liquid. The viscosity of a clear, thick liquid such as glycerin can be found by dropping ball bearings through a tall cylinder full of the liquid. The ball bearings quickly reach their terminal constant velocity which can be measured. The radius of the ball bearing also affects this terminal velocity.

The viscosity of the liquid η is given by:

$$\eta = \frac{kr^2}{v}$$

The radius of the ball bearing is r, v is the terminal velocity and k is a constant of known value.

The value of η can be found by plotting a suitable linear graph. The equation for a straight line is y = mx + c or, if it passes through the origin, y = mx.

To decide what graph to plot, the first step is to re-arrange the viscosity relationship so that v is the subject of the equation and compare this to the equation for a straight line.

$$v = \frac{kr^2}{\eta}$$

$$y = mx$$

Plotting a graph of v on the y axis and r^2 on the x–axis will give a straight line, the gradient of which equals $\frac{k}{\eta}$.

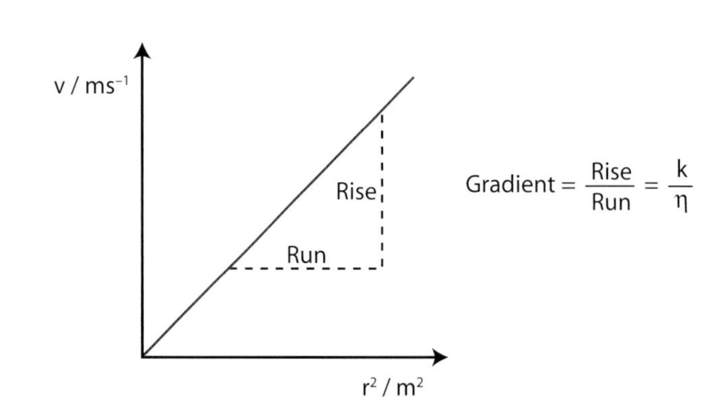

Worked Example

An obstructed pendulum

An obstructed pendulum is one where the swing is obstructed by an object in the path of the swing. The arrangement is shown in the diagram on the right.

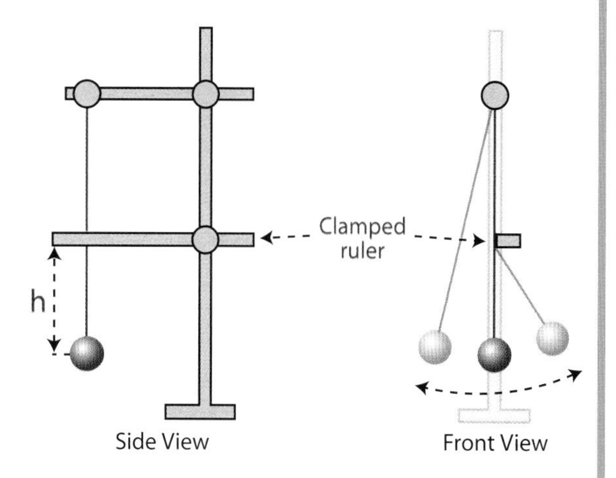

In this experiment the length h is altered and the period time T for different values of h is measured.

The acceleration due to gravity g can be determined from this experiment. The relationship between g, T and h is shown below. A is a constant.

$$g = \frac{\pi^2 h}{(T - A)^2}$$

This relationship needs to be re-arranged so that T becomes the subject of the equation. In other words, we need to have it in a form that satisfies the general equation for a straight line y = mx + c.

$$g(T - A)^2 = \pi^2 h$$

Taking the square root of each side and re-arranging:

$$T - A = \frac{\pi\sqrt{h}}{\sqrt{g}}$$

Finally we have:

$$T = \pi\sqrt{\frac{h}{g}} + A$$

This also has the form y = mx + c. A graph of T (y–axis) against $\sqrt{h}$ (x–axis) will give a straight line. The gradient gives $\frac{\pi}{\sqrt{g}}$ and the intercept on the y axis gives the constant A.

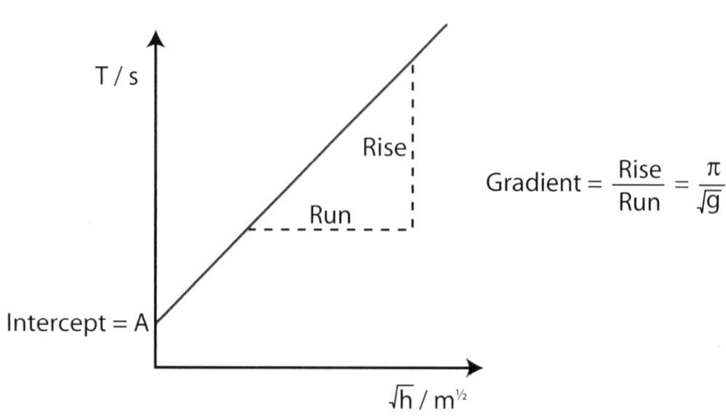

Exercise 50

1 In the study of atomic line spectra is often convenient to use the wavenumber of the spectral line. The wavenumber is calculated as $1/\lambda$, where λ is the wavelength in mm. The wavenumbers for some of the spectral lines emitted by sodium are shown below.

Spectral line number N	3	4	5	6	8	9
Wavenumber/mm^{-1} W	3500	3730	3850	3930	4010	4140

The wavenumbers W of the spectral lines are related by $W = A - \frac{B}{(N+1)^2}$.

A and B are both constants.

(a) What linear graph should you plot in order to find the constants A and B?

(b) Explain how both constants are found from the graph.

(c) On graph paper, plot the points and draw the best straight line through them.

(d) Find the values of A and B and give the appropriate unit for each.

[CCEA 2004]

2 The resonance of sound waves in a tube closed at one end can be used to determine the speed of sound in air. The diagram below shows the apparatus. The frequency f of the sound emitted by the loudspeaker is set at a particular value and the length L of the air column varied until resonance occurs. This is detected when the sound increases considerably. This process is repeated for a range of frequencies.

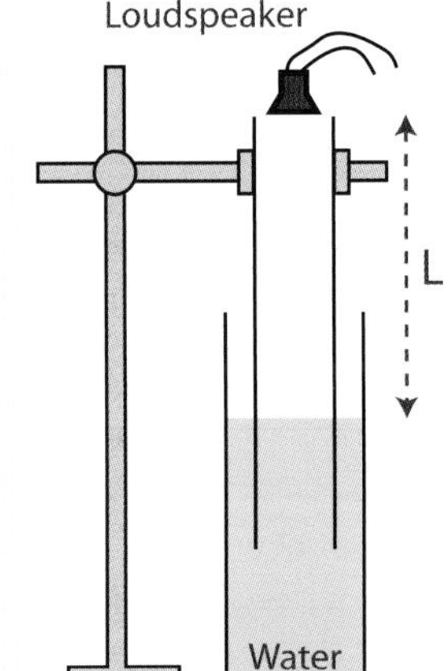

The velocity of sound is given by the relationship:

$v = 4Lf$

where L is the length of air column, in m, at which resonance is detected

f is the frequency of the sound, in Hz

(a) Re-arrange the equation into the form a straight line with L and f as the variables.

(b) What graph should be plotted? State what should be plotted on the x– and y–axes.

(c) Describe how you would use the graph to find the velocity of sound in air.

3 The the relationship between the resistance of a metal R in Ω and its temperature θ in °C is given by the expression:

$$R = R_0(1 + \alpha\theta)$$

where R_0 is the resistance of the metal at 0°C

α is a constant known as the temperature coefficient of resistance

(a) Re-arrange the expression so that a linear graph can be obtained from the measured values of R and θ.

(b) Sketch the graph that would be obtained, labelling each axis.

(c) Explain how R_o and α can be obtained from the graph, giving the units of each.

4 The characteristic lines in the X-ray spectra of elements may be analysed using a Bragg spectrometer. This instrument uses a crystal as a diffraction grating for the X-rays.

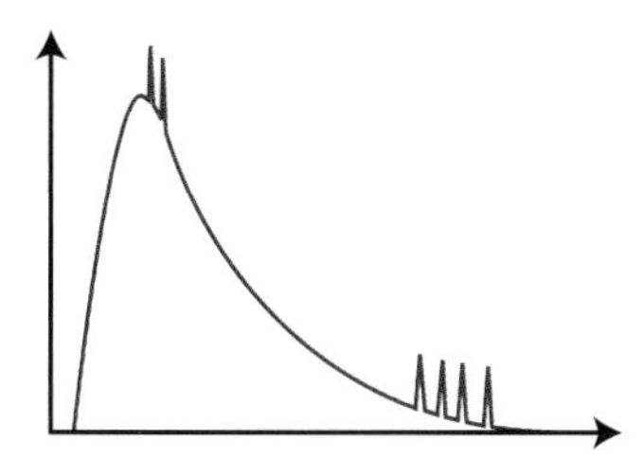

The Bragg diffraction law for these lines is expressed by:

$$2d \sin \theta = \lambda$$

where d is the spacing of the atomic planes in the crystal
θ is the observed angle of diffraction
λ is the wavelength of the X-ray emission line

An experimenter recorded values of θ for a number of elements as shown below.

Element	chromium	molybdenum	tin	europium	tungsten
θ/°	23.72	7.260	5.080	3.170	2.280

(a) State the number of significant figures to which the data is recorded.

(b) The vernier protractor used to measure the angles produces an uncertainty of ±0.01°. However the experimenter has recorded all the angles to the same number of significant figures. Explain why this procedure is not meaningful considering the uncertainty in each measurement of the angles.

(c) State the element(s), the value(s) of which require modification.

[CCEA 2006]

Use of Logarithms

A logarithm is defined as the power to which a base must be raised to produce a given number. Two bases are common: base 10 and base e (2.71828). The notation **lg** or **lg_{10}** is taken to mean logarithm to the base 10, and the notation **ln** is taken to mean the base e.

For example:

$\lg 1000 = 3.0$ since $10^3 = 1000$

$\lg 0.01 = -2.0$ since $10^{-2} = 0.01$

$\ln 5 = 1.6094$ since $2.71828^{1.6094} = 5$

If the relationship under investigation involves a power then the use of logarithms becomes an essential tool in the analysis of the data.

The diagram represents a loaded cantilever: a metre rule clamped at one end with a mass **M** attached at the other end. When the free end is displaced and released the metre rule oscillates.

The periodic time T of these oscillations is given by a relationship of the form $T = kM^n$, where k and n are constants.

Since the only variables in this expression are the mass M and the period T clearly the investigation involves varying the mass M and measuring the corresponding period T. To process the data to determine the value of the power n requires us to take the logarithms ($\log_{10}$) of each side of the expression:

$$T = k \quad M^n$$
$$\lg T = \lg k + n \lg M$$
$$\uparrow \qquad \uparrow \qquad \uparrow \; \uparrow$$
$$y = c + m \; x$$

Comparing this with the general equation for a straight line you can see that lg **T** plotted on the y–axis against lg **M** on the x–axis will give a straight line, while the intercept of the y–axis will give lg **k** and the gradient will give the power **n**.

Worked Example

CCEA 2007 (modified)

The aerodynamic drag force experienced by motorcycles is an important factor when considering their design. The factors affecting the amount of drag on a motorcycle can be investigated by carrying out experiments in the controlled environment of a wind tunnel. Wind tunnel experiments have shown that the aerodynamic drag force F_D on an object depends on a number of variables, according to the equation below:

$$F_D = \frac{\rho A C_D}{2} v^n$$

where C_D is the drag coefficient
A is the frontal area of the motorcycle
ρ is the density of the air in the wind tunnel
v is the speed of the air relative to the object
n is the a constant

In a wind tunnel experiment, the variation in drag force with speed was investigated. The results of the wind tunnel experiment are shown on the next page.

1 One of the values in the column headed F_D has been quoted to an inconsistent number of significant figures. How should this value have been recorded?

v/ms^{-1}	F_D/N		
10	17.0		
15	38.50		
20	68.3		
25	106		
30	154		
35	209		

Solution

All the velocity values have been recorded to two significant figures. All but one of the drag force values have been recorded to three significant figures, i.e. apart from 38.50.

This should be recorded as **38.5.**

2 Process the expression for the drag force F_D so that a linear graph can be obtained from which the value of n can be determined. State what should be plotted on the y–axis and on the x–axis.

Describe how the value of n can be found from this graph.

Solution

Taking logarithms to the base 10 of each side of the above expression and comparing the result with the general equation for a straight line graph gives:

$$\begin{array}{ccccc} \lg_{10} F_D & = & \lg_{10}\left(\dfrac{\rho A C_D}{2}\right) & + n & \lg_{10} v \\ \uparrow & & \uparrow & \uparrow & \uparrow \\ y & = & c & + m & x \end{array}$$

$\lg_{10}F_D$ should be plotted on the y–axis and $\lg_{10}v$ plotted on the x–axis. The gradient of the straight line gives the value of n.

3 Insert suitable column headings in the blank columns of the table. Complete the columns, quoting the data to three significant figures.

Solution

$\lg_{10}30 = 1.47712$

but to 3 significant figures it becomes 1.48

v/ms^{-1}	F_D/N	$\lg_{10} v$	$\lg_{10} F_D$
10	17.0	1.00	1.23
15	38.5	1.18	1.59
20	68.3	1.30	1.83
25	106	1.40	2.03
30	154	1.48	2.19
35	209	1.54	2.32

4 Plot the appropriate graph and use it to find the value of n in the above expression for the drag force F_D.

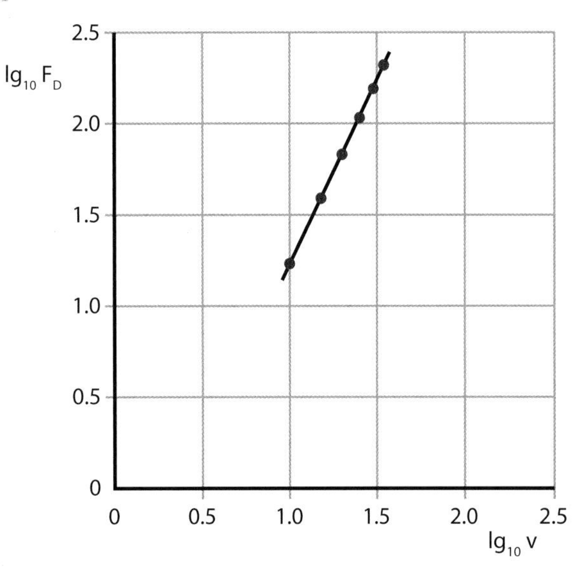

The gradient of the line = 2

The value of n in the expression for the drag force is therefore **2**

Note!

When plotting a log quantity it is simply a number; it does not have a unit. Notice that in the graph above units are **not** shown for the x and the y axis.

Exercise 51

1 A light dependent resistor (LDR) is a semiconductor device, the resistance of which varies as the light falling on it varies. The variation of resistance of an LDR with illumination is given by:

$$R = R_0 E^{-n}$$

where R is the intensity measured in Ω

E is the illumination measured in a unit called lux

R_o and n are constants

(a) Using logarithms arrange the above expression so that a linear graph can be drawn.

(b) Sketch the graph, label each axis and explain how R_o and n can be obtained from the graph.

[CCEA 2008]

2 Rayleigh scattering is the scattering of light or other electromagnetic radiation by particles much smaller than the wavelength of the light. It can occur when light travels in transparent solids and liquids, but is most prominently seen in gases. Rayleigh scattering of sunlight in clear atmosphere is the main reason why the sky is blue.

For particles of a specific size the intensity I of the scattered light depends on the wavelength λ of the light. The relationship between I and λ is given below.

$$I = A\lambda^n$$

(a) Using logarithms arrange the above expression so that a linear graph can be drawn.

(b) Sketch the graph, label each axis and explain how **A** and **n** can be obtained from the graph.

Exponential or Natural Logarithms

Exponential or natural logarithms are to the base e. The quantity e is associated with systems that grow or decay exponentially and continuously. Examples are radioactive decay, capacitor charge and discharge. The value of e is 2.71828.

Worked Example

When a large number of identical particles are suspended in a liquid, they tend to settle in the way shown in the diagram. There are many particles at the bottom of the liquid column, but progressively fewer towards the top of the column.

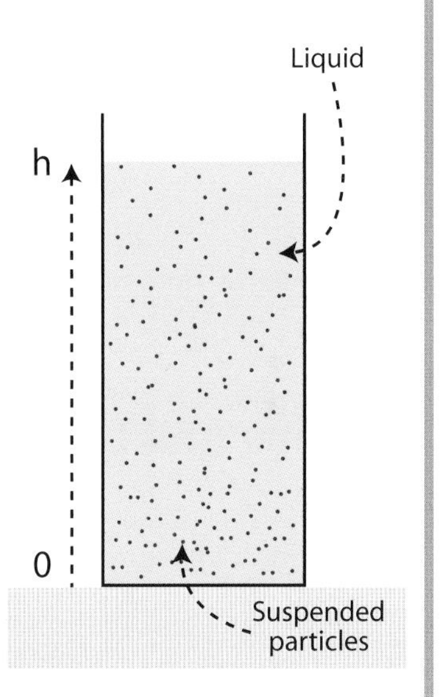

The number density n of particles at a height h above the bottom of the liquid column is given by:

$$n = n_0 e^{-\frac{Wh}{kT}}$$

where n_0 is a constant
W is the weight of the particle
k is Boltzmann's constant
T is the temperature in Kelvin

(a) What name is given to the mathematical function represented by the equation above?

Solution Exponential

(b) The quantity n is the number density of the particles. Explain what this means.

Solution The number of particles per m^3.

(c) What is the physical interpretation of the constant n_0 in the above equation?

Solution $n = n_0$ when h=0, so n_0 is the number density if all the particles were at the base of the column.

(d) Using exponential (natural) logarithms arrange the above equation so that a linear graph can be drawn.

Solution

$$\ln n = \ln n_0 - \frac{W}{kT} h$$

$$\uparrow \quad \uparrow \quad \uparrow \uparrow$$

$$y = c + m \, x$$

Note

$\ln e = 1$ so: $\ln(e^{-\frac{Wh}{kT}}) = -\frac{Wh}{kT}$

(e) Sketch the graph, label each axis.

Solution

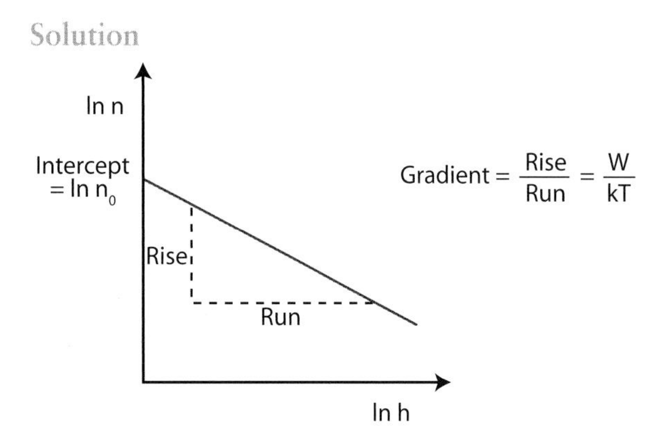

(f) What will the gradient of this graph provide?

Solution $\frac{W}{kT}$

Planning and Design

In this question in the assessment of Practical Techniques (A2, 3) you will be expected to identify and define a practical problem using information provided in the question from your knowledge of Physics.

You will be expected to show that you can choose effective and safe procedures after consideration of appropriate qualitative and quantitative methods for collecting data.

You will also be expected to demonstrate that you can select suitable apparatus for the measurement of a physical quantity. The list of apparatus you are expected to be familiar with is: mass (spring and top-pan balances), length (rule, vernier scale, micrometer, callipers), liquid volume (graduated cylinder), angle (protractor), time (clock, stopwatch, calibrated timebase of cathode ray oscilloscope), temperature (thermometer), electric current (ammeter), potential difference (voltmeter, cathode ray oscilloscope), and you should be aware of the use of ICT for data capture.

Worked Example

CCEA June 2008

Traditionally, bronze 2p British circulation coins were made from an alloy of copper, tin and zinc. However, in 1992 the Royal Mint changed the material of these coins to copper-plated steel.

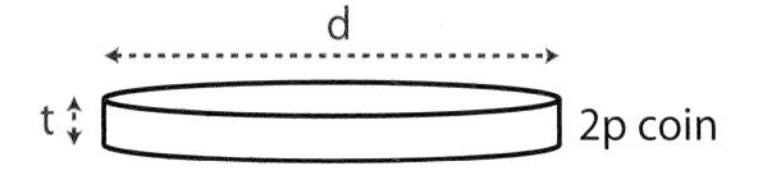

In this question you are to discuss a plan to find the density of a pre-1992 2p coin and a post-1992 2p coin.

The density of a material may be calculated using:

$$\text{density} = \frac{\text{mass}}{\text{volume}} \quad \textbf{Equation 1}$$

It is proposed to measure the mass of each coin with a top-pan balance.

1 The volume V of a disk of diameter d and thickness t can be determined using:

$$V = \tfrac{1}{4}\pi d^2 t \quad \textbf{Equation 2}$$

(a) Suggest an appropriate instrument which may be used to measure the thickness of the coins with precision. Give the uncertainty in measurements made with this instrument.

Solution **Instrument** – A micrometer gauge is best suited to such a thickness

Uncertainty – The thickness can be measured to an accuracy of ± 0.01 mm

(b) (i) What feature of a coin makes the measurement of the thickness uncertain?

Solution The coins have a pattern embossed on the coins as shown in the photograph. This means they are not of uniform thickness.

(ii) Suggest how you would take this feature into account when taking measurements.

Solution It is always a good scientific method to take a number of measurements and find their average, and the same applies here.

(c) The vernier calliper is an appropriate instrument for the measurement of the diameter of this coin.

(i) State the uncertainty in the measurement taken with this instrument.

Solution The uncertainty in the use of the vernier calliper is ± 0.1 mm

(ii) Explain why the measurement of this quantity will have a greater influence on the error in the calculation of density.

Solution In equation 2 the diameter d is squared, this means that the uncertainty in d^2 is twice that in d.

(d) Briefly describe another procedure to measure the volume of a coin. State **one** source of uncertainty. State also how this method may be improved upon.

Solution This is the displacement method, using the displacement of water in a graduated cylinder. Water is first of all placed in the graduated cylinder and the volume recorded. Next add the 2p coin and note the new volume. The difference between the two readings gives the volume of the coin.

The source of uncertainty is governed by the size of the divisions on the graduated cylinder. If the smallest division is 2 cm^3 then you could you improve the situation by using a smaller graduated cylinder with smaller divisions, say 1 cm^3.

Alternatively you could use many 2p coins, this will produce a much greater difference in the volumes and finally dividing by the number of coins to give the volume of one coin.

2 Suppose that a top-pan balance is not available for measuring the mass of the coins. It is proposed that the mass of the coins should be determined using the concept of centre of

gravity, the principle of moments and the following apparatus: several coins of each type, a metre rule of known mass and a knife edge pivot.

(a) State the principle of moments.

Solution At equilibrium, the total clockwise moment equals the total anticlockwise moment, when taken about the same point. Remember that "at equilibrium" must be stated since it is the only situation in which the resultant moment is zero.

(b) Draw a labelled diagram and describe a procedure using the principle of moments and the apparatus listed above to find the mass of the coins.

Solution

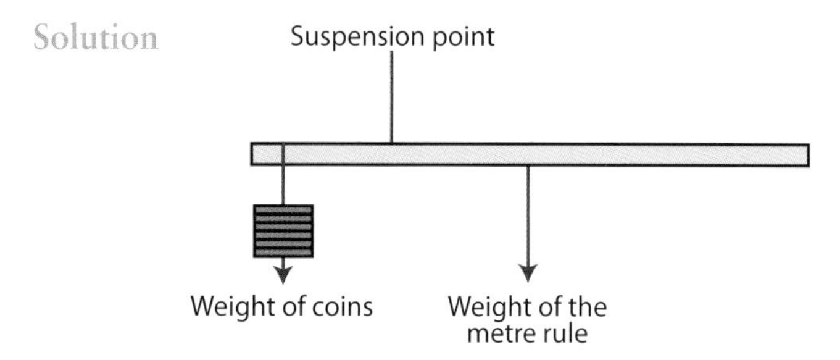

Since you are not provided with other masses, such as 50g or 100g you have to use the weight of the metre rule to provide a clockwise moment to balance the anticlockwise moment due to the weight of the coins. This means suspending the metre at a position other than the 50 cm mark, which is its centre of gravity.

Measure the distance between the suspension point (pivot) and the 50 cm mark (i.e. the centre of gravity of the metre rule). Move the coins toward or away from the suspension point until equilibrium is achieved. Measure the distance between the thread supporting the coin and the suspension point when you have equilibrium. At equilibrium the clockwise moment due the weight of the metre rule equals the anticlockwise moment due to the weight of the coins.

3 With the change in material came a change in density of the coins. The post-1992 coins are 9% less dense than the pre-1992 coins. For sorting and counting purposes at banks, it is important that all 2p coins in circulation have the same mass and diameter. Hence the thickness of the coins had to change when the material they were made from changed. The pre-1992 coins have a thickness of 1.85 mm. What thickness must the post-1992 coins be to keep the mass and diameter constant?

Solution

where ρ is the mass of the pre - 92 coins.

The mass of a coin = density × volume.

$$\text{Mass}_{\text{post92}} = 0.91\rho \times \frac{\pi d^2 t_{\text{post92}}}{4} \text{ and } \text{Mass}_{\text{pre92}} = \rho \times \frac{\pi d^2 t_{\text{pre92}}}{4}$$

To keep the mass and diameter constant the ratio $\frac{\text{mass}}{\text{diameter}^2}$ must also be constant.

Re - arranging the above and simplifying we can write:

$$t_{\text{post92}} = \frac{t_{\text{pre92}}}{0.91} = \frac{1.81}{0.91} = 2.03 \text{ mm}$$

Thus the thickness of post-1992 coin = 2.03 mm

4 Copper-plated coins have the same mass, colour, diameter and design as those made of bronze and circulate alongside them. Apart from differences in density and thickness, there is one other notable difference which makes the two types easy to separate by a simple item of school laboratory apparatus.

What is this item, and why can it be used to separate pre-1992 and post-1992 coins?

Solution

The post 1992 coins are steel coated with copper so they will be attracted by a magnet.

Exercise 52

Planning and design question – CCEA June 2007

Introduction

In this question you are asked to discuss the details of an experiment to find a value for the viscosity η of a clear, thick liquid called glycerin by dropping ball bearings through a transparent cylinder full of the liquid.

Viscosity is the internal property of fluids that makes flow difficult – it is fluid friction. The maximum velocity attained by an object in free fall is called the terminal velocity and is strongly affected by the viscosity of the fluid through which it is falling. When terminal velocity is attained, the body experiences no acceleration, so the forces acting on the body are in equilibrium.

In this experiment, the state of terminal velocity is very quickly attained. However, the speed of descent does not remain constant near the bottom of the cylinder.

(a) Theory

The viscosity η of a fluid is given by

$$\eta = \frac{kr^2}{v}$$

where r is the radius of the ball bearing

v is the terminal velocity of the ball bearing falling through the glycerin

k is a known constant

(i) In the experiment you will need to determine values for r and v.
By plotting a suitable graph, you will then find the value of η.
Sketch the graph you would expect to obtain and label the axes.

(ii) Explain how you would use your graph to find a value for η.
Remember, k is a known constant.

(iii) Use your answer to (a)(ii) to obtain a unit for η. The SI unit for k is kg $m^{-2}s^{-2}$.

(b) Planning and design procedure

You are provided with five ball bearings of different sizes all made from the same material. Their diameters vary between about 2 mm and 12 mm.

(i) Name a suitable measuring instrument and explain how you will obtain a precise and reliable value for the radius r of each ball bearing.

(ii) Write an account of the experiment you would carry out to measure the terminal velocity of the ball bearings as they fall through the glycerin. Remember, the terminal velocity is attained very quickly and the speed is not constant near the bottom of the cylinder.

Your account should include (1) a description of the procedure, (2) a list of readings to be taken, and (3) how the readings are used to obtain a value for terminal velocity.

You may use a diagram to illustrate your answer.

(c) Interpretation

The viscosity of a fluid is temperature dependent. At 20°C the viscosity of glycerin is 1420 units. At 40°C the viscosity of glycerin is 280 units.

(i) State qualitatively what happens to the time taken by a ball bearing to fall through the glycerin as the temperature of the glycerin increases. Explain your answer.

(ii) Assuming a linear relationship between viscosity and temperature in the range stated above, calculate the numerical value of the viscosity change per degree Celsius.

Answers

Exercise 1

1 (i) $p = mv = 0.015 \times 250 = 3.75$ Ns

(ii) Momentum before collision = Momentum after collision

$$3.75 = 3.015v$$

$$v = 3.75 \div 3.015 = 1.244 \text{ ms}^{-1}$$

(iii) $KE = \frac{1}{2}mv^2 = \frac{1}{2} \times 3.015 \times 1.244^2 = 2.32$ J

(iv)

$$mgh = KE$$

$$3.015 \times 9.81 \times h = 2.32$$

$$h = 7.9 \text{ cm}$$

2 (i) Friction is a force external to the system of colliding bodies. The Principle of Conservation of Linear Momentum only applies in the absence of external forces. So to verify the principle for these colliding bodies, the track must be friction-free.

(ii) In an elastic collision kinetic energy is conserved. In an inelastic collision kinetic energy is **not** conserved.

(iii) Momentum before collision = Momentum after collision

$$0.5 \times 0.18 = 2.0 \times v$$

$$v = 0.045 \text{ ms}^{-1} \text{ to the right}$$

(iv)

$$a = \frac{\Delta v}{t} = \frac{(0 - 0.045)}{0.15} = -0.3 \text{ ms}^{-2}$$

$$F = ma = 2 \times (-0.3) = -0.6 \text{ N}$$

The minus sign shows the force is opposing the motion.

3 (a) Momentum before spring expansion = Momentum after spring expansion

$$0 = mv + 3mu$$

$$v = -3u$$

The minus sign shows the speed is to the left

(b) (i) Since the collision with the fixed spring is elastic, the speed of recoil from the spring is 3u to the right. So from $t = t_1$ to $t = t_2$, trolley A travels at speed 3u to the right.

(ii) At the instant of the collision:

B is at a distance from the fixed spring of: $3ut_1 + ut_2$

A is at a distance from the fixed spring of: $3u(t_2 - t_1)$

So, $3ut_1 + ut_2 = 3ut_2 - 3ut_1$ which simplifies to $2ut_2 = 6ut_1$

So, $t_2 = 3t_1$

(c) Momentum before collision = Momentum after collision

$$m \cdot 3u + 3m \cdot u = 4m \cdot v$$

$$v = 1.5u \text{ (to the right)}$$

(d) Total momentum is not conserved because the spring provides an external force.

4 (a) See table on page 9

(b) $mu = mv + MV$ and $\frac{1}{2}mu^2 = \frac{1}{2}mv^2 + \frac{1}{2}MV^2$

(c)
$$R = \frac{KE_{after}}{KE_{before}} = \frac{\frac{1}{2}mv^2}{\frac{1}{2}mu^2} = \left(\frac{v}{u}\right)^2 = \left(\frac{m-M}{m+M}\right)^2$$

(d) If m = M, R = 0 and both particles stop.

(e) As $M \to \infty$, $R \to 1$

(f) The best moderator has the least value of $\frac{v}{u}$:

With carbon: $\frac{v}{u} = \left(\frac{m-M}{m+M}\right) = \frac{1-12}{1+12} = \frac{-11}{13}$ so, $v \approx -0.846u$

With lead: $\frac{v}{u} = \left(\frac{m-M}{m+M}\right) = \frac{1-206}{1+206} = \frac{-205}{207}$ so, $v \approx -0.990u$

So there is greater speed reduction with carbon than with lead. Hence carbon is a better moderator than lead.

Exercise 2

1 PV has units $Nm^{-2} \times m = N \times m = J$

2 (a) (i) PV = constant, so $100 \times 24 = 50 \times V_2$ so V_2 = 48 litres

(ii) PV = constant, so $100 \times 24 = 150 \times V_3$ so V_3 = 16 litres

(b) (i) PV = constant, so $100 \times 24 = P_2 \times 4.8$ so P_2 = 500 kPa

(ii) PV = constant, so $100 \times 24 = P_2 \times 8$ so P_2 = 300 kPa

(iii) PV = constant, so $100 \times 24 = P_2 \times 12$ so P_2 = 200 kPa

3 (a) isothermals (b) isobars

Pressure, P in MPa	0.5	1.0	1.5	2.0	2.5	3.0
Volume, V in cm³	80.0	40.0	26.7	20.0	16.0	13.3
P × V in J	40.0	40.0	40.0	40.0	40.0	40.0
P^{-1} in $(MPa)^{-1}$	0.20	1.00	0.67	0.50	0.40	0.33

Graph of P against V is a hyperbola. At higher temperature curve is further from origin.

Graph of PV against P is a horizontal straight line. At higher temperature line is further from origin.

Graph of V against P^{-1} is a straight line of positive slope through (0,0) origin. At higher temperature slope is greater.

4 Oil of low vapour pressure is essential as otherwise we are compressing an oil vapour and air mixture, whose oil content will change with pressure. This in turn would cause the mass of the material being compressed to change.

It is essential to keep the temperature of the air constant because we are investigating the variation of volume with pressure. A change in air temperature will also change the volume – so we will be unable to say whether the volume change arose from a change in temperature or pressure.

One technique to ensure that the air is at constant temperature is to wait a minute or so after the air pressure has increased before recording the volume. This is to allow the compressed air to cool to room temperature.

Exercise 3

1 The graph of volume against Celsius temperature does not pass through the (0,0) origin.

2 $\frac{V_1}{T_1} = \frac{V_2}{T_2}$ so $\frac{12}{(273+27)} = \frac{V_2}{(273+127)}$

$\frac{12}{300} = \frac{V_2}{400}$ so $V_2 = \frac{12 \times 400}{300} = 16$ litres

3 $\frac{V_1}{T_1} = \frac{V_2}{T_2}$ so $\frac{25}{(273+77)} = \frac{40}{T_2}$

so $T_2 = \frac{40 \times 350}{25} = 560\text{ K} = 287°\text{C}$

4 $\frac{V_1}{T_1} = \frac{V_2}{T_2}$ so $\frac{12}{373} = \frac{1.2}{T_2}$

so $T_2 = 37.3\text{ K} = -235.7°\text{C}$

Exercise 4

1 Assuming a constant temperature:

$p_1V_1 = p_2V_2$ so $3 \times 2 = 1 \times V_2$ so $V_2 = 6\text{ cm}^3$

2 $\frac{p_1V_1}{T_1} = \frac{p_2V_2}{T_2}$

Substituting gives $\frac{(1 \times 70)}{280} = \frac{(P_2 \times 30)}{300}$

so $P_2 = 2.5$ atmospheres

Exercise 5

1 75°C

2 47 cm^3

3 (i) The volume of a fixed mass of an ideal gas at constant pressure is directly proportional to its temperature in Kelvin (ii) See text relating to Charles' Law.

4 (i) The pressure of a fixed mass of an ideal gas at constant pressure is directly proportional to its temperature in Kelvin. (ii) Straight line of positive slope crossing the vertical axis at a positive, non-zero pressure. (iii) 402.5 Pa $°\text{C}^{-1}$ (or 402.5 Pa K^{-1})

Exercise 6

(a) 2.01×10^{-3} kg (b) 2.24×10^{-2} m^{-3} (c) 0°C

Exercise 7

1 474 ms^{-1}

2 1.2 mol

3 (i) 0.18 kg m^{-3} (ii) 1300 ms^{-1}

4 313 ms^{-1}

Exercise 8

1 (i) 514 ms^{-1}

(ii) Straight line graph of positive slope with positive value of $\langle c^2 \rangle$ at 0°C. (Hint: Remember $\frac{1}{2}m\langle c^2 \rangle = \frac{3}{2}kT$, where T is the Kelvin temperature)

2 295 K (22°C)

3 (i) $\langle c^2 \rangle$ represents the mean of the squares of the speeds of the molecules.

(ii) $\langle c^2 \rangle$ is a statistical term. It is only meaningful if the molecules are relatively far apart and moving with a full range of velocities. This in turn means that the number of gas molecules must be large, and they must be relatively far apart so that there are no inter-molecular forces.

(iii) If they were inelastic, kinetic energy would be lost in each collision so the average KE of the gas molecules would fall. The average KE is linked to temperature through the equation $\frac{1}{2} m\langle c^2 \rangle = 3/2\ kT$. So if the average KE decreased, the temperature of the gas would also decrease.

(iv) See text above on the internal energy of a gas.

(v) Zero Kelvin is sometimes called absolute zero because it represents the temperature at which the translational kinetic energy of gas molecules is zero. There is therefore no temperature lower than zero Kelvin.

(vi) The molecules would no longer be moving.

4 See text on Charles' Law.

5 (a) Pressure increases when temperature rises; speed of molecules increase; this produces more collisions per second with container walls; this results in greater momentum change in each collision; this produces greater force (and hence greater pressure) in the walls.

(b) (i) 3.55×10^{22} molecules (Hint: Use $pV = NkT$)

(ii) 1.08×10^{5} Pa (Hint: Use Pressure Law)

(iii) 7.35 J (Hint: Use $\frac{1}{2}m\langle c^2 \rangle = \frac{3}{2}kT$)

Exercise 9

1 $J\ kg^{-1}\ K^{-1} = Nm\ kg^{-1}\ K^{-1} = kg\ ms^{-2}\ m\ kg^{-1}K^{-1} = m^2s^{-2}K^{-1}$

Exercise 10

1 $Q = mc\theta = 4 \times 385 \times (80 - 25) = 4 \times 385 \times 55 = 84700 \text{ J}$

2 Temperature rise $\Delta\theta = \frac{Q}{mc} = \frac{IVt}{mc} = \frac{2 \times 12 \times 720}{3 \times 500} = \frac{17280}{1500} = 11.5°C$

Final temperature = initial temperature + rise in temperature = 20 + 11.5 = 31.5°C

3 $KE = \frac{1}{2}mv^2 = \frac{1}{2} \times 1400 \times 30^2 = 630\,000 \text{ J}$

Temperature rise $\Delta\theta = \frac{Q}{mc} = \frac{630\,000}{104 \times 600} = \frac{630\,000}{62400} = 10.1°C$

It is assumed that all of the car's kinetic energy is transformed into thermal energy in the brakes.

4 Heat transferred per minute = mass entering per minute × c × $\Delta\theta$

$= 35 \times 4200 \times (375 - 285) = 35 \times 4200 \times 90 = 13\,230\,000 \text{ J}$

$= 13.23 \text{ MJ}$

Exercise 11

1 (i) The specific heat capacity of a material is the quantity of heat energy needed to raise the temperature of 1 kg of the material by 1 K.

(ii) Heat required $Q = mc\Delta\theta = 0.24 \times 10^{-3} \times 450 \times 1 = 0.108 \text{ J}$

Compressing and releasing the spring produces 10% of 4 mJ as heat

$= 0.4 \text{ mJ} = 4 \times 10^{-4} \text{ J}$

Number of times spring must be compressed and released

$= \frac{0.108 \text{ J}}{4 \times 10^{-4} \text{ J}} = 270$ times

2 (a) $E = IVt = 2.50 \times 230 \times 15 = 8625 \text{ J}$

$\Delta\theta = \frac{Q}{mc} = \frac{0.55 \times 8625}{0.65 \times 380} = 19.2°C$

(b) The temperature rise calculated in (a) is greater than what occurs in practice. This is because heat is lost to the drill bit and to the atmosphere, resulting in a lower rise in temperature than would otherwise be expected.

3 (a) $Q = IVt = 12.4 \times 3.74 \times (5 \times 30) = 13900 \text{ J}$

$c = \frac{Q}{m\Delta\theta} = \frac{13900}{0.126 \times (43 - 19)} = 4590 \text{ J kg}^{-1} \text{ °C}^{-1}$

(b) Agitate the milk using the stirrer. Stirring encourages an even temperature throughout the milk and ensures that the milk at bottom is heated.

(c) The inner metal calorimeter absorbs heat from the heater causing its temperature to rise. Calculate the heat energy absorbed by this calorimeter using $Q = m_c \times c_c \times \Delta\theta$

Subtract this heat absorbed by the calorimeter from the total heat energy supplied to obtain the heat supplied to the milk.

Exercise 12

1 (a) The centripetal force is provided by the horizontal component of the tension in the string. The weight is balanced by the vertical component of the tension in the string.

(b) Curve showing a positive minimum tension at P, maximum at R and equal, non-zero values at Q and S.

(c) $T_{min} = mr\omega^2 - mg$

$= (0.135 \times 0.320 \times 8.50^2) - (0.135 \times 9.81) = 1.80\text{ N}$

2 (a) (i) Tangent drawn to circle at the ball with arrowhead pointing "north".

(ii) If the angular velocity is to remain constant, the speed of the ball must not change. Therefore, there can be no component of the force parallel to the instantaneous velocity. So, the force must be perpendicular to the velocity. Its direction is inwards, because the circle curves inwards.

(b) (i) 14.2 N (ii) 5.9°

3 (a) (i) Angular velocity is the angle swept out by the radius vector every second.

(ii) The centripetal force is the force at right angles to the velocity vector – it forces the object to change direction and move in a circular path.

(iii) Towards the centre of the circle.

(b) (i) $F = mr\omega^2 = 2.50 \times 1.20 \times 1.35^2 = 5.47\text{ N}$

(ii) Speed $= r\omega = 1.20 \times 1.35 = 1.62\text{ ms}^{-1}$

Velocity change = final velocity at Y – initial velocity at X

$= 1.62\text{ ms}^{-1}$ upwards – 1.62 ms^{-1} downwards

$= 1.62 - (-1.62) = 3.24\text{ ms}^{-1}$

4 (a) (i) 11.17 radians per second (ii) 111.7 cms^{-1}

(b) (i) Straight line of positive slope through the origin

(ii) 0.152 m

(c) 14.5 radians per second

5 (a) (i) Angular velocity, ω, must decrease with increasing radius

because $\omega = \dfrac{v}{r}$ and v is constant

(ii) $\omega = 14.6 \times 2\pi = 91.73\text{ rad s}^{-1}$

$r = \dfrac{v}{\omega} = \dfrac{3.84}{91.73} = 0.042\text{ m} = 4.2\text{ cm}$

(b) Friction provides the centripetal force.

Centripetal force $F = \dfrac{mv^2}{r}$

Since linear velocity (v) of DVD is higher than that of CD, then a larger centripetal force is needed; so dust particles more likely to stick to a CD.

Exercise 13

1 (a) 1. the acceleration is proportional to its displacement from a fixed point and
2. the direction of the acceleration is always towards that fixed point.

(b) $a_{max} = \omega^2 A$, so: $3.01 = \left(\frac{2\pi}{T}\right)^2 \cdot A = \left(\frac{2\pi}{0.625}\right)^2 \cdot A$

$$A = 3.01 \times \left(\frac{0.625}{2\pi}\right)^2 = 0.02978 \approx 0.030 \text{ m}$$

(c) $x = A\cos\omega t = 0.030\cos\left(\frac{2\pi t}{0.625}\right)$

$= 0.030\cos(10.05\,t)$

2 (a) (i)

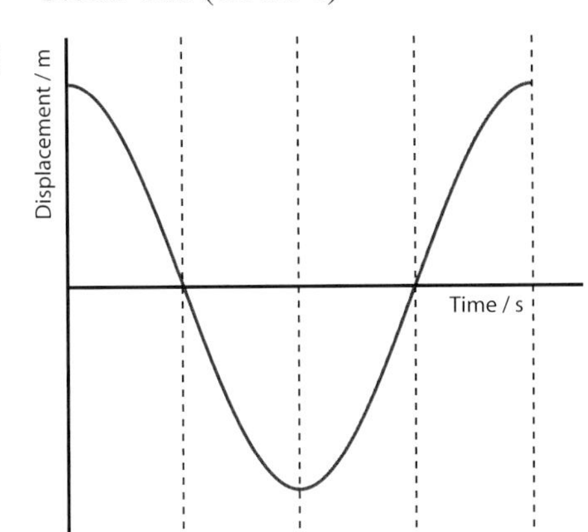

(ii)

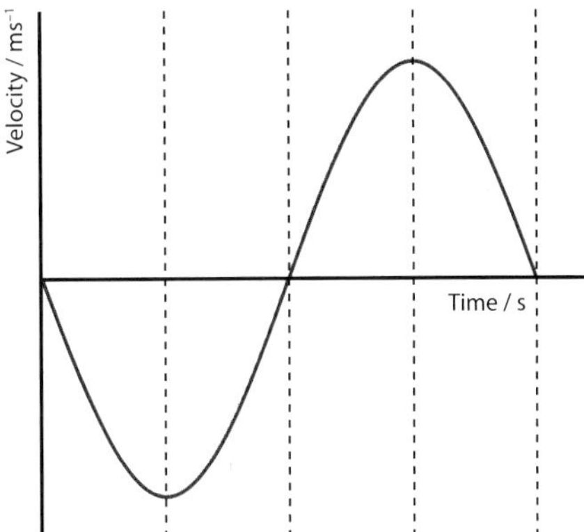

Acceleration leads velocity by 90° (or $\pi/2$ radians or a quarter period)

(iii)

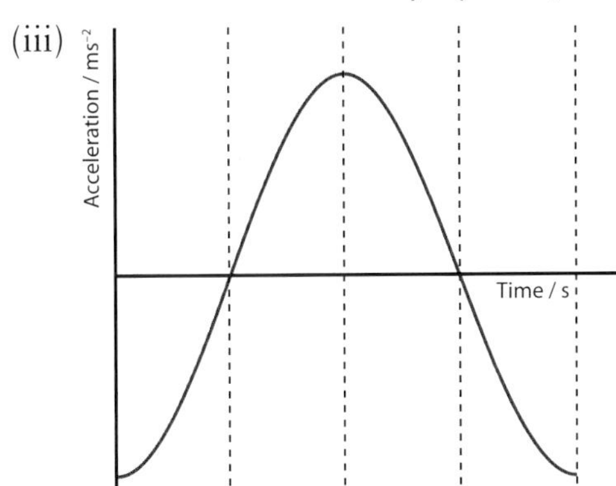

Acceleration and displacement are π radians or 180° out of phase

(b) (i) Amplitude $= \frac{1}{2}(5.6 - 2.4) = 1.6$ m

(ii) $\omega = \frac{2\pi}{T} = \frac{2\pi}{12.5} = 0.5027$ rad hour^{-1}

Assuming time t = 0 corresponds to high tide (as in the displacement graph above), then we require to find the water depth (6.25 + 2.4) or 8.65 hours after high tide. The level of the water midway between high and low tides is 2.4 + 1.6 = 4.0 metres. Height of water above sea bed $= 4.0 + 1.6\cos\omega t = 4.0 + \cos(0.5027 \times 8.65)$

$= 4.0 - 0.57 = 3.43$ m

3 (a) Force is directly proportional to the displacement and occurs in the opposite direction.

(b) Since $a_{max} = -\omega^2 A$, then when the acceleration is half its maximum value, the displacement is half the amplitude, so $x = \frac{1}{2}A$

4 (a) Work done in compressing the spring $= \frac{1}{2}kx^2$

$$25 \times 10^{-3} = \tfrac{1}{2}k\left(75 \times 10^{-3}\right)^2 = k \times 2.8125 \times 10^{-3}$$

$$k = 25 \times 10^{-3} \div 2.8125 \times 10^{-3} = 8.89\ \text{Nm}^{-1}$$

(b) From the definition of SHM, $a = -\omega^2 x$

Ignoring the minus sign, $\omega^2 = \dfrac{a}{x} = \dfrac{3.0}{7.5 \times 10^{-3}} = 40\ \text{s}^{-2}$ (since the minus sign merely indicates the direction).

The general equation for the period, $T = \dfrac{2\pi}{\omega}$ so $T^2 = \dfrac{4\pi^2}{\omega^2}$

Since $T = 2\pi\sqrt{\dfrac{m}{k}}$ then $T^2 = 4\pi^2\dfrac{m}{k}$

Comparing this to the equation for T^2 above gives: $\dfrac{m}{k} = \dfrac{1}{\omega^2}$

So: $m = \dfrac{k}{\omega^2} = \dfrac{8.89}{40} = 0.22\ \text{kg}$

Exercise 14

1 Graph of f^2 in s^{-2} (vertical axis) against $\dfrac{1}{l}$ in m^{-1} (horizontal axis) is a straight line through the origin and has a gradient of $\dfrac{g}{4\pi^2}$

2 (a) See text.

(b) $T = 2\pi\sqrt{\dfrac{m}{k}} = 2\pi\sqrt{\dfrac{0.02}{8.8}} = 0.2995\ \text{s}$

$f = \dfrac{1}{T} = \dfrac{1}{0.2995} = 3.34\ \text{Hz}$

3 (a) A = amplitude; possible unit: metres

ω = angular frequency possible unit: radians per second

(b)

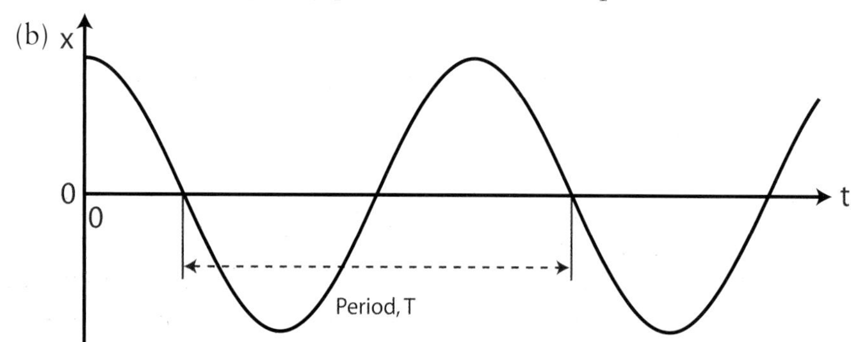

Use the scale on the time axis to determine the period, T of the vibration as illustrated above. The value of ω is then calculated using $T = \dfrac{2\pi}{\omega}$

Exercise 15

1-3 See text.

4 (a) (i) Force = mass × acceleration, so graph of acceleration against displacement similar to force against displacement.

The graph shows a straight line through origin, so acceleration and force are both proportional to displacement.

The graph has a negative gradient, so acceleration and force are in opposite direction to displacement.

(ii) Amplitude, A = maximum displacement = 5 mm = 0.005 m

$$\text{Maximum acceleration} = \omega^2 A = 7.5\ \text{ms}^{-2}$$

$$\omega^2 = \frac{7.5}{0.005} = 1500\ \text{s}^{-2}$$

$$\text{Period, T} = \frac{2\pi}{\omega} = \frac{2\pi}{\sqrt{1500}} = 0.16\ \text{s}$$

(b) • Oscillating system undergoing forced vibrations.

• Vary forcing frequency, observe amplitude of response.

• Typical frequency response curve is shown in sketch below.

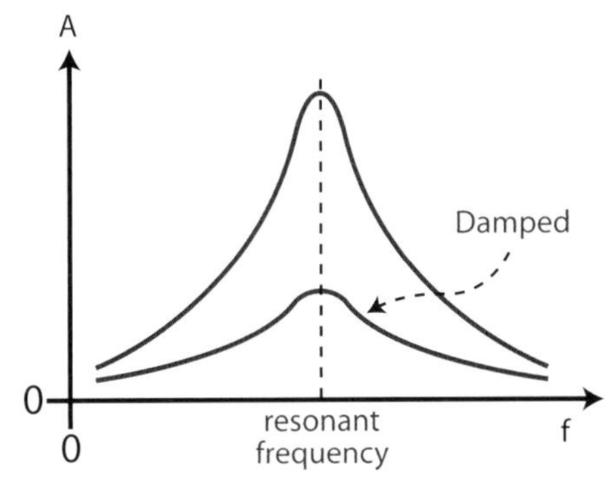

• The resonant frequency is that at which the amplitude is a maximum.

• Damping effect is to reduce the maximum amplitude at resonance as shown in sketch.

Exercise 16

1 (a) (i) r = radius of nucleus; r_o = proportionality constant and radius of a single nucleon

A = mass number (nucleon number)

(ii) 1 fm

(b) (i) From text: $\rho = \frac{3m}{4\pi r_0^3} = \frac{3 \times 1.66 \times 10^{-27}}{4\pi \times 1 \times 10^{-45}} = 3.96 \times 10^{17}\ \text{kg m}^{-3}$

(ii) The huge density of nuclear matter in comparison with everyday matter is a reflection of the fact that in ordinary matter there is a great deal of empty space between the nucleus and the orbiting electrons. There is no "empty space" between the particles inside the nucleus.

2 (i) Volume of a sphere, $V = \frac{4\pi r^3}{3}$, so $r = \sqrt[3]{\frac{3V}{4\pi}} = \sqrt[3]{\frac{3 \times 4.6 \times 10^{-43}}{4\pi}}$

$$= 4.78876 \times 10^{-15} \approx 4.79 \times 10^{-15}\ \text{m}$$

(ii) $A = \left(\frac{r}{r_0}\right)^3 = \left(\frac{4.79 \times 10^{-15}}{1.2 \times 10^{-15}}\right)^3 = 64$ (to nearest integer)

3 (i) At A = 150, r = 4.7 fm

(ii) Nuclear mass $= 150 \times 1.66 \times 10^{-27}$ kg $= 2.49 \times 10^{-25}$ kg

$$\text{Nuclear volume} = \frac{4\pi r^3}{3} = \frac{4\pi \times (4.7 \times 10^{-15})^3}{3} = 4.35 \times 10^{-43}\ \text{m}^3$$

$$\text{Density} = \frac{\text{Mass}}{\text{Volume}} = \frac{2.49 \times 10^{-25}}{4.35 \times 10^{-43}} \approx 5.7 \times 10^{17}\ \text{kg m}^{-3}$$

(iii) Vertical axis: r^3 Horizontal axis: Nucleon number, A

4 (i) Almost all of the gold atom consist of empty space

(ii) There exists a tiny nucleus at the centre of the atom where almost all its mass is concentrated. This central, tiny nucleus is positively charged.

(iii) The vacuum is essential as collisions with air molecules would deflect the α–particles and give erroneous results.

5 (i) 13 protons and 14 neutrons in nucleus surrounded by 13 orbiting electrons

(ii) $r = r_0 A^{1/3} = 1.2 \times 27^{1/3} = 1.2 \times 3 = 3.6$ fm $= 3.6 \times 10^{-15}$ m

Exercise 17

1 (a) (i) In α-decay, mass number decreases by 4 and atomic number decreases by 2.

(ii) In β-decay, mass number is unchanged and atomic number increases by 1.

(b)

Element (symbol)	Atomic number	Mass number	Decays by emitting	Leaving element
U	92	238	α	Th
Th	90	234	β	Pa
Pa	91	234	β	U
U	92	234	α	Th
Th	90	230	α	Ra
Ra	88	226	α	Rn
Rn	86	222	α	Po
Po	84	218	α	Pb
Pb	82	214	β	Bi
Bi	83	214	β	Po

(d) γ-ray emission changes neither Atomic Number nor Mass Number so it cannot be shown on a graph of mass number against atomic number.

(e) (i) Uranium-238 and Uranium-234 and (ii) Thorium-234 and Thorium-230

2 $^{238}_{92}U \rightarrow ^{234}_{90}Th + ^{4}_{2}He$ $\quad ^{14}_{6}C \rightarrow ^{14}_{7}N + ^{0}_{-1}\beta$

Exercise 18

1 (a) (i) ${}^{14}_{6}C \rightarrow {}^{14}_{7}N + {}^{0}_{-1}e$ (ii) ${}^{238}_{92}U \rightarrow {}^{234}_{90}Th + {}^{4}_{2}He$

(b) (i) $\lambda = \dfrac{0.693}{T_{1/2}} = \dfrac{0.693}{8} = 0.086625 \text{ days}^{-1}$

$N = N_0 e^{T_{1/2}} = 5 \times 10^{14} \times e^{-0.086625 \times 29}$

$= 4.0548 \times 10^{13} \approx 4.1 \times 10^{13}$ undecayed atoms

(ii) $\lambda = \dfrac{0.086625}{24 \times 3600} = 1.0026 \times 10^{-6} \text{ s}^{-1}$

$A = \lambda N = 1.0026 \times 10^{-6} \times 4.0548 \times 10^{13}$

≈ 40.6 MBq

2 (a) (i) The half-life of a radioactive material is the time taken for the activity of that material to fall to half of its original value.

(ii) $A = A_0 e^{-\lambda t}$ but from the definition of half - life $t = T_{1/2}$:

$\dfrac{A_0}{2} = A_0 e^{-\lambda T_{1/2}}$ divide both sides by A_0 and take natural logs:

$\ln 1 - \ln 2 = -\lambda T_{1/2}$ but since $\ln 1 = 0$, this gives:

$\ln 2 = \lambda T_{1/2}$ or $T_{1/2} = \dfrac{\ln 2}{\lambda} = \dfrac{0.693}{\lambda}$ as required.

(b) $T_{1/2} = 4.5 \times 10^{9}$ years $= 4.5 \times 10^{9} \times 3.2 \times 10^{7}$ seconds

$= 1.44 \times 10^{17}$ seconds

$\lambda = \dfrac{0.693}{T_{1/2}} = \dfrac{0.693}{1.44 \times 10^{17}} = 4.8125 \times 10^{-18} \text{ s}^{-1}$

$A = \lambda N = 4.8125 \times 10^{-18} \times 3.0 \times 10^{21}$

$\approx 14\,400$ Bq

Exercise 19

1 (a) Mass defect $= (8 \times \text{mass}_{\text{proton}} + 8 \times \text{mass}_{\text{neutron}}) - \text{mass}_{\text{oxygen nucleus}}$

$= (8 \times 1.0078 + 8 \times 1.0087) - 15.9905$ u

$= 16.132 - 15.9905 = 0.1415$ u

Binding energy $= \Delta mc^2 = 0.1415 \times 1.66 \times 10^{-27} \times (3 \times 10^{8})^2$

$= 2.11401 \times 10^{-11}$ J

$= 2.11401 \times 10^{-11} \div 1.6 \times 10^{-13}$ MeV ≈ 132.13 MeV

(b) Average BE per nucleon $= 132.13 \div 16 \approx 8.26$ MeV / nucleon

2 (a) Mass difference $= 227.97929 - 223.97189 - 4.00151 = 0.00589$ u

(b) Mass difference $= 0.00589 \times 1.66 \times 10^{-27}$ kg $= 9.7774 \times 10^{-30}$ kg

(c) Using $E = \Delta mc^2$

(i) $E = 9.7774 \times 10^{-30} \times (3 \times 10^8)^2 = 8.79966 \times 10^{-13}$ J

(ii) $E = (8.79966 \times 10^{-13}) \div 1.6 \times 10^{-13}$ MeV ≈ 5.5 MeV

(d) The energy appears mainly as the kinetic energy of the α particle, and, to a lesser extent, the kinetic energy of the radium nucleus.

3 Mass difference $= 28.97330 - (28.96880 + 0.000549) = 3.951 \times 10^{-3}$ u

$= 3.951 \times 10^{-3} \times 1.66 \times 10^{-27}$ kg $= 6.55866 \times 10^{-30}$ kg

Energy released $E = \Delta mc^2 = 6.55866 \times 10^{-30} \times (3 \times 10^8)^2 = 5.902794 \times 10^{-13}$ J

$= 5.902794 \times 10^{-13} \div 1.6 \times 10^{-13}$ MeV ≈ 3.69 MeV

Exercise 20

1 BE of lanthanum = 146 × 8.41 = 1227.86 MeV

BE of bromine = 87 × 8.59 = 747.33 MeV

Total BE after fission = 1227.86 + 747.33 = 1975.19 MeV

BE of uranium–236 = 236 × 7.59 = 1791.24 MeV

Energy released = Increase in Binding Energy = 1975.19 – 1791.24 = 183.95 MeV ≈ **184 MeV**

2 (i) 1 kg of fuel contains 30 grams of uranium–235, which contains $30\,N_A \div 235$ uranium–235 atoms. So, 1 kg of fuel contains $(30 \times 6.02 \times 10^{23}) \div 235 = 7.69 \times 10^{22}$ uranium–235 atoms.

(ii) Energy released by fission of all uranium–235 nuclei is
$7.69 \times 10^{22} \times 3 \times 10^{-11}$ J $= 2.307 \times 10^{12}$ J
Time = energy ÷ power $= (2.307 \times 10^{12}) \div 500000 = 4\,614\,000$ s = 53.4 days

3 Combined mass of 3 protons and 4 neutrons = 3 × 1.008 + 4 × 1.009 = 7.060 u

Mass defect = Mass of separate nucleons – Mass of lithium nucleus = 7.060 – 7.018 = 0.042 u

Binding energy = 0.042 × 931 MeV = 39.1 MeV

Exercise 21

1 (a) (i) 3.59×10^{-12} J $= 3.59 \times 10^{-12} \div 1.6 \times 10^{-13}$ MeV ≈ 22.4 MeV

(ii) Since 8 nucleons of fuel are used, energy released per nucleon = 22.4 ÷ 8 = 2.8 MeV

(iii) Fission releases only 200 ÷ 235 = 0.85 MeV/nucleon so there is a smaller yield per nucleon of fuel with fission.

(b) With fusion there is none of the toxic and highly radioactive waste associated with fission. With fusion there are readily available and virtually inexhaustible supplies of fuel (seawater contains 1 atom of deuterium for every 5000 of ordinary hydrogen) – there are limited supplies of uranium ore for fission.

2 Mass of reactants = 2.014102 + 3.016030 = 5.030132 u
Mass of products = 4.002604 + 1.008665 = 5.011269 u
Mass difference = 0.018863 u $= 0.018863 \times 1.66 \times 10^{-27}$ kg
$= 3.131258 \times 10^{-29}$ kg
$E = mc^2 = 3.131258 \times 10^{-29} \times (3 \times 10^8)^2 \approx 2.82 \times 10^{-12}$ J

3 Mass of reactants = 6.0151 + 2.0141 = 8.0292 u
Mass of products = 2×4.0026 = 8.0052 u
Mass difference = 0.0240 u = $0.0240 \times 1.66 \times 10^{-27}$ kg
= 3.984×10^{-29} kg

$$E = mc^2 = 3.984 \times 10^{-29} \times (3 \times 10^8)^2 \approx 3.59 \times 10^{-12} \text{ J}$$

4 Mass of reactants = 2×2.014102 = 4.028204 u
Mass of products = 3.016030 + 1.008665 = 4.024695 u
Mass difference = 0.003509 u = $0.003509 \times 1.66 \times 10^{-27}$ kg
= 5.82495×10^{-30} kg

$$E = mc^2 = 5.82495 \times 10^{-30} \times (3 \times 10^8)^2 \approx 5.242446 \times 10^{-13} \text{ J}$$

$$E = 5.242446 \times 10^{-13} \div 1.6 \times 10^{-13} \text{ MeV} \approx 3.28 \text{ MeV}$$

Exercise 22

1 See text.

2 (i) uranium–235 (ii) graphite (iii) carbon dioxide gas (iv) boron

3 (a) Total mass of fuel rods = 2000×14 = 28 000 kg
Mass of uranium - 235 = 3% of 28 000 kg = 840 kg

(b) Total heat = heat from 1 kg uranium - 235 × mass of fuel
= $1 \times 10^{14} \times 840 = 8.4 \times 10^{16}$ J

(c) $\text{Time} = \dfrac{\text{Energy}}{\text{Power}} = \dfrac{8.4 \times 10^{16}}{2.4 \times 10^9} = 3.5 \times 10^7 \text{ s} \approx 1.2 \text{ years}$

Exercise 23

1 (a) For nuclei to overcome electrostatic repulsion they must be heated to around 800 million kelvins. It is difficult to achieve a plasma at a temperature of around 800 million kelvins and then to confine the plasma for long enough for fusion to take place.

(b) Charged plasma particles circulate in helical paths around circular magnetic field lines in an evacuated chamber. The magnetic field is produced by water-cooled, current-carrying toroidal field coils.

2 Mean kinetic energy = $\frac{3}{2}kT$
= $1.5 \times 1.38 \times 10^{-23} \times 8.0 \times 10^8$
= 1.656×10^{-14} J $\approx 1.7 \times 10^{-14}$ J

3 (a) Plasma is the 4th phase of matter. When a gas is heated to a sufficiently high temperature all the electrons break free of the atoms, and the gas becomes a mixture of free-moving electrons and ions. This is the plasma state.

(b) Inertial confinement involves using intense ion or laser beams directed at a solid fuel pellet. The beams provide the energy to heat the material to the temperature required for fusion. The idea is to produce fusion for long enough to extract the energy before the plasma escapes.

4 (a) ${}^1_1H + {}^1_1H \rightarrow {}^2_1H + {}^0_1\beta^+ + \upsilon$

(b) No long-lived, dangerous radioactive waste and almost limitless fuel supply (deuterium) from seawater

(c) Magnetic and inertial

(d) Mean kinetic energy = $\frac{3}{2}kT$ so:

$$T = \frac{2 \times \text{mean kinetic energy}}{3k}$$

$$T = \frac{2 \times 110 \times 1.6 \times 10^{-16}}{3 \times 1.38 \times 10^{-23}} = 850 \times 10^{6}\ \text{K}$$

Exercise 24

1 Rearranging $F = G\frac{m_1m_2}{r^2}$ gives $G = F\frac{r^2}{m_1m_2}$ and substituting the units for force (N), distance (m) and mass (kg) gives the unit for G as **N m² kg⁻²**

From F = ma, the newton in SI base units is the kg ms^{-2}

Hence the unit for G in SI base units is kg ms^{-2} m^2 kg^{-2} which simplifies to kg^{-1} m^3 s^{-2}

2

$$\text{Centripetal force} = \frac{mv^2}{r} = \frac{9.11 \times 10^{-31} \times (2.2 \times 10^{6})^2}{150 \times 10^{-12}}$$

$$= 8.82 \times 10^{-8}\ \text{N}$$

$$\text{Gravitational force } F = G\frac{m_1m_2}{r^2}$$

$$= \frac{6.67 \times 10^{-11} \times 1.66 \times 10^{-27} \times 9.11 \times 10^{-31}}{(50 \times 10^{-12})^2} = 4.03 \times 10^{-47}\ \text{N}$$

Comment: The gravitational force is much too small to provide the centripetal force to make the electron orbit the proton in a hydrogen atom.

3 $\text{Field strength } g = \frac{F}{m} = \frac{180}{5} = 36\ \text{N kg}^{-1}$

4 An inverse square law means halving the separation from 2 m to 1 m quadruples the force. So the answer to (a) is 4 x 36 = 144 N. Tripling the distance from 1 m to 3 m causes the force to decrease by a factor of 3^2. So the answer to (b) is $\frac{144}{9} = 16\ \text{N}$

5 Gravitational force has increased by a factor of 35/14 or 2.5. So, the field strength on Xenon is 2.5 times greater than on Earth.

Hence, $g_{xenon} = 2.5 \times g_{earth} = 2.5 \times 9.81 = 24.53\ \text{N kg}^{-1}$

Exercise 25

1 $\frac{Mv^2}{R} = Mg_{moon}$

and cancelling the mass M of the landing craft on both sides gives

$$v = \sqrt{Rg_{moon}} = \sqrt{2 \times 10^{6} \times 1.23} = 1.57 \times 10^{3}\ \text{ms}^{-1}$$

$$\text{Momentum} = Mv = 17500 \times 1.57 \times 10^{3} = 2.74 \times 10^{7}\ \text{Ns}$$

Direction: Tangential (to orbital path at any point)

2 (a) (i) Any **two** from:

- Orbit is in the equatorial plane; i.e. plane of the orbit passes through the centre of the earth.
- Orbital period = 24 hours **or** Remains in fixed position relative to earth **or** Same angular speed as earth.
- Moves in same direction as earth.

(ii) $T^2 = \dfrac{4\pi^2 r^3}{GM_E}$ so: $86400^2 = \dfrac{(4 \times 3.14^2 \times r^3)}{6.67 \times 10^{24}}$

$r^3 = 7.565 \times 10^{22}$ so: $r = 4.23 \times 10^7$ m

(b) Since $g = \dfrac{GM_E}{r^2}$ where r is the radius of the Earth

Since G and r are constants, and ΔM_E is the total increase in the Earth's mass,

Increase in gravitational field strength $\Delta g = \dfrac{G \cdot \Delta M_E}{r^2}$

$\Delta g = 0.001 \times 9.81 = \dfrac{6.67 \times 10^{-11} \times \Delta M_E}{(6.38 \times 10^6)^2}$

$\Delta M_E = 0.001 \times 9.81 \times \dfrac{(6.38 \times 10^6)^2}{6.67 \times 10^{-11}}$

$\Delta M_E = 6.0 \times 10^{21}$ kg to increase the gravitational field strength by 0.1%

Time needed = total increase in mass ÷ increase in mass per year

Time $= 6.0 \times 10^{21} \div 7.9 \times 10^7 = 7.6 \times 10^{13}$ years

3 (a) (i) The gravitational force between two point masses is proportional to the product of their masses and inversely proportional to the square of their separation.

(ii) Gravitational fields are always attractive, electric fields can be attractive or repulsive. Objects can be shielded from electric fields but not from gravitational fields.

(b) Gravitational force between Sun and planets provides the centripetal force to keep them in orbit. In the case of satellites its is the gravitational force between the earth and the satellite that provides the centripetal force.

(c) (i) See page 110.

(ii) For the Earth in orbit around the Sun we use $T^2 = \dfrac{4\pi^2 r^3}{GM_{sun}}$

$T = 1 \text{ year} = 3.156 \times 10^7$ seconds

r = radius of the earth's orbit $= 1.5 \times 10^{11}$ m

$$M_{sun} = \frac{4\pi^2 r^3}{T^2 G} = \frac{4 \times 3.14^2 \times (1.5 \times 10^{11})^3}{(3.156 \times 10^7)^2 \times 6.67 \times 10^{-11}} = 2 \times 10^{30} \text{ kg}$$

Exercise 26

1 (a) To ensure the units are correct for each side of the equation and to fit properties to medium.

(b) Vertical force $= QE_1 = 2.0 \times 10^{-6} \times 6.0 = 12 \times 10^{-6}$ N
Vertical force $= QE_2 = 2.0 \times 10^{-6} \times 4.5 = 9 \times 10^{-6}$ N
By Pythagoras, resultant force $= \sqrt{(122 + 92)} \times 10^{-6} = 15 \times 10^{-6}$ N

Direction $= \tan^{-1}\left(\frac{12 \times 10^{-6}}{9 \times 10^{-6}}\right) = 53°$ above the horizontal

(c) By Coulomb's Law, $F = (4\pi\varepsilon_0)^{-1} \cdot \frac{q_1 q_2}{r^2}$
Force on +2.0 μC charge from 25.0 μC

$= \frac{(9 \times 10^9 \times 25 \times 2 \times 10^{-6})}{10^2} = 4.5 \times 10^{-3}$ N to the right

However this is greater than the resultant force $(3 \times 10^{-3}\ \text{N})$
So force due to X is $4.5 \times 10^{-3} - 3 \times 10^{-3} = 1.5 \times 10^{-3}$ N to the left
Hence X is negative to reduce it

By Coulomb's Law, $F = (4\pi\varepsilon_0)^{-1} \cdot \frac{q_1 q_2}{r^2}$ so $q_X = F \cdot \frac{r^2 (4\pi\varepsilon_0)}{q_1}$

$q_X = \frac{1.5 \times 10^6 \times 6^2}{9 \times 10^9 \times 2.0 \times 10^{-6}} = 3.0 \times 10^{-6}\ \text{C} = 3.0\ \mu\text{C}$
$q_X = -3.0\ \mu\text{C}$

2 (a) (i) By Coulomb's Law, $F = (4\pi\varepsilon_0)^{-1} \cdot \frac{q_1 q_2}{r^2}$

$F = \frac{9 \times 10^9 \times 3 \times 10^{-6} \times 6 \times 10^{-6}}{(30 \times 10^{-3})^2} = 180$ N

(ii) Opposite charges since they move towards each other.

(b) Electric field strength is uniform between parallel plates, but around isolated point charges the field is radial and decreases as $1/r^2$. Diagrams the same as p115-6.

3 (a) A field of force is not an area, which has two dimensions – it is a region a space, which has three dimensions. It is **only** an electric field in which a charge experiences a force. In general, a field will exert a force on a body only if that body has a specific property. For example, a gravitational field will exert a force only on a mass and a magnetic field will exert a force only on a moving charge. There is no requirement to have a unit property, such as a unit of charge, when defining what is meant by a field of force.

(b) The strength of an electric field, E, at a point is defined as the force which would be produced on a test charge of +1 C at that point. The direction of an electric field at a point is in the direction of decreasing potential, i.e. it is in the direction in which that test charge would move if it were free to do so.

4 (a) (i) The force is repulsive and decreases as 1/distance^2

(ii) The force is attractive and 4 times larger, at any distance, than the force in (i), it also decreases as 1/distance^2

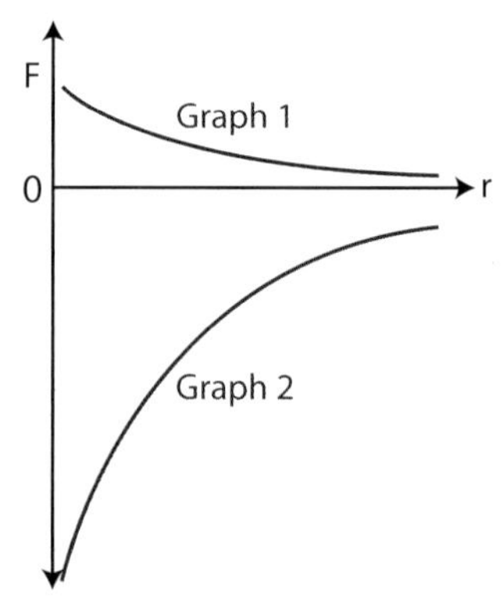

(b) (i) At X ←, at Y ←, at Z ← See the diagrams on page 115.

(ii) At Y, the two electric fields act in the same direction at Y and Y is the closest point to both charges.

(c) (i) The electric field strength between the plates is given by E = V/d

For a constant potential difference V, $E \propto \frac{1}{d}$ this produces a linear graph that passes through the origin.

(ii) The direction of the electric field is taken as that in which a positive charge would experience a force, so the top plate is negative.

Exercise 27

1 (a) All three capacitors in series $\frac{1}{C} = \frac{1}{4} + \frac{1}{4} + \frac{1}{4} = \frac{3}{4}$ so: $C = \frac{4}{3} = 1.33\ \mu F$

(b) Two in series gives $\frac{1}{C} = \frac{1}{4} + \frac{1}{4} = \frac{1}{2}$ so: $C = 2\ \mu F$ and with the third in parallel with both of these gives C = 4 + 2 = 6 μF

2 YZ has 6 μF and 4 μF in series: $\frac{1}{C} = \frac{1}{6} + \frac{1}{4} = \frac{5}{12}$ so: $C = \frac{12}{5} = 2.4\ \mu F$

XY has 2 μF and 6 μF in series: $\frac{1}{C} = \frac{1}{2} + \frac{1}{6} = \frac{4}{6}$ so: $C = \frac{6}{4} = 1.5\ \mu F$

XZ has 2 μF and 4 μF in series: $\frac{1}{C} = \frac{1}{2} + \frac{1}{4} = \frac{3}{4}$ so: $C = \frac{4}{3} = 1.3\ \mu F$

Maximum capacitance is obtained between terminals Y and Z.

3 (a) Between X and P the two 10μF capacitors in series giving capacitance of 5μF. This is repeated in the lower branch of the circuit. Effectively we have two 5 μF capacitors in parallel giving a value of 10μF. This is replicated between P and Y. This means we can regard the circuit as two 10μF capacitors in series giving a total capacitance of 5 μF.

(b) The circuits are not joined at P. Between X and Y we have four 10 μF capacitors in series which is equivalent to 2.5 μF. This is repeated in the other branch of this circuit. Effectively this means we have two 2.5 μF capacitors in parallel giving a total capacitance of 5 μF.

4 Each branch of the circuit has a capacitance of ½ C. Since the three branches are in parallel the total capacitance is 3 × ½ C, 1.5 C = 33 μF so C = 22 μF

Exercise 28

1 (a) $C = \frac{Q}{V}$ re-arranging gives: $V = \frac{Q}{C}$

For 1 μF capacitor, $V = \frac{1.5 \times 10^{-9}}{1.0 \times 10^{-6}} = 1.5 \times 10^{-3}\ V$

For 100 pF capacitor, $V = \frac{1.5 \times 10^{-9}}{100 \times 10^{-12}} = 15\ V$

(b) $E = \frac{Q^2}{2C}$

For 1 μF capacitor: $E = \frac{(1.5 \times 10^{-9})^2}{2 \times 1 \times 10^{-6}} = 1.125 \times 10^{-12}$ J

For 100 pF capacitor: $E = \frac{(1.5 \times 10^{-9})^2}{2 \times 100 \times 10^{-12}} = 1.125 \times 10^{-8}$ J

2 (a) 3.6×10^{-3}C .

Explanation: the initial charge on C_1 is $Q = CV = 12 \times 10^{-6} \times 400 = 4.8 \times 10^{-3}$ C

When connected to the 4μF, the total capacitance is 16μF storing the same charge. Since the capacitances are in the ratio 3 to 1, the 12μF capacitor stores 3.6×10^{-3}C and the 4 μF capacitor stores 1.2×10^{-3} C

(b) 3 times.

Explanation: each time C_2 is connected to C_1 it takes ¼ of the charge remaining on C_1.

First time C_2 takes 1.2×10^{-3}C leaving C_1 with 3.6×10^{-3}C

Second time C_2 takes ¼ of 3.6×10^{-3}C leaving C_1 with 2.7×10^{-3}C

Third time C_2 takes ¼ of 2.7×10^{-3}C leaving C_1 with 2.025×10^{-3}C

3 (a) 5μ and 25μ are connected in parallel

(b) 5μF and 25μ in parallel have an effective capacitance of 30μF

30μF and 10μF in series have a total capacitance of

$\frac{1}{C} = \frac{1}{10} + \frac{1}{30} = \frac{4}{30}$ $\quad C = \frac{30}{4} = 7.5$ μF

(c) $E = \frac{1}{2}CV^2 = \frac{1}{2} \times 7.5 \times 10^{-6} \times 12^2 = 5.4 \times 10^{-4}$ J

4 (a) $E = \frac{1}{2}CV^2 \qquad 1.44 \times 10^{-4} = \frac{1}{2}C \times 12^2$

Total capacitance C = 2 μF

(b) The capacitors are in series so $\frac{1}{2} = \frac{1}{4} + \frac{1}{C} + \frac{1}{12} \qquad \frac{1}{C} = \frac{1}{6} \qquad C = 6$ μF

Exercise 29

1 (i) The potential difference across the capacitor decreases exponentially. The graph on page 131 shows how the current in the circuit decreases exponentially, the potential difference behaves in the same way.

(ii) The time constant is the period of time that elapses before the charge or current or potential difference falls to 0.37 of its starting value. See page 132.

(iii) Time constant $\tau = CR = 47 \times 10^{-12} \times 22 \times 10^{6} = 1.03 \times 10^{-3}$ s

$V = V_0 e^{\frac{-t}{\tau}} \qquad \frac{V}{V_0} = 0.14 \quad$ so: $\quad 0.14 = e^{\frac{-t}{1.03 \times 10^{-3}}}$

Taking natural logs gives: $\ln(0.14) = \frac{-t}{1.03 \times 10^{-3}}$, $-1.97 = \frac{-t}{1.03 \times 10^{-3}}$

$t = 2.03 \times 10^{-3}$ s

2 (a) The charge on the capacitor rises quickly at the start but then slows finally levelling off at a constant value. The graph on page 130 shows how the potential difference across the capacitor varies with time for two values of resistance, the charge on the capacitor shows the same type of variation with time.

(b) The current drawn from the battery starts of large and gradually falls. This exponential decrease is also shown on page 130.

3 (a) The circuit diagram required is shown on page 130, the only missing component is the milliameter which should be placed in series with the capacitor.

(b) To find the initial current drawn from the battery use Ohm's law

$$I = \frac{V}{R} = \frac{20}{1\times10^4} = 2\text{ mA}$$

(c) $\tau = CR = 470\times10^{-6} \times 1\times10^4 = 4.7$ s

(d) The current decreases exponentially given by $I = I_0 e^{\frac{-t}{\tau}}$

$$\frac{I}{I_0} = 0.25 = e^{\frac{-t}{4.7}}, \quad \ln(0.25) = \frac{-t}{4.7}, \quad -1.386 = \frac{-t}{4.7}, \quad t = 6.51$$

4 (a) Plot ln(current) on y–axis and time on x–axis,

(b) Gradient = $1/\tau = 1/CR$ $C = 8 \times 10^{-6}$ F giving $R = 2.1 \times 10^6\ \Omega$

Exercise 30

1 $E = \frac{1}{2}CV^2 = \frac{1}{2} \times 50\times10^{-6} \times V^2$ so: $V = 2000$ V

2 Charge stored on the capacitor is $Q = CV = 50\times10^{-6} \times 2000 = 0.1$ C

$$I = \frac{Q}{t} = \frac{0.1}{3} = 33\text{ mA}$$

3 Energy stored = 100 J Power = energy/time = 100/3 = 33.3 W

4 9 capacitors needed, arranged as groups of three in series with three such groups in parallel. Explanation: three in series means that 1500 V is divided equally between the three, 500 V across each capacitor. Each group of three in series has a total capacitance of 50/3 = 16.67μF. To have a total capacitance of 50 μF needs three of these series groups in parallel.

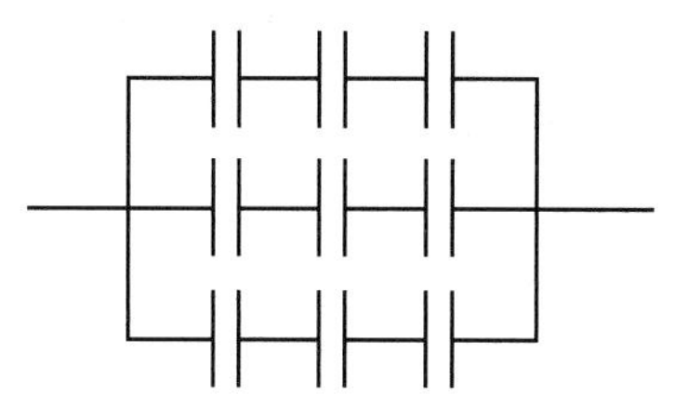

Exercise 31

1 Energy stored = average power × time = $1.5\times10^4 \times 20\times10^{-6} = 0.3$ J

Energy stored $E = \frac{1}{2}CV^2$, $0.3 = \frac{1}{2} \times 47\times10^{-6} \times V^2$ so: $V = 113$ V

Exercise 32

1 (a) The energy stored $E = \frac{1}{2}CV^2$, the graph can be seen on page 127

(b) $E = \frac{1}{2}CV^2 = \frac{1}{2} \times 200\times10^{-6} \times 1.5^2 = 2.25\times10^{-4}$ J

(c) The two capacitors when joined in parallel have a total capacitance of 600μF

The 400μF capacitor initially stored a charge $Q = CV = 200 \times 10^{-6} \times 1.5 = 3 \times 10^{-4}C$

The conservation of electric charge tells us that the same charge is now stored in the new capacitance. $Q = CV$, $3 \times 10^{-4} = 600 \times 10^{-6} \times V$ giving $V = 0.5$ V

(d) Stored energy $E = \frac{1}{2}CV^2 = \frac{1}{2} \times 600 \times 10^{-6} \times 0.5^2 = 7.5 \times 10^{-5}$ J

2 (a) Capacitance is defined as the charge stored per volt. See page 122.

(b) (i) $Q = CV = 47 \times 10^{-6} \times 6 = 2.8 \times 10^{-4}C$ (ii) $E = \frac{1}{2} CV^2 = 8.4 \times 10^{-4}J$

(c) (i) Electric charge is conserved so the new combination holds $2.8 \times 10^{-4}C$.

$Q = CV$ $2.8 \times 10^{-4} = C \times 4.1$ giving a total capacitance of 68.3 μF

The two capacitors are connected in parallel so the value of C is 68.3 – 47 = 21.3μF

(ii) $E = \frac{1}{2}QV = \frac{1}{2} \times 2.8 \times 10^{-4} \times 4.1 = 5.74 \times 10^{-4}$ J

(iii) The charge flow onto the uncharged capacitor constitutes an electric current and heat is generated due to the resistance of the connecting wires.

3 (a) (i) To transfer electrons between the plates requires work to be done against the repulsive force of the electrons already on that plate. See page 126.

(ii) No charge means there is not a repulsive force.

(iii) work done = charge × potential difference $\Delta W = V\Delta q$

(iv) The work done the total charge transferred × average voltage = $Q \times \frac{1}{2}V = \frac{1}{2}QV$.

(b) (i) 8μF and 32μF in parallel have an effective capacitance of 40μF and this in series with 20μF has value given by $\frac{1}{C} = \frac{1}{40} + \frac{1}{20} = \frac{3}{40}$ so: $C = \frac{40}{3} = 13.3$ μF

(ii) Total charge $Q = CV = 13.3 \times 10^{-6} \times 12 = 1.6 \times 10^{-4}$ C

Since the 20μF is in series with the 40μF the same charge resides on it namely $1.6 \times 10^{-4}C$. Using $V = Q/C$ gives the potential difference across the 20μF as $1.6 \times 10^{-4} / 20 \times 10^{-6} = 8V$

4 (a) Exponential decrease with time. (b) Exponential increase from zero.

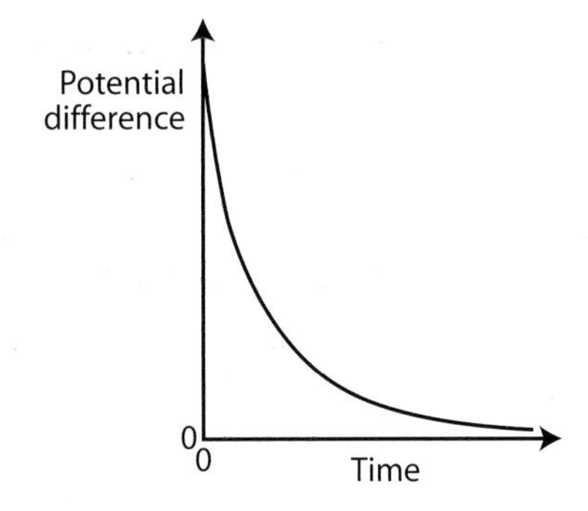

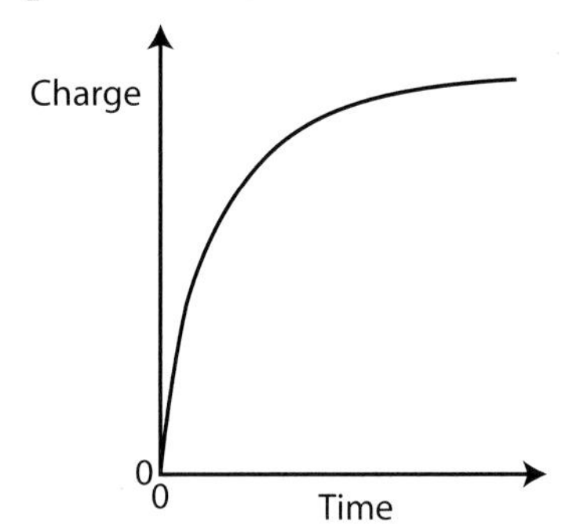

Exercise 33

1 (a) To cause the wire to lift there must be an upward force. Applying Fleming's Left Hand rule gives the direction of the current as X to Y.

(b) The upward force must equal the weight of the wire which is $0.8 \times 10^{-3} \times 9.81 = 7.85 \times 10^{-3}$ N. This upward force can be calculated using the relationship $F = BIL$.

$7.85 \times 10^{-3} = 2 \times 10^{-5} \times I \times 1.2$ give a current $I = 327$ A

2 (a) The interaction of the current flowing from B to C and the uniform magnetic field produces an upward force which reduces the reading on the balance.

(b) The upward force is the difference in the balance reading converted to a force.
Difference in the balance reading is 50.57 – 50.035 = 0.535g converted to a force we have $0.535 \times 10^{-3} \times 9.81 = 5.25 \times 10^{-3}$ N

The upward force is calculated using F = BIL, $5.25 \times 10^{-3} = B \times 2.8 \times 50 \times 10^{-3}$

This gives a value for B = 0.0375 T.

(c) Use Fleming's Left Hand rule the direction of the magnetic field is into the page.

(d) When the current in BC is reversed the force on the length BC is also reversed. The downward acting force now increases the reading on the balance by 0.535g to a value of 51.105g

Exercise 34

1 The induced e.m.f. is equal to the negative of the rate of change of magnetic flux linkage, this is the negative of the gradient of the graph shown.

Time 0s to 2s emf $= \frac{-2}{2} = -1$ V   Time 2s to 3s emf = 0

Time 3s to 4s emf $= -\left(\frac{-4}{1}\right) = +4$ V   Time 4s to 7s emf = 0

Time 7s to 8s emf $= \frac{-6}{1} = -6$ V   Time 8s to 9s emf = 0

Time 9s to 13s emf $= -\left(\frac{-4}{4}\right) = +1$ V

2 (a) Electromagnetic induction is the production of an e.m.f. across a conductor, such as a coil, placed in a changing magnetic field or by moving it through a magnetic field.

(b) (i) When the current in the solenoid falls to zero the magnetic flux linking it also changes and an e.m.f. is produced which lights the bulb.

(ii) The e.m.f. generated is proportional to the rate of change of magnetic flux linkage. In this case the rate of change is large enough to produce an e.m.f. which is greater than the e.m.f. of the battery. However it lasts a short period of time.

3 (a) Farady's law of electromagnetic induction states that the magnitude of the induced e.m.f. is proportional to the rate of change of magnetic flux linking the circuit.

(b) Pushing the magnet into the coil with a N pole facing the coil generates an e.m.f. which in turn causes a current to pass around the coil so that the end of the coil nearest the N pole of the magnet is also a N pole, in accordance with Lenz's law. Work has to be done against this repulsive force, kinetic energy of the moving magnet is being converted to electrical energy. The current passing through the resistor generates heat energy. The faster the magnet is moved the larger the induced em.f. and current according to Faraday's law.

4 (a) The movement of the stylus means that the magnetic flux linking the coil is changing and this change of magnetic flux linkage induces an e.m.f. in the coil.

(b) At these times the gradient of the curve is a maximum which indicates that magnetic flux linkage through the coil has its maximum rate of change and so the induced e.m.f. is a maximum at these times.

Exercise 35

1 (a) A suitable input voltage, applied to the primary coil, is an alternating one. The alternating current produces a changing magnetic field. This means that the magnetic flux linkage through the secondary is continually changing so an e.m.f. is induced in this coil.

(b) A step down transfer is used. To reduce the output voltage to half the input voltage the number of turn on the secondary coil must be half those on the primary coil.

The turns ratio gives the relationship between the number of turns and the voltages

$$\frac{N_p}{N_s} = \frac{V_p}{V_s} = \frac{2}{1}$$

2 (a) Using the turns ratio $\frac{N_p}{N_s} = \frac{V_p}{V_s} = \frac{11\ \text{kV}}{275\ \text{kV}} = 0.04$

(b) Assuming a 100% efficiency means input power = output power

Input power = 176×10^6 W = $11 \times 10^3 \times I_p$ giving I_p = 16000 A

Output power = 176×10^6 W = $275 \times 10^3 \times I_s$ giving I_s = 640 A

(c) 2% of the power generated = 3.52×10^6 W.

Power loss = I^2R, where R is the resistance of the transmission lines.

$3.52 \times 10^6 = 640^2 \times R$ giving R = 8.6 Ω

(d) Power losses result from: Resistance of the windings of both coils. Not all the magnetic flux from the primary coil passes through the secondary coil. Repeatedly reversing the direction of magnetisation of the core generates heat as large currents (eddy currents) are induced in the core, which heats up the coils and increase their resistance so causing even more energy loss.

Exercise 36

1 (a) The time spent in the electric field depends only on the length of the plates and the initial horizontal velocity of the ions, it does not depend on the mass of the ion.

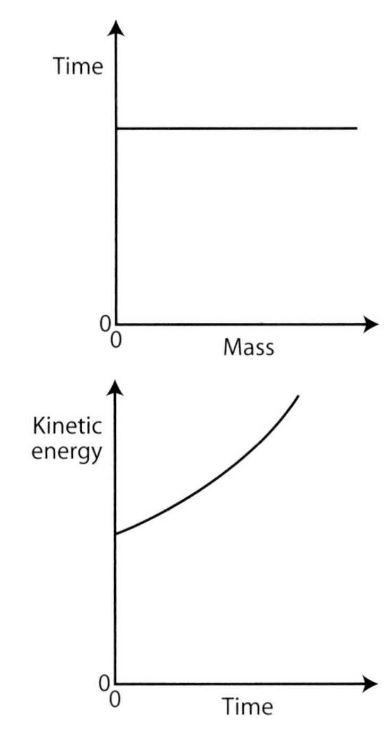

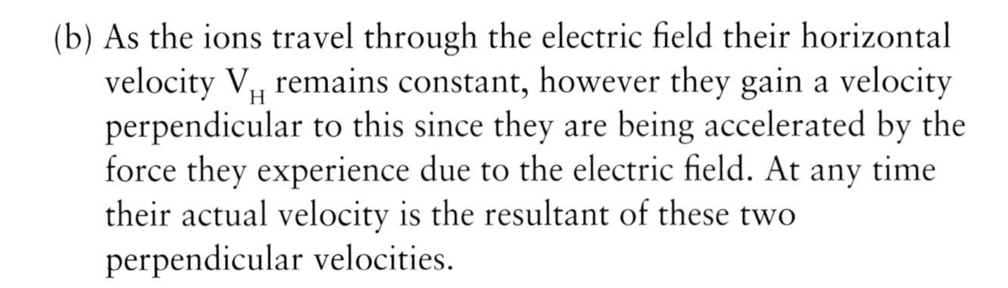

(b) As the ions travel through the electric field their horizontal velocity V_H remains constant, however they gain a velocity perpendicular to this since they are being accelerated by the force they experience due to the electric field. At any time their actual velocity is the resultant of these two perpendicular velocities.

$$v_{Res}^2 = v_H^2 + v_V^2 = v_H^2 + (at)^2 \quad \text{where a = acceleration} = \frac{F}{m} = \frac{eV}{md}$$

$$v_{Res}^2 = v_H^2 + \left(\frac{eV}{md}t\right)^2 \quad \text{where } v_H\text{, e, V, m and d are all constants}$$

Therefore we can simplify the above to:

$v_{Res}^2 = A + Bt^2$ where A and B are constants

The kinetic energy is proportional to v_{Res}^2 so it varies with time in the same way

This is the equation of a parabola, that does not pass through the origin.

2 (a) $F = ma = 6.64 \times 10^{-26} \times 1.15 \times {}^{11} = 7.64 \times 10^{-15}$ N

(b) $F = mg = 6.64 \times 10^{-26} \times 9.81 = 6.51 \times 10^{-25}$ N

(c) Electric force F = qE, where E is the electric field strength

$$E = \frac{7.64 \times 10^{-15}}{2.39 \times 10^{-19}} = 3.1 \times 10^4 \text{ Vm}^{-1} \text{ or NC}^{-1}$$

3 (a) Gain of kinetic energy = loss of electric potential energy

$$\tfrac{1}{2}mv^2 = eV \text{ , rearrange to: } v^2 = \frac{2eV}{m} \quad \text{so} \quad v = 4.19 \times 10^6 \text{ ms}^{-1}$$

(b) $\text{Time} = \dfrac{\text{distance}}{\text{average velocity}}$ Average velocity $= \tfrac{1}{2}(0 + 4.19 \times 10^6)$

$$\text{Time} = \frac{20 \times 10^{-3}}{2.095 \times 10^6} = 9.55 \times 10^{-9} \text{ s}$$

4 Gain of kinetic energy = loss of electric potential energy $\tfrac{1}{2}mv^2 = qV$

$$V = \frac{mv^2}{2q} = \frac{8.35 \times 10^{-27} \times (5.75 \times 10^4)^2}{2 \times 3.2 \times 10^{-19}} = 43.1 \text{ V}$$

5 (a) The top plate is positive since it attracts the negatively charged electrons.

(b) 2000 Vm^{-1}, a uniform electric field has the same field strength throughout.

(c) $F = eE = 1.6 \times 10^{-19} \times 2000 = 3.2 \times 10^{-16}$ N.

(d) The electrons experience uniform acceleration in a direction perpendicular to the line RS. The deflection in this direction can be obtained from $s = ut + \tfrac{1}{2}at^2$, where t is the time since they first entered the electric field and u = 0; remember that they enter without a velocity in this direction. The electrons have a constant velocity along RS, the time to reach Q is 3/8 of the time to reach S. The deflection at S is 16 mm, the deflection at Q will be $(3/8)^2$ of this or 9/64 of 16 = 2.25 mm.

Exercise 37

1 (a) Fleming's Left Hand Rule (see page 173)

(b) Into the page.

(c) The force is always perpendicular to the velocity.

(d) $F = \dfrac{mv^2}{r} = Bqv \text{ , } \quad r = \dfrac{mv}{Bq}$

(e) Specific charge means that the value of $q/m = 5 \times 10^8$. Substitution of the values in the expression for the radius gives r = 0.39 m

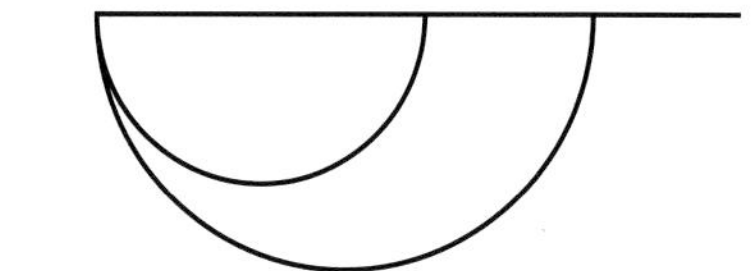

(f) As the velocity of the ions increases so does the radius of the semi circular path they follow.

2 (a) $F = Bev = 150 \times 10^{-3} \times 1.6 \times 10^{-19} \times 2.5 \times 10^{6} = 6 \times 10^{-14}$ N

(b) Out of the page. Fleming's Left Hand Rule is used see page 173. Remember that the direction of electron flow is opposite to the direction of conventional current.

3 (a) Into the field at the bottom, out of the field at the top, again Fleming's Left Hand Rule is used to obtain this.

(b) Momentum is a vector, so its direction is continually changing as the ions move in a semi circle inside the magnetic field.

(c) The centripetal force $\frac{mv^2}{r}$ is provided by the magnetic force Bqv

$\frac{mv^2}{r} = Bqv$, rearranging gives mv = Bqr

Exercise 38

1 (a) A horizontal scale of 6 cm represents a total time of 6 cm × 250μs cm^{-1} = 1500μs

A voltage of frequency 1000 Hz has a period of 1000μs. The CRO would display 1 ½ complete waves. The height of the wave, i.e. from the centre of the screen to the peak, would be:

$\frac{3V}{2}$ Vcm^{-1} = 1.5 cm

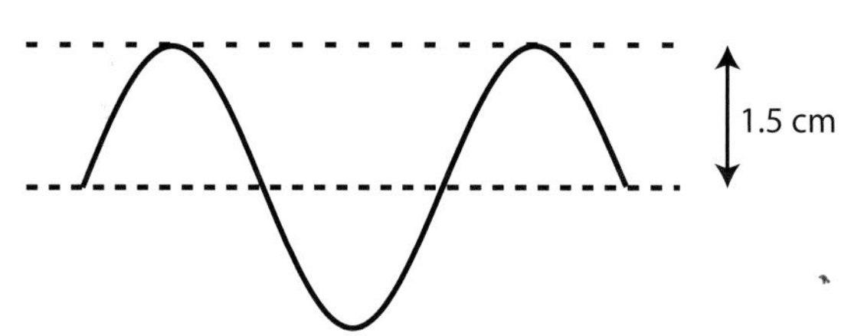

(b) A steady voltage of 3 V would simply produce a horizontal trace on the screen. This line would be deflected $\frac{3V}{2}$ Vcm^{-1} = 1.5 cm upwards from the centre of the CRO screen.

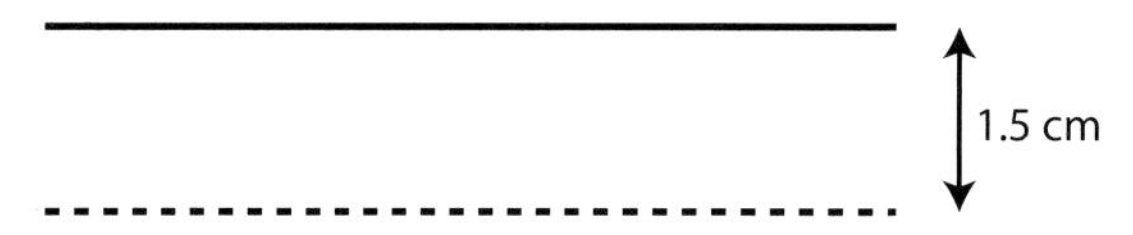

(c) When both signals are applied at the same time to the CRO the sinusoidal trace is seen but it is displaced upwards 1.5 cm due to the addition of the steady voltage.

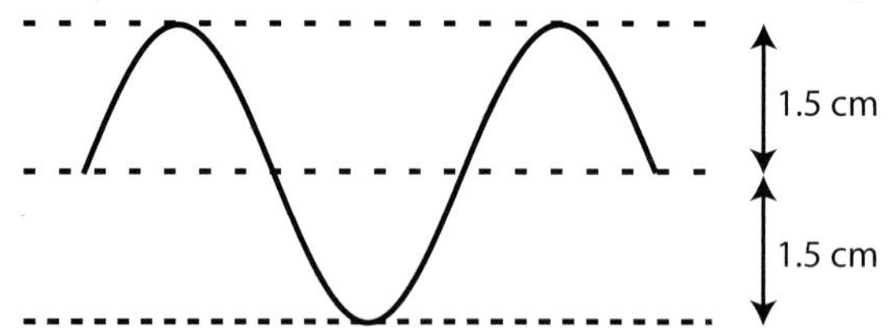

2 (a) Peak voltage = 3 cm × 5 mV cm^{-1} = 15 mV

(b) 1 cycle = 4 cm, Period T = 4 × 50 μs cm^{-1} = 200 μs

$$\text{Frequency} = \frac{1}{\text{Period}} = \frac{1}{200 \times 10^{-6}} = 5000 \text{ Hz}$$

(c) 1 complete wave lasts for 200 μs. There are 10 cm across the screen so each cm must equal 20 μs. The time base setting needed is therefore 20 μs cm^{-1}.

Exercise 39

1 (a) See page 189

(b) (i) In the cyclotron the magnetic field exists over the total area of the accelerator. In the synchrotron the magnetic field only exists at the orbit of the charged particles.

(ii) The magnetic field increases in strength as the protons are accelerated. This keeps the particles in an orbit of fixed radius.

(iii) The frequency of the applied voltage increases since the time for the particles to complete an orbit decreases as they are accelerated.

(iv) The magnetic force on the protons provides the centripetal force so $\frac{mv^2}{r} = Bev$

Rearranging gives $B = \frac{mv}{re} = \frac{1.66 \times 10^{-27} \times 1.5 \times 10^{7}}{(50 \times 1.6 \times 10^{-19})} = 3.11 \text{ mT}$

2 (a) On the diagram drift tubes 1, 3 and 5 are connected to one side of the voltage supply and drift tubes 2 and 4 are connected to the other side.

(b) The frequency of the voltage supply remains constant.

(c) Inside the drift tubes the protons move with constant velocity. Remember, there is no electric field inside the tubes so no acceleration of the protons takes place.

(d) Time spent inside the drift tubes = half the period of the applied alternating voltage. The relationship is maintained by increasing the length of successive drift tubes.

(e) L = vt, where t is the time spent inside the drift tube.

For synchronous acceleration t = T/2, where T is the period of the applied voltage.

T = 1/f, where f is the frequency of the applied voltage, t = 1/2f. L = v/2f.

3 (a) The second tube being 20% longer means that the velocity of the protons has increased by 20%, it is now 1.2v, where v was the velocity in the first drift tube. The kinetic energy is proportional to v^2, so the kinetic energy is now $1.2^2 \times 500$ keV = 720 keV.

(b) The kinetic energy has increased by 220 keV, so the potential difference between the electrodes is 220kV.

(c) Acceleration occurs across the gaps of the linear accelerator, so 43 drift tubes means 42 gaps. At each crossing of a gap the energy increases by 220 keV. After 42 gaps the final energy will be 500 + 220 × 42 = 9740 keV or 9.74 MeV.

Exercise 40

1 (a) The protons spiral outwards from the centre of the cyclotron, see page 188.

(b) The magnetic field always acts at right angles to the plane of the dees. Fleming's left hand rule is used to determine the direction of the magnetic field, in or out of the plane of the paper will depend on the initial direction of movement of the protons. The splitter shows the protons spiral clockwise, so the field is out of the plane of the paper.

(c) Acceleration occurs across the gap between the dees, inside the dees the protons move with constant speed. When the arrive at the gap the alternating potential difference ensures that they are attracted across the gap by a negatively charged dee.

(d) The centripetal force is provided by the magnetic force on the protons, $Bev = \frac{mv^2}{r}$ this gives $B = \frac{mv}{er}$. To involve frequency you need to eliminate the velocity v,

$$v = \frac{\text{circumference of the orbit}}{\text{orbital period}} = \frac{2\pi r}{T}$$

$$\text{Frequency} = \frac{1}{T} \quad \text{so} \quad B = \frac{2\pi mf}{e}$$

(e) Re-arranging the above gives $f = \frac{Be}{2\pi m}$

$$f = \frac{1.5 \times 10^{-3} \times 1.6 \times 10^{-19}}{2 \times 3.14 \times 1.66 \times 10^{-27}} = 2.3 \times 10^4 \text{ Hz}$$

2 (a) In the synchrotron the radius of the particles is constant whereas in the cyclotron the radius increases. In the synchrotron the magnetic field used to keep the particles in their fixed radius orbit increases whereas in the cyclotron the magnetic field is of constant strength.

(b) The centripetal force needed to keep the particles moving in a circle is provided by the magnetic force on the charged particle.

$\frac{mv^2}{r} = Bqv$, re-arranging gives $r = \frac{mv}{Bq}$

(c) (i) An antiparticle is one with the same mass but opposite charge to the particle, e.g. the electron is the particle and the positron is its anti-particle.

(ii) An event known as annihilation happens, the mass of the electron and the positron when they collide is converted to two gamma ray photons which fly off in opposite directions to conserve momentum.

(iii) From part (b) the velocity of the accelerated particles is given by $v = \frac{Bqr}{m}$.

The LEP accelerator has a larger radius than the SPS accelerator, 4.25 km compared to 1.1 km, so the LEP can produce particles with much greater velocities.

3 (a) (i) Annihilation (see page 195)

(ii) It happens when particles meet their anti-particles. Anti-particles are not part of normal matter so particles such as positrons are very rare.

(b) The linear accelerator is described on pages 184 to 186. This is a past paper question and examiners would expect an answer to contain some of the following:

- a diagram showing source of the charged particles,
- a number of drift tubes arranged in a straight path,

- details of how the drift tubes are connected to a.c. voltage supply,
- description of accelerating of particles takes place across the gaps,
- particles repelled by the tube they leave and attracted by the one they enter,
- particles travelling at constant speed inside the drift tubes,
- explanation of why the drift tubes get longer as the particles accelerate,
- explanatioin of what is meant by synchronous acceleration and how it is achieved.

Exercise 41

1 The antiproton p^- consists of three quarks, a negative charge of –1e and has a baryon number of B = –1. The proton is made of the u u d quarks so the antiproton p^- is made of the corresponding antiquarks $\bar{u}\,\bar{u}\,\bar{d}$. In terms of charge we have $-\frac{1}{3}e + -\frac{1}{3}e + -\frac{1}{3}e = -1e$.
The baryon number of each of these antiquarks is $-\frac{1}{3}$ give a baryon number for the antiproton of –1.

2 Mesons are made up from 2 quarks.
$\pi^+ = u\bar{d}$, the charge $= +\frac{2}{3}e + \frac{1}{3}e = +1e$, and the baryon number $= \frac{1}{3} + \left(-\frac{1}{3}\right) = 0$
$\pi^- = \bar{u}d$, the charge $= -\frac{2}{3}e + \left(-\frac{1}{3}e\right) = -1e$, and the baryon number $= -\frac{1}{3} + \frac{1}{3} = 0$

3 (a) The W^- particle, see page 207.
(b) $\bar{\nu}_e$, the antielectron neutrino. Lepton number is conserved, we have 1 = 1 – 1 + 1 = 1. See page 202 for the lepton numbers.

4 (a) The baryon number of p^+ = 1 and p^- (antiproton = –1) so the RHS has B = 0
Mesons are not baryons so they have B number of 0.
(b) Annihilation.
(c) Annihilation results in the formation of two gamma ray photons which in this case then form 8 mesons. This is an example of matter turning into energy and then energy turning back into matter.

5 (a) Baryon number B is not conserved, $1 \neq -1 + 0 + 0$
Lepton number L is not conserved, $0 \neq 0 + (-1) + (-1)$
(b) $p^+ = n + e^+ + \nu_e$

6 (a) Hadrons = p^+, n, π^0 (b) Leptons = e^+ and μ^- (c) p^+ and n (d) π^0

Exercise 42

1 (a) A particle that is not made up of smaller particles.
(b) Neutron, pi-meson, proton
(c) Leptons are fundamental particles, the hadrons are not or you could state that leptons are not affected by the strong nuclear force and hadrons are.
(d) (i) Baryons and mesons.
(ii) Baryons are made up of 3 quarks, mesons are made of only two quarks.
(iii) proton = uud, neutron = udd,

2 (a) The nucleus contains 4 protons and 5 neutrons giving a total of 9 baryons. It is a neutral atom so it has 4 electrons (same number of protons) which are leptons. It has no mesons: these are too short lived to be a part of a stable atom.

(b) The neutron and proton have a baryon number of 1 each, so the baryon number is conserved. The electron is a lepton so that particle X must also be one. The left hand side of the reaction has a total charge of 0 so the right hand side must also have a total charge of 0 so the particle X must have no charge. It is a neutrino.

3 (a) Baryon - proton or neutron
Lepton - electron or neutrino or muon or a tau
A pion is an example of a meson
Antiparticle = positron, or p^-

(b) Quark structure of proton = uud. The charge is $+\frac{2}{3}e + \frac{2}{3}e + \left(-\frac{1}{3}e\right) = +1e$

(c) (i) $d \rightarrow u + W^-$ this is followed by $W^- \rightarrow e^- + \nu_e$ See page 207

(ii) Weak nuclear force

Exercise 43

1 7.27 cm

2 4.03 cm

3 9.18 cm

Exercise 44

1 3.56 cm

2 5.80 cm

3 7.72 cm

Exercise 45

1 16.3 ± 6 Ω

Exercise 46

1 (a) 5 (b) 3 (c) 5 (d) 4 (e) 4

2 (a) 645.6 (b) 0 (c) 1700 (d) 1.2

Exercise 47

1 (a) 1.36×10^4 (b) 3.0×10^8 (c) 1.9×10^{-5}.

Exercise 48

1 (a) 25 (b) −10 (c) 2

Exercise 49

1 3

2 0.707, 0.500, 0.365, 0.333

3 (a) 3 or 4 (b) 251 nm = 0.000251 mm wavenumber = 3984

Exercise 50

1 (a) W (y axis) against $1/(N+1)^2$ (x axis)

(b) The intercept on the y axis gives A and the gradient gives B

(d) A = 4180 mm^{-1} B = -1.12×10^{-4} mm^{-1}

2 (a) L = v/4f (b) L (y axis) against 1/f (x axis) (c) Gradient equals v/4

3 (a) $R = R_o + R_o\alpha\theta$

(b) Plot R on the y axis and θ on the x–axis, a straight line with a intercept on the y–axis

(c) R_o is the intercept on the y–axis and α is equal to the gradient/intercept

4 (a) 4

(b) Giving the angle for molybdenum as 7.260 implies that the vernier protractor can be read to ± 0.001°. This is wrong. The protractor can be only read to ± 0.01°.

(c) molybdenum, tin, europium and tungsten all need modification to 7.26, 5.08, 3.17 and 2.28 respectively.

Exercise 51

1 (a) $\log R = \log R_o - n\log E$

(b) Straight line graph of negative slope with intercept on the y–axis. The intercept on the y–axis will give log R_o and the gradient will give n.

2 (a) $\log I = \log A + n\log\lambda$

(b) Straight line graph of positive slope with intercept on the y–axis. The intercept on the y–axis will give log A and the gradient will give n.

Exercise 52

(a) Theory

(i) Axes, labelled v against r^2. The graph is a straight line through the origin.

(ii) Gradient = k/η

(iii) Units of η = **kg $m^{-1}s^{-1}$**

(b) **Planning and design procedure**

(i) Use a micrometer screw gauge, measure several diameters, take the average and the radius is half this value.

(ii) A suitable measurement region is not far below the surface and you would avoid selecting a region near the bottom. The apparatus you need would be: metre rule and stopclock. Measure two marks a distance apart, time the ball bearing to move between the marks, repeating and taking an average. The terminal velocity = distance/time.

(c) **Interpretation**

(i) The time taken decreases because v is proportional to $1/\eta$

(ii) –57 (units)

Other Resources

ISBN: 978 1 904242 43 7

Price: £25

Author: Pat Carson & Roy White

Physics for CCEA AS Level – Revised Specification

This important resource has been developed in co-operation with CCEA to cover the revised AS level Physics Specification and follows the same headings as the specification The resource is presented in full colour with several hundred illustrations and diagrams, as well as numerous worked examples.

The book contains 47 exercises, and answers are provided at the back for all numerical questions. Sample examination questions are also provided throughout.

Included is a most useful section on the practical techniques required for best practice in physics experimentation, such as precision, accuracy, analysis and interpretation.

Resource CD for AS Level

ISBN: 978 1 906578 61 9

Price: £39.99 + VAT

Resource CD for A2 Level

ISBN: 978 1 906578 62 6

Price: £39.99 + VAT

Resource CDs

These resources contain all the diagrams from the textbooks in jpeg format for use with interactive whiteboards.

Contact Colourpoint Educational at:

Tel: 028 9182 6339 Fax: 028 9182 1900
Web: www.colourpointeducational.com

All orders to MiMO Distribution:

Tel: 028 9182 0505 E-mail: sales@mimodistribution.co.uk

Colourpoint Books, Colourpoint House, Jubilee Business Park,
Jubilee Road, Newtownards, Co Down, BT23 4YH